13th International Conference on Electrostatics 2011

AA002681

Journal of Physics: Conference Series Volume 301

Bangor, Wales, United Kingdom
10-14 April 2011

Editors:

D. Martin Taylor

ISBN: 978-1-61839-004-2
ISSN: 1742-6588

Printed from e-media with permission by:

Curran Associates, Inc.
57 Morehouse Lane
Red Hook, NY 12571

Some format issues inherent in the e-media version may also appear in this print version.

Printed by Curran Associates, Inc. (2011)

For permission requests, please contact the Institute of Physics
at the address below.

Institute of Physics
Dirac House, Temple Back
Bristol BS1 6BE UK

Phone: 44 1 17 929 7481
Fax: 44 1 17 920 0979

techtracking@iop.org

Additional copies of this publication are available from:

Curran Associates, Inc.
57 Morehouse Lane
Red Hook, NY 12571 USA
Phone: 845-758-0400
Fax: 845-758-2634
Email: curran@proceedings.com
Web: www.proceedings.com

13th International Conference on Electrostatics 2011

Journal of Physics: Conference Series Volume 301

Bangor, Wales, United Kingdom
10-14 April 2011

ISBN: 978-1-61839-004-2
ISSN: 1742-6588

IOP Institute of Physics

TABLE OF CONTENTS

BILL BRIGHT MEMORIAL LECTURE

SECTION A: ATMOSPHERIC AND PLANETARY ELECTROSTATICS

SECTION B: APPLICATIONS

SECTION C: DISCHARGES

SECTION D: ENVIRONMENT AND HEALTH

SECTION E: ELECTROHYDRODYNAMICS

SECTION F: HAZARDS

SECTION G: MEASUREMENTS

SECTION H: ESD, ELECTRONIC DEVICES AND MEMS

SECTION I: MODELLING

SECTION J: STANDARDS

SECTION K: TRIBOELECTRIFICATION

PREFACE

Electrostatics 2011 was held in the city of Bangor which is located in North West Wales in an area of outstanding natural beauty close to the Snowdonia mountain range and bordering the Irish Sea. The history of the area goes back into the mists of times, but a continuous technological thread can be traced from the stone- and bronze-age craftsmen, who inhabited the area several thousand years ago, via the civil engineering and fortifications of the Romans and Edward I of England, through Marconi's long-wave trans-Atlantic transmitter near Caernarfon to the conference host. The School of Electronic Engineering at Bangor University has contributed much to the discipline of Electrostatics not only in teaching and research but also in supporting industry. It was a great pleasure for me, therefore, to have the pleasure of welcoming the world's experts in Electrostatics to Bangor in April 2011.

In my preface to the Proceedings of Electrostatics 1999, I reported that almost 90 papers were presented. Interestingly, a similar number were presented in 2011 testifying to the importance and endurance of the subject. The all-embracing nature of electrostatics is captured in the pictorial depiction used for the conference logo: a hand-held plasma ball with its close link to gaseous discharges and the superimposed Antarctic aurora highlighting the featured conference themes of atmospheric, planetary and environmental electrostatics. Leading these themes were three invited contributions, the first by Giles Harrison who delivered the Bill Bright Memorial Lecture 'Fair weather atmospheric electricity', Carlos Calle on 'The electrostatic environments of Mars and the Moon' and Istvan Berta on 'Lightning protection – challenges, solutions and questionable steps in the 21st century'. Leading other key sessions were invited papers by Atsushi Ohsawa on 'Statistical analysis of fires and explosions attributed to static electricity over the last 50 years in Japanese industry' and Antonio Ramos on 'Electrohydrodynamic pumping in microsystems'.

Of the papers submitted for publication 69 passed through the thorough review process and I take this opportunity to warmly thank the reviewers for their constructive criticism and rapid turnaround which has allowed the Proceedings to be delivered to the publisher on time. It is a pleasure also to thank members of the International Advisory Panel, and the Organizing and Programme Committees for their guidance and suggestions and especially Claire Garland and her team at the Institute of Physics for their support, all of which ensured a successful and enjoyable conference. Special thanks are due to Jeremy Smallwood for organising the pre-conference workshop, to Tom Jones, Martin Glor and Dave Swenson for their highly informative and educational contributions at the workshop, to CST for organising the simulation workshop, and to CST and JCI Chilworth for their much appreciated sponsorship of the conference.

I am sure you will enjoy reading this record of Electrostatics 2011, covering as it does the wide range of subjects upon which static electricity impinges. Especially important is the development of electrostatic-based methods for reducing atmospheric pollution. In this context it is interesting to see how Masuda's work on the surface-discharge-based Boxer charger, first reported over 30 years ago, has now developed into dielectric barrier discharge (DBD) systems for the removal of noxious molecules from industrial and vehicle exhaust gases.

Thanks to our hard working conference chairman, Paul Holdstock, the conference retained its now well-established reputation for providing a friendly, sociable atmosphere for discussing the newest developments in this important scientific area. Finally, my sincere thanks go to all the presenters and to all those who attended and contributed to another successful conference.

Professor D. Martin Taylor
Proceedings Editor
Bangor, May 2011

Fair weather atmospheric electricity

R.G. Harrison

Department of Meteorology, University of Reading, P.O. Box 243, Earley Gate, Reading RG6 6BB

r.g.harrison@reading.ac.uk

Abstract. Not long after Franklin's iconic studies, an atmospheric electric field was discovered in "fair weather" regions, well away from thunderstorms. The origin of the fair weather field was sought by Lord Kelvin, through development of electrostatic instrumentation and early data logging techniques, but was ultimately explained through the global circuit model of C.T.R. Wilson. In Wilson's model, charge exchanged by disturbed weather electrifies the ionosphere, and returns via a small vertical current density in fair weather regions. New insights into the relevance of fair weather atmospheric electricity to terrestrial and planetary atmospheres are now emerging. For example, there is a possible role of the global circuit current density in atmospheric processes, such as cloud formation. Beyond natural atmospheric processes, a novel practical application is the use of early atmospheric electrostatic investigations to provide quantitative information on past urban air pollution.

The material summarised here was presented in the Bill Bright memorial lecture at Electrostatics 2011.

1. Historical perspective

Benjamin Franklin's findings opened a new era in electrostatics, of central importance in establishing the science of atmospheric electricity. Discovering [1] the similarity between electricity in thunderclouds and that generated by machines by kite and rod experiments began the study of thunderstorm electricity, but, with it, an awareness also emerged that electrification remained apparent despite the absence of thunderstorms, even, to use an early description, in "serene" (*i.e.* fine) weather conditions.

Early direct observations of terrestrial fair weather electrification were made in summer 1753, no doubt inspired by Franklin's work. For example, at the Château de Maintenon, Abbé Mazeas established a long wire antenna system for electrical atmospheric measurements [2],

> *"On the 14th of June I accompanied the Marcehal de Noailles to his castle of Maintenon. At my arrival, I set up an apparatus, which consisted of an iron wire 370 feet long, raised to 90 feet above the horizon. It came down from a very high room in the castle, where it was fastened to a silken cord six feet long, and was carried from thence to the steeple of the town; where it was likewised fastened to another silken cord of eight feet long, and shelter'd from rain…"*

finding that

> *"...in weather void of storms, the electricity of a piece of sealing-wax of two inches long was about twice as strong as that of the air."*

Experiments on thundercloud electrification using a rain-shielded electrometer were undertaken in the UK from May 1753 by John Canton [3], which led him to conclude [4], specifically for fine conditions rather than just storm-free conditions, that

> *"The air without-doors I have sometimes known to be electrical in clear weather."*

Several other series of atmospheric electricity experiments are known to have been made through the late eighteenth and nineteenth centuries, notably the heroic continuous observations of Giambatista Beccaria in Piedmont, north-west Italy [5], which inspired Lord Kelvin to develop an "incessant" automatic photographic recording system for his water dropper electrometer [6]. Kelvin's well calibrated instrumentation allowed quantitative observations of the atmospheric electric field's vertical component, E_z, to be obtained, conventionally recorded as the Potential Gradient F, where $F = - E_z$. Converted to modern units [7], Kelvin's determination of the fair weather PG near Aberdeen, Scotland at 0800 on 14th September, 1859 was +137 Vm^{-1}.

2. Synthesis of observations in the global atmospheric electric circuit

Although the first century of atmospheric electricity observations confirmed the continuous presence of an electric field in fair weather conditions, its origin remained a key geophysical problem. Further discoveries around the beginning of the twentieth century, such as the vertical profile of the ionization generated by cosmic rays, and confirmation, by radio, of a conducting region in the upper atmosphere allowed C.T.R. Wilson to propose the existence of an atmospheric electrical circuit. This synthesis drew heavily on the established electrostatic and geophysical observations made since 1753, together representing "central tenets" of the global circuit [8], Table 1.

Century	*Observation*	*Investigators*
18th	Finite conductivity of atmospheric air	Coulomb
19th	Fair weather positive Potential Gradient	Kelvin
	Continual atmospheric ionisation	Wilson
20th	Upper atmosphere highly conductive	Heaviside, Kennelly, Marconi
	Vertical current flow in fair weather	Wilson
	Cosmic ray atmospheric ionization profile	Hess
	Finite conductivity of earth's surface	Chapman

Table 1. "Central tenets" of the global atmospheric electrical circuit.

Wilson's conceptual model linked current generation in disturbed weather zones (thunderstorms and shower clouds acting as "generators") with the fair weather zones, by an electrical connection through the conductivity of the upper atmosphere and the surface (Figure 1a).

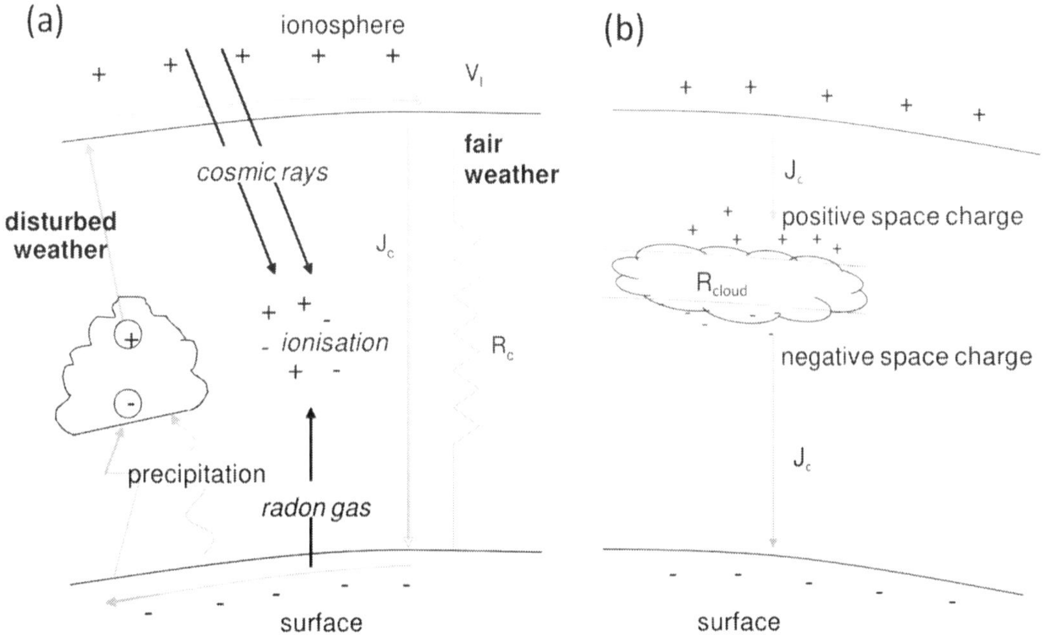

Figure 1. (a) The global atmospheric electric circuit. Charge generated in disturbed weather zones drives current through the ionosphere and back through the surface, with intermediate current flow through the finite air conductivity in fair weather zones. The ionospheric potential $V_{i,}$ (~ +250 kV), conduction current density J_c (~2 pA m^{-2}) and unit area columnar resistance R_c are approximately related by Ohm's Law. (b) Vertical current flow in the fair weather zone of the global circuit, through an extensive horizontal resistive cloud layer.

3. Fair weather measurements

Subsequent studies have acted to confirm the global circuit concept [8], such as agreement between the diurnal variations in the global thunderstorm generator, the PG in clean air, and the ionospheric potential. Many additional measurements are now available, but the fair weather region of the global circuit has probably received less attention than the disturbed weather region. In part this may be due to the difficulty in correcting local site-dependent factors, and the technical challenge in maintaining high integrity insulation for reliable measurements of small currents. PG measurements are the most common, made straightforward by environmentally robust field mill technology [6, 9]. Table 2 summarises PG measurements made at Reading under different weather conditions. The field mill was standardised to a measurement at 1m by calibration against a long wire antenna.

The raw data from which Table 2 was derived are presented as histograms in Figure 2, with the classification of the different weather conditions used briefly described. (Solar radiation measurements were used to infer cloud.) These histograms clearly indicate that the PG varies with weather conditions, showing distributions skewed negative in rain, and strongly positive in fog. Fair weather PG values, as long recognized, are positive, as are the PG values determined during non-raining overcast conditions. It is evident from Table 2 that the PG in overcast conditions is slightly reduced over that obtained in clear conditions. This would be associated with an increase in the columnar resistance, as a result of the increased resistance of the cloud layer over clear air, although the relative contribution to the columnar resistance will vary with the depth and height of the cloud layer present.

Weather conditions	Median PG (Vm^{-1})	PG variability (IQR/1.349) (Vm^{-1})	hours of data
all conditions	80.6	42.9	8447
snow	191.6	36.1	22
fog	170.6	105.3	139
clear sky	93.5	32.2	177
broken cloud on dry days	91.1	30.9	349
overcast on dry days	81.4	33.1	479
heavy rain	-4.9	133.4	174

Table 2. Summary of Reading Observatory PG measurements in 2010. (The variability is expressed in terms of the Inter-Quartile Range IQR, scaled to be equal to one standard deviation if the distribution were normal.)

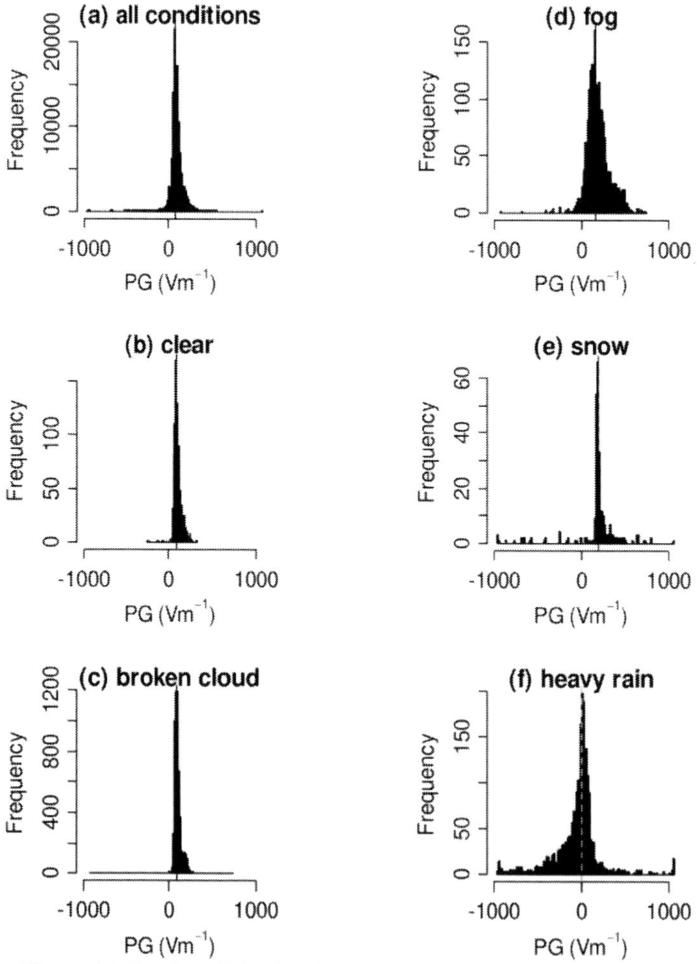

Figure 2. Histograms of Potential Gradient (PG) (5 minute averages) made at Reading Observatory during 2010, for (a) all weather conditions, (b) clear days, (c) broken cloud on dry days, (d) fog (wet bulb and dry bulb temperatures equal, with relative humidity > 90%), (e) snow (recorded by observer) and (f) heavy rain (> 4mm/hour).

A succinct visual summary of the PG data in Table 2 can be made by plotting the variability against the median for the different weather types, as shown in figure 3. This representation shows the distinction in electrical conditions between rain, fog and snow, which can be difficult to separate using some automated monitoring systems. Atmospheric electricity data may therefore aid classification of weather conditions for automatic weather stations.

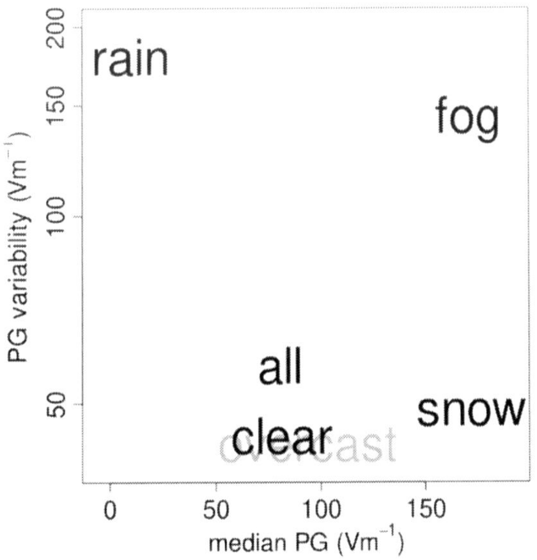

Figure 3. Representation of the PG data obtained during a variety of weather conditions at Reading during 2010, as summarized in Table 2.

Observations of the sensitivity of atmospheric electrical properties to weather changes seems very likely to have formed the basis for the famous comment [10] of Lord Kelvin concerning weather forecasting:

> *"There can be no doubt but the electric indications, when sufficiently studied, will be found important additions to our means for prognosticating the weather; and the speaker hoped soon to see the atmospheric electrometer generally adopted as a useful and convenient weather-glass."*

4. Applications and outstanding problems

Because of its linkage between different regions of the planet, the global circuit potentially provides a monitoring framework able to provide corroborating information relevant to climate change. Changes in thunderstorm activity for example, may lead to a change in the ionospheric potential and associated global circuit variables, but the details of such changes are complex and poorly known [11]. On the more local scale, an application for the many urban measurements of the PG is in reconstructing past air pollution burdens. This is because the PG is sensitive to smoke pollution [12], which provides quantitative relationship allowing smoke concentrations in some past urban situations to be inferred from atmospheric potential measurements. For example, using eighteenth century electrostatic measurements, it has been possible to establish that a two maxima per day existed in London's smoke pollution before the advent of motor traffic [13].

Beyond the traditional fair weather regime of an electrically quiescent atmosphere, the intermediate "semi-fair weather conditions" raises many new science questions. Firstly, charge generated in lofted atmospheric particles, such as the ash produced by the April 2010 eruption of Eyjafjallajökull [14] provides an opportunity to use low cost detection technologies. Secondly, the sensitivity of the global circuit to surface ionization may allow observation of widespread geological changes in radon emission, which could be a precursor phenomenon for some earthquakes [15]. Finally, when the vertical conduction current density of the global circuit encounters horizontal layers of dust or clouds, the abrupt reduction of air conductivity in the transition from clear air to cloudy air leads to space charge on the cloud boundary (this process is illustrated in Figure 1(b)). This may in turn influence the properties of the clouds and, perhaps, climate. In some respects this leads back to the historical aspects discussed in section 1, and particularly to the fashions of early electrostatics in speculating on cloud phenomena. This era is epitomized in the writings of the pioneer meteorologist Luke Howard, who described the shower cloud *Nimbus* as one

> *"in which minute drops constituting cloud...are by a change in their electrical state made to coalesce, and descend in drops of Rain"*. [16]

References

[1] Berger G and Ait Amar S 2009 The noteworthy involvement of Jacques de Romas in the experiments on the electric nature of lightning. *J. Electrostat.* **67**, 531-535 doi:10.1016/j.elstat.2009.01.033

[2] Parsons J and Mazeas W 1753 Observations upon the Electricity of the Air, Made at the Chateau de Maintenon, during the Months of June, July, and October, 1753. *Phil Trans* **48**, (1753 - 1754) (377-384)

[3] Canton J 1753 Electrical Experiments, with an Attempt to Account for Their Several Phaenomena; Together with Some Observations on Thunder-Clouds *Phil Trans* **48**, (1753 - 1754) (350-358)

[4] Canton J 1754 A Letter to the Right Honourable the Earl of Macclesfield, President of the Royal Society, concerning Some New Electrical Experiments. *Phil Trans* **48**, 780-785.

[5] Beccaria G 1775 *Della elettricite terrestre atmosferica a cielo serno*, Turin.

[6] Bennett A J and Harrison R G 2007 Atmospheric electricity in different weather conditions, *Weather*, **62**, 10, 277-283.

[7] Bennett A J and Harrison R G 2007 A simple atmospheric electrical instrument for educational use *Adv. Geosci.* **13**, 11-15

[8] Aplin K L, Harrison R G and Rycroft M J 2008 Investigating Earth's atmospheric electricity: a role model for planetary studies *Space Science Reviews* **137**, 11-27

[9] Chubb J N 1990 Two new designs of"field mill" type fieldmeters not requiring earthing of rotating chopper. *IEEE Trans. Ind. Appl.* **26**, 6, 1178–1181.

[10] Thomson W 1872 *Reprint of papers on Electrostatics and Magnetism*, Macmillan, London, (p233)

[11] Price C 2009 Will a drier climate result in more lightning? *Atmos Res* **91,** 479–484

[12] Harrison R G 2006 Urban smoke concentrations at Kew, London, 1898-2004 *Atmos Environ* **40**, 18, 3327-3332

[13] Harrison R G 2009 Two daily smoke maxima in eighteenth century London air, *Atmos Environ* **43**, 1364-1366

[14] Harrison R G, Nicoll K A, Ulanowski Z and Mather T A 2010 Self-charging of the Eyjafjallajökull volcanic ash plume, *Environ Res Lett* **5** 024004 http://stacks.iop.org/1748-9326/5/024004

[15] Harrison R G**,** Aplin K L and Rycroft M J 2010 Atmospheric electricity coupling between earthquake regions and the ionosphere *J. Atmos Solar Terr Physics* **72**, 376-381

[16] Howard L 1837 Lecture 5 in *Seven Lectures on Meteorology* (2nd ed. Harvey and Dalton, 1843).

Atmospheric ionisation in Snowdonia

K L Aplin[1,2], J H Williams[3]

1. Department of Physics, University of Oxford, Denys Wilkinson Building, Keble Road, Oxford OX1 3RH UK

2. RAL Space, R25, Harwell Oxford, Didcot OX11 0QX UK

3. Envirodata-Eyri, Bryn Goleu, Penmaen Park, Llanfairfechan, Gwynedd LL33 0RL UK

k.aplin1@physics.ox.ac.uk

Abstract Atmospheric ionisation from natural radioactivity and cosmic rays has been measured at several sites in Snowdonia from 2005-present. The motivation for this project was a combination of public engagement with science, and research into the effects of ionisation on climate. A four-component atmospheric radiometer instrument is co-located with the ionisation detectors and the data is remotely logged and displayed on the Web. Atmospheric ionisation from natural radioactivity varies with local geology, and the cosmic ray ionisation component is modulated by solar activity and altitude. Variations due to all these effects have been identified and are described.

1. Introduction

Motivated by a possible relationship between atmospheric ionisation and climate [1], we have been measuring atmospheric ion production [2], and the long and short wave radiation at mountain locations in Snowdonia since 2005. We have also been displaying the data, updated daily, on the Web, to increase public understanding of atmospheric and particle physics in the North Wales area and beyond[1].

The experiment has been located at two mountain sites, Snowdon Summit (1085m) (2004-2006, with ionisation measurements from 2005), Marchlyn Mawr Reservoir (660m) (2009-present) and briefly at a low-level local site, Llanberis (120m) (2005 and 2009), all within 8 km of each other in Snowdonia. Additionally, the equipment underwent an extended outdoor test at a second low-level site in Gloucestershire (2008-2009). The basic instrumentation comprises a Kipp and Zonen CNR1 four-component radiometer to measure atmospheric radiation, and ZP1442 Geiger counters, which are used to measure ionisation from both local radioactivity and cosmic rays. Ionisation rate is measured, as it is an effective proxy for ion concentration in clean air [3]. (Instruments that measure atmospheric ions directly need more maintenance than is possible at a remote mountain site.) The Geiger counters were located with the data logger inside an IP68 sealed box, which was in turn placed inside a secure, sealed box containing the battery. Drainage holes drilled in the bottom of the boxes prevent catastrophic flooding of the equipment.

[1] http://snowdon.physics.ox.ac.uk/Snowdon/Cosmic_Ray_Station.html

This range of sites has enabled the equipment to be optimised for the harsh mountain conditions and has also permitted measurements at different altitudes and climates. Solar activity has also changed over the duration of the experiment. The sun's magnetic field deflects cosmic rays away from Earth, and also varies on an 11-year cycle, with a deep minimum, and associated cosmic ray maximum, in 2009, and a more active Sun (fewer cosmic rays) when the measurements began in 2005.

Table 1 Summary of Welsh mountain measurements, including an extended low-level test. Local radioactivity was measured using Geiger counters, cosmic rays using a Geiger telescope, broadband atmospheric radiation with a Kipp and Zonen CNR1 net radiometer and narrowband radiation with a filter radiometer.

Location (lat/long N/W, altitude)	**Dates**	**Measurements**	**Power and data logging**
Llanberis (53.1167, 4.11959, 120m)	July 2005 July 2009	Local radioactivity and cosmic rays	Battery, Campbell 21X (2005) and Campbell CR3000 (2009)
Marchlyn Mawr (53.1367, 4.0691, 660m)	July 2009 – present	Local radioactivity, cosmic rays, down and up-welling short- and longwave radiation	Mains power to charge 90Ah lead acid battery Campbell CR3000 logger, data retrieved by mobile phone
Snowdon Summit (53.0689, 4.0756, 1085m)	May 2004 – September 2006	Local radioactivity, downwelling short- and long-wave radiation, cosmic rays (29th July 2005 - 25th March 2006)	60W solar panel to charge 90Ah dryfit battery Campbell 21X logger, data retrieved manually over RS-232 interface
Cheltenham (51.8929, 2.1300, 30m)	July 2008 - May 2009	Local radioactivity, cosmic rays, downwelling short- and longwave radiation, narrowband longwave radiation [1]	80W solar panel to charge 90Ah lead acid battery Campbell CR3000 logger, data retrieved by mobile phone or manually over RS-232 interface

2. Atmospheric ionisation measurements

Near the Earth's surface, cluster-ions are formed by both natural radioactivity and cosmic rays. Natural radioactivity is principally alpha and gamma radiation emitted from radon isotopes and the soil [4], and cosmic rays are energetic ionising radiation from outer space. In the lower troposphere most cosmic ray ionisation is from muons with mean energy 4GeV, half of which is lost to ionisation [5]. The instrumentation used for ionisation rate measurements are two Geiger counters, both independently responsive to beta and gamma radiation, stacked vertically. Coincident triggering of both counters is usually due solely to high-energy particles, and this enables the cosmic ray ionisation to be measured. This coincidence configuration is known as a cosmic ray telescope [2]. At the surface, the typical ionisation rate q is usually quoted as 10 $cm^{-3}s^{-1}$ (although this can be highly variable) with 20% from cosmic radiation, 40% alpha radiation from radon and 40% beta and gamma radiation from radon decay [4]. In clean air, q is related to the ion concentration n by the ion balance equation with $q = \alpha n^2$ where α is a self-recombination coefficient [3].

2.1 Ion production from natural radioactivity

The ZP1442 Geiger counters used are sensitive to beta and gamma radiation. The individual count rates of each counter are therefore expected to represent about 40% of q. The Geiger tube is calibrated in terms of a known dose of gamma radiation from a ^{137}Cs source of 10 μGy/hr, corresponding to 1.14mR/hr in air where 1Roentgen=208230 ions m^{-3} [6] and giving the number of ions produced in air for one Geiger count as 66 cm^{-3}. Natural radioactivity measurements are shown in table 2 below. For the low altitude test site, the mean count rate is 11.4 cpm corresponding to (12.4±1.4) ions $cm^{-3}s^{-1}$, within the range reported in the literature [4]. It is also possible to estimate the radon concentration if the background gamma dose rate is assumed to be linearly proportional to the ^{222}Rn concentration in the air [7]. From the Geiger tube calibration, which relates dose to count rate, the mean count rate at Cheltenham corresponds to a radon concentration of 1.3±0.4 atoms $cm^{-2}s^{-1}$. (This is likely to be an over-estimate, since beta radiation will also contribute to the measured count rate.) The European radon inventory, based on interpolated gamma dose rate measurements, gives the radon concentration for the Cheltenham grid square to be 0.70 atoms $cm^{-2}s^{-1}$, consistent with our measurements. In North Wales the underlying geology is different, leading to higher radon concentrations [7] and increased ion production rates compared to Cheltenham.

Table 2 Summary of Geiger counter measurements. Energetic particles are recorded in counts per hour (cph), and local radioactivity in counts per minute (cpm) with standard errors (se) indicated.

Location (height asl)	**Date**	**Mean Energetic Particle Rate (cph±se)**	**Top Geiger Count Rate (cpm±se)**	**Bottom Geiger Count Rate (cpm±se)**
Low altitude test site				
Cheltenham	19/8/08-12/11/08 12590 5 minute averages, summed energetic particles	12.46±0.11	11.51±0.01	11.38±0.01
High altitude sites				
Marchlyn Mawr (660m)	28/07/09-25/11/09 32,621 5 minute averages, summed energetic particles	14.50±0.07	16.93±0.01	16.97±0.01
Snowdon Summit (1085m)	26/07/05-27/10/05 2229 hourly averages, summed energetic particles	15.08±0.02	18.88±0.93	16.76±0.02

2.1 Ion production from cosmic rays

Cosmic ray ionisation increases with altitude, and, as expected, the highest telescope count rates were measured at Snowdon Summit, with significantly lower high-energy radiation detected at Marchlyn and Llanberis. The energetic particle count rates at Llanberis in 2005 and 2009 were 10.3±1.0 and 12.4±1.0 counts per hour, a statistically significant difference. This is likely to be related to solar activity, for which a simple proxy is sunspot number. There were (40±34) sunspots on average in July 2005 compared to (3±5) in July 2009 [8], consistent with the more active Sun deflecting cosmic rays away from Earth in July 2005, and preventing them from reaching the atmosphere.

The cosmic ray particles measured by the telescope are almost all muons (heavy electrons moving at relativistic velocities). The typical muon surface flux is a few hundred particles $m^{-2}s^{-1}sr^{-1}$ [3], but the count rates recorded by our insensitive detectors are much lower. The ionisation caused by muons is $21cm^{-1}$ [9], but converting our count rates to volumetric ionization rates is not straightforward due to pressure and temperature related variations in muon production. The planned installation of a barometer and thermometer will permit further work on this.

The upper Geiger counter of the telescope always records a slightly (but statistically significant) higher count rate than the lower one, which persists when the tubes are swapped, Figure 1. This is unlikely to be related to natural radioactivity, since the two Geiger tubes are close together, and may be from lower-energy cosmic rays without sufficient energy to trigger both Geiger counters. The "soft component" is mostly electrons, and comprises 10% of the total cosmic rays at the surface [5].

3. Atmospheric ionisation and downwelling longwave radiation

Laboratory experiments showed that atmospheric cluster-ions can absorb radiation in the 9-12 μm range of the infra-red (IR) spectrum [10]. However, it has not been possible to detect any relationship

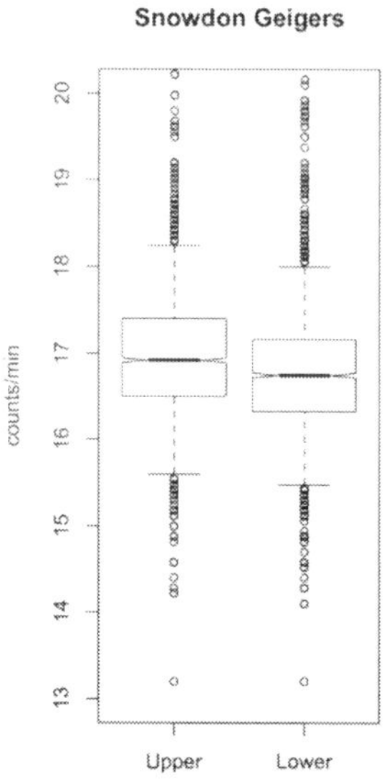

Figure 1 Box plot showing differing Geiger counter rates at Snowdon (2005-2006).

between atmospheric ionisation and the downwelling broadband IR radiation in clear sky at the Snowdon Summit site, due to atmospheric temperature effects modulating both the muon data and downwelling IR. Further measurements with a filter radiometer "tuned" to the ion absorption band did show a relationship between ionisation and atmospheric IR radiation [1].

4. Conclusions

Atmospheric ionisation measurements have been made at several sites in Snowdonia and in the South of England and have shown the variations expected with local geology, altitude and solar cycle. There was no clear link between ionisation and downwelling broadband radiation.

Beyond the scientific research aspect of this experiment, the Marchlyn data is made available to the general public on the Web. We intend to develop this website, including a Welsh translation, and produce data files so that school students can investigate simple aspects of atmospheric and particle physics for themselves. The equipment will be moved to the new Snowdon Summit building this summer to increase both the cosmic ray count rate and the impact of the measurements on the general public, who visit Snowdon in their thousands every year. The public engagement aspect of this work is currently funded by the Royal Astronomical Society and the work was previously funded by NERC and STFC. Technical assistance from J G Firth, C A Eley (RAL Space) and N T Clifford (Oxford University) is gratefully acknowledged.

References

[1] Aplin K L and McPheat R A 2008 An infra-red filter radiometer for atmospheric cluster-ion detection *Rev. Sci. Instrum.* **79**, 106107

[2] Aplin K L and Harrison R G 2010 Compact cosmic ray detector for unattended atmospheric ionisation monitoring *Rev. Sci. Instrum.* **81**, 124501

[3] Aplin K L and Harrison R G 2000 A computer-controlled Gerdien atmospheric ion counter *Rev. Sci. Instrum.* **71**, 8, 3037-3041

[4] Dhanorkar S and Kamra A K 1994 Diurnal variation of ionisation rate close to ground, *J. Geophys. Res.* **99**, D9, 18,523–18,526

[5] Nakamura K et al (Particle Data Group), Cosmic rays, *J Phys G* **37**, 075021 (2010)

[6] Geiger Muller Tubes databook, Centronic

[7] Szegvary T, Cohnen F and Ciais P 2009 European ^{222}Rn inventory for applied atmospheric studies, *Atmos. Env.*, **43**, 1536-1539

[8] http://solarscience.msfc.nasa.gov/SunspotCycle.shtml

[9] Danforth W E and Ramsey W E 1936 The specific ionisation of cosmic ray particles as determined by Geiger-Muller counter efficiency *Phys. Rev.* **49**, 854

[10] Aplin K L and McPheat R A 2005 Absorption of infra-red radiation by atmospheric molecular cluster-ions, *J. Atmos. Sol.-Terr. Phys.*, **67**, 8-9, 775-783 doi:10.1016/j.jastp.2005.01.007

Charge measurements in stratiform cloud from a balloon based sensor

K.A. Nicoll, R.G. Harrison

Department of Meteorology, University of Reading, Earley Gate, Reading, Berkshire, UK, RG6 6BB

Email: k.a.nicoll@reading.ac.uk, r.g.harrison@reading.ac.uk

Abstract. The electrification of stratiform clouds has is little investigated in comparison with thunderstorms and fair weather atmospheric electricity. Theory indicates that, at the upper and lower horizontal boundaries of layer clouds, charging will arise from vertical flow of cosmogenic ions in the global atmospheric electric circuit. Charge is transferred to droplets and particles, affecting cloud microphysical processes such as collision and droplet activation. Due to the lack of in-situ measurements, the magnitude and distribution of charge in stratiform clouds is not well known. A sensitive, inexpensive, balloon borne charge sensor has been developed to make in-situ measurements of edge charging in stratiform cloud using a standard meteorological radiosonde system. The charge sensor has now been flown through over 20 stratiform clouds and frequently detected charge up to 200 pC m^{-3} near cloud edges. These results are compared with measurements from the same sensor used to investigate charge in particle layers, such as volcanic ash from the Eyjafjallajökull eruption, and Saharan dust in the Cape Verde Isles.

1. Introduction

Although most previous research into cloud electrification has been concerned with thunderstorms, other cloud types can also be electrified. Stratiform clouds in particular are thought to be charged as theory predicts that the upper and lower boundaries of these clouds should become charged through vertical current flow in the global atmospheric electric circuit [1]. Cloud droplets and aerosol particles in these regions will become charged, for which implications are expected in cloud microphysical processes, particularly in respect to collision and scavenging processes (e.g. [2]-[4]). However, very few electrical measurements have been made in-situ in stratiform clouds, and consequently very little is known about the magnitude of the charge.

2. Charge generation on stratiform cloud edges

2.1. Theory

The electrical characteristics of Earth's atmosphere are conveniently summarised in the conceptual framework of the Global Electric Circuit, GEC [5]. In the GEC, a vertical conduction current, J_c, flows between the ionosphere and Earth's surface, as a result of the large difference in potential between these two regions. J_c is typically considered in terms of the fair weather atmosphere (i.e. no clouds or significant weather), however in the presence of cloud, J_c is predicted to charge the upper and lower edges of stratiform clouds of large horizontal extent. This occurs due to the gradient in conductivity, σ_t, between the high conductivity clear air outside the cloud, and the low conductivity air inside the cloud (due to ion-droplet attachment). The space charge, ρ, predicted to occur at the upper and lower cloud edges is given by Gauss' law,

$$\rho = \varepsilon_0 J_c \frac{d}{dz}\left(\frac{1}{\sigma_t}\right) \quad (1)$$

where $\varepsilon_0 = 8.85 \times 10^{-12}$ F m^{-1}, z is vertical distance, $d/dz\ (1/\sigma_t)$ is the vertical gradient in conductivity, and both z and J_c are positive downwards. The vertical conductivity gradient near stratiform cloud edges is therefore associated with a positive layer of space charge at cloud tops (resistivity $(1/\sigma_t)$ increasing in the direction of positive z and J_c) and negative charge at cloud base.

2.2. Instrumentation

To measure stratiform cloud charging in-situ, a Cloud Edge Charge Detector (CECD) has been specially developed for use on weather balloons [6]. The CECD consists of a spherical electrode, connected to a sensitive electrometer which measures the voltage on the electrode. Changes in the electrode voltage result from the presence of charge, via induction and impaction effects. The CECD electrode voltage is reset every 60 seconds to prevent prolonged saturation of the electrometer in highly charged areas. The CECD flies alongside a standard meteorological radiosonde which measures pressure, temperature, relative humidity, and GPS position. A specially developed data acquisition system, similar to the one described in [7], is used to transmit the CECD data over the radio link of the radiosonde, synchronous with the meteorological data. Measurements are made at 1Hz, giving a typical vertical sampling resolution of 5m (for a 5ms^{-1} ascent rate).

2.3. Measurements

The CECD has been flown on over twenty balloon flights through stratiform cloud from a variety of locations around the UK. Figure 1. shows results from one of these flights, at the base of an extensive layer of stratocumulus cloud over the Southern UK. Figure 1(a) shows cloud droplet number concentration, (b) the voltage on the CECD electrode as it descended through the cloud layer, and (c) the resulting space charge (derived from the rate of change of CECD electrode voltage and ascent rate [6]. The cloud base is at ~3.30km, where the cloud droplet number concentration (and thus the air conductivity) begins to change. This is also the region where the CECD voltage changes, denoting the presence of space charge.

From the small sample of stratiform clouds performed during this study, it has been found that the majority of stratiform cloud edges (two-thirds of those measured) are charged, with typical mean charges of ~20pC m^{-3} and maximum charges of ~200pC m^{-3}. In addition, a case study of the CECD flight detailed in Figure 1. shows that the measured cloud edge space charge agrees closely with that expected from theory [8], supporting the hypothesis of charging from the vertical conduction current.

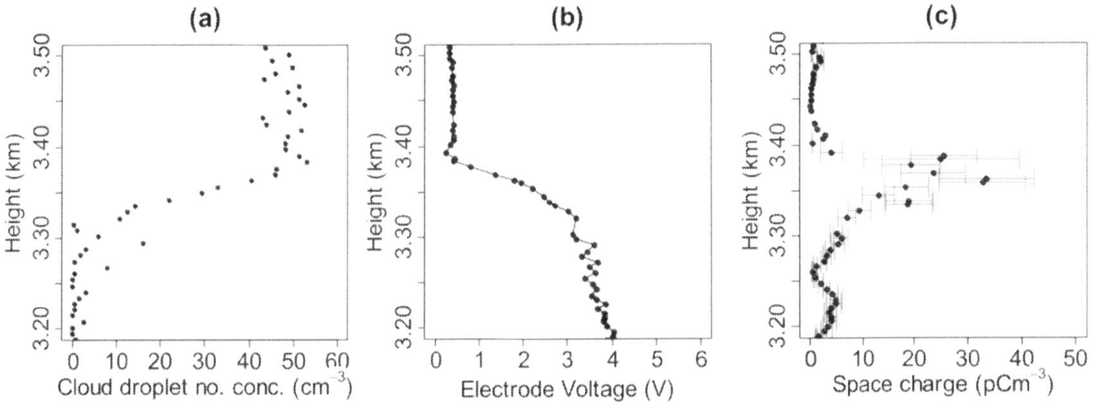

Figure 1(a) Vertical profile of cloud droplet number concentration through an extensive layer of stratocumulus cloud, measured by a Cloud Droplet Probe on the FAAM research aircraft, flying on the same day. (b) CECD electrode voltage measured on radiosonde descent through the same cloud layer. (c) Magnitude of space charge derived from rate of change of CECD electrode voltage and sonde ascent rate. Adapted from [8].

3. Charge associated with other atmospheric phenomena

Space charge is separated whenever there is a gradient in conductivity, such as at boundaries between clear and droplet/particle laden air, therefore charging by this mechanism is expected to occur in stratified dust and pollution layers, as well as in stratiform cloud. The CECD has thus been used to investigate charging in volcanic plumes and Saharan dust layers, in which charging may partly arise by this mechanism. During flights though ash and dust the CECD was flown alongside a specially developed aerosol particle counter [9], which measures particles of diameter 0.6 to 10.6μm using a light scattering method. Results from flights through a volcanic ash plume and elevated Saharan dust layer will be described in subsequent sections.

3.1. Volcanic ash plume charging

Radiosondes are well suited to making measurements in areas that are hazardous to manned aircraft, hence the CECD and aerosol particle counter were used to sample the Eyjafjallajokull volcanic ash plume during the flight ban in April 2010 [10]. A layer of positively charged ash of ~0.5pC m^{-3} was detected between 3.8 and 4.5km on a sounding made from Stranraer, Scotland on 19th April. Possible explanations for the charge generation include charging at the eruption site - volcanic lightning was observed on 17th April, when the plume was emitted from the volcano - however charge decay considerations suggest that any residual charge from the eruption should have decayed within 10km of the vent (Stranraer is ~1200km from the eruption site). A second explanation is charging of the upper and lower edges of the plume from vertical current flow in the GEC as described in section 2.1, however the observed charging was not confined to the plume edges, and was unipolar, suggesting an additional charge generation mechanism.

3.2. Saharan dust layer charging

Several balloon measurements in elevated layers of dust were made from the Cape Verde Isles, ~600km off the west coast of Africa, which experience frequent Saharan dust outbreaks. Two balloon soundings instrumented with the aerosol particle counter and CECD detected dust particles up to 4km altitude, with increased levels of charge inside the dust layers. Charges of up to 25pC m^{-3} were detected inside the dust layers, which, like the charges observed in the volcanic ash plume, are not expected and again suggest that a continuous charge generation mechanism is present. Charge generation during the initial dust lofting stage is likely to occur due to particle-particle collisions and collisions with the surface, however, the large distances between the lofting region and measurement site suggest that this mechanism cannot account for the observed charging. Edge charging of dust layers, from vertical current flow through the layers provides an alternative generation mechanism, however, modelling the space charge on the layer edges produces charges two orders of magnitude smaller than those observed.

4. Conclusions

Theory predicts that charge should be generated at the boundaries between clear and droplet/particle laden air, due to flow of the vertical conduction current in the Global Electric Circuit. Such charge layers have been observed at the upper and lower edges of stratiform clouds by a specially developed balloon borne charge sensor. The sensor has also detected small amounts of charge in volcanic ash plumes, and Saharan dust layers far from their point of origin [11]. In the stratiform cloud case, the measured charges agree with that predicted by edge charging theory via the conduction current, however this is not so in the volcanic ash and Saharan dust layer cases, suggesting the presence of additional charge generation mechanisms, which are, at present, incompletely understood.

References

[1] Tinsley B A 2000 Influence of Solar Wind on the Global Electric Circuit, and Inferred Effects on Cloud Microphysics, Temperature, and Dynamics in the Troposphere *Space Sci. Rev.* **94**(1-2), 231-258

[2] Tinsley B A, R P Rohrbauch, M Hei, and K V Beard 2000 Effects of image charges on the scavenging of aerosol particles by cloud droplets and on droplet charging and possible ice nucleation processes *J. Atmos. Sci.* **57**, 2118-2134.

[3] Khain A, M Arkhipov, M Pinsky, Y Feldman, and Y Ryabov 2004 Rain enhancement and fog elimination by seeding with charged droplets. Part1: theory and numerical simulations *J. Appl. Meteorol.* **43** 1513-1529

[4] Harrison R G and M H P Ambaum 2008 Enhancement of cloud formation by droplet charging *Proc. R. Soc. A* **464**, 2561-2573

[5] Rycroft M J, S Israelsson and C Price 2000 The global atmospheric electric circuit, solar activity and climate change *J. Atmos. Sol.-Terr. Phys.* **62** 1563

[6] Nicoll K A and R G Harrison 2009 A lightweight balloon-carried charge sensor *Rev. Sci. Instrum.* **80** 014501-4

[7] Harrison R G 2005 Inexpensive multichannel digital data acquisition system for a meteorological radiosonde *Rev. Sci. Instrum.* **76** 026103

[8] Nicoll K A and R G Harrison 2010 Experimental determination of layer cloud edge charging from cosmic ray ionisation *Geophys. Res. Lett.* **37** L13802

[9] Ulanowski Z, E Hirst ,P H Kaye,R G Harrison, K A Nicoll and G Rogers 2010 Radiosonde aerosol counter for vertical profiling of atmospheric dust layers *EGU General Assembly (Vienna, Austria)* 13512.

[10] Harrison R G, K A Nicoll, Z Ulanowski and T A Mather 2010 Self-charging of the Eyjafjallajokull volcanic ash plume *Environ. Res. Lett.* **5** 024004 doi:10.1088/1748-9326/5/2/024004

[11] Nicoll K A, R G Harrison and Z Ulanowski, 2011 Observations of Saharan dust electrification *Environ. Res. Lett.* **6** 014001

Electrostatic phenomena in volcanic eruptions

S J Lane, M R James and J S Gilbert

Lancaster Environment Centre, Lancaster University, Lancaster LA1 4YQ, UK

s.lane@lancaster.ac.uk

Abstract. Electrostatic phenomena have long been associated with the explosive eruption of volcanoes. Lightning generated in volcanic plumes is a spectacular atmospheric electrical event that requires development of large potential gradients over distances of up to kilometres. This process begins as hydrated liquid rock (magma) ascends towards Earth's surface. Pressure reduction causes water supersaturation in the magma and the development of bubbles of supercritical water, where deeper than c. 1000 m, and water vapour at shallower depths that drives flow expansion. The generation of high strain rates in the expanding bubbly magma can cause it to fracture in a brittle manner, as deformation relaxation timescales are exceeded. The brittle fracture provides the initial charge separation mechanism, known as fractoemission [1]. The resulting mixture of charged silicate particles and ions evolves over time, generating macro-scale potential gradients in the atmosphere and driving processes such as particle aggregation. For the silicate particles, aggregation driven by electrostatic effects is most significant for particles smaller than c. 100 μm. Aggregation acts to change the effective aerodynamic behaviour of silicate particles [2], thus altering the sedimentation rates of particles from volcanic plumes from the atmosphere. The presence of liquid phases also promotes aggregation processes [3] and lightning.

1. Introduction

The presence of electrical phenomena in volcanic eruptions, specifically explosive eruptions, has long been known and is exemplified by lightning [see review by Mather and Harrison, 4]. Electric charge is highly heterogeneous at the scale of atoms and molecules, but to have a macroscopic effect this heterogeneity must increase its scale through a series of physical and chemical processes. In the case of volcanic lightning this scale is of order kilometers, an increase of c. 10^{13}.

2. Magma fragmentation

Volcanoes are the product of molten rock (magma) emerging at Earth's surface. The molten rock is produced by partial melting at depths of 10s to 100s of km, resulting in a melt phase that, at high pressure, has water and other volatile species dissolved in it. The melt ascends buoyantly in Earth's crust and, when pressure is reduced to the extent that water is supersaturated in the magma, bubbles nucleate and grow. This process causes expansion, reduces density and drives magma to the surface. If the magma is of medium to high viscosity (> c. 10^4 Pa s) then some form of explosive event involving magma fragmentation [5] by brittle failure is likely.

2.1. Fractoemission

The fracturing of solid materials necessarily involves disruption of the atomic-scale electric fields that bind solids together. If fracture propagation speeds are sufficiently rapid that relaxation of the perturbed electrical structure cannot decay within the solid, then a range of phenomena can occur. These include the emission of photons, electrons and positive ions as well as the creation of charged fragments of the solid; this is the process of fractoemission [1].

The magma fragmentation process occurs at a range of depths below and heights above the Earth's surface. Initial fragmentation can take place several 100s m below ground level, with the clasts generated continuing to fragment during collisions [6] and continuing gas expansion. The fragmentation process results in clasts generally ranging from metres to less than a micron in size, representing a very large surface area of fresh fracture surface. The collapse of lava domes to produce avalanches of pyroclasts produces similarly high degrees of fragmentation. It is thought that fractoemission begins the process of increasing the scale of charge heterogeneity, and is the dominant pathway for the initial generation of charged particles in particulate volcanic flows.

2.2. Atmospheric interaction

On interaction with the atmosphere the large surface area resulting from the small particle sizes promotes rapid heat exchange with cold entrained air. Should the average density of the hot mixture of rock particles and air be less than that of the surrounding atmosphere, a thermally buoyant volcanic plume develops and can range in height from a few 100 meters to 10s of kilometers, depending on the thermal (mass) flux of the eruption.

The processes involved in plume development can act to further separate the charged plume components. There is field [7] and experimental [8] evidence that gas and/or small particles take an average positive charge, whilst larger particles are negative on average. The origin of this process may lie in the partitioning of charge between ions and silicate fragments during the fragmentation process and from interactions with the atmosphere.

3. Electrostatic aggregation

The products of magma fragmentation are silicate particles, water vapour and other gases such as SO_2 and halogens. The larger sizes of hot silicate particles (pyroclasts) are not coupled to the gas phase and follow ballistic trajectories (Figure 1). The small particles (ash) couple more strongly to the gas phase and are lofted in the buoyant plume. The presence of unbalanced electric charge on the surface of silicate particles < c. 100 μm in size promotes aggregate formation [2] (Figure 2a), with the presence of a liquid phase (Figure 2b) enhancing this process [3]; note that > 50 % by mass of explosive volcanic products are in this size range.

Particle aggregation has a profound effect on sedimentation from volcanic plumes, acting to increase the effective fall velocity of any particle within the aggregate [9]. Hence, small ash particles are more rapidly removed from the atmosphere than if they were transported as individuals. This process is implicated in the development of bimodal particle size distributions and in ash deposits that increase in thickness with distance from the volcanic source.

4. Atmospheric potential gradients

Close to the volcanic source, the charge associated with volcanic ejecta can dominate the electrical environment of the local atmosphere and perturbs the fair-weather atmospheric potential gradient. The potential gradient can be used to imply charge distributions and magnitudes within the volcanic plume [4] and models suggest that one charge tends to dominate the upper reaches of the plume, whereas the lower reaches are dominated by the opposite polarity. The details of how charge separates remains poorly constrained, but possibilities include gas and small particles taking one charge, and large particles another.

The fallout of volcanic ash at measurement localities indicates that larger particles often have a net negative charge, but this polarity may vary between eruptions and volcanoes. Laboratory experiments have shown that the fracture of pumice generally produces ions of a net positive charge and ash particles of a net negative charge [7][8]. It is hypothesized that, over time, the large surface area presented by small particles scavenges more ions, leaving large particles more negatively charged than small particles on average. Gravitational separation then acts to produce the observed atmospheric potential gradients.

5. Volcanic lightning

Volcanic lightning is the result of potential gradients exceeding the electrical breakdown strength of the atmosphere, a process that may also take place on other solar system bodies [10]. McNutt and Williams [11] provide a contemporary review of volcanic lightning. The mechanisms of electrification in plumes 1 - 4 km high may be attributed to fragmentation processes, whereas lightning in plumes 7 - 12 km high may also involve electrification mechanisms seen in thunderclouds. Evidence for the involvement of freezing processes in the generation of lightning detectable by long range meteorological observation comes from the recent Eyjafjallajökull eruption of 2010 [12]. Thus there appear to be at least two mechanisms leading to lightning, namely magmatic mechanisms giving discharges near the vent and at temperatures above c. -20°C, and atmospheric mechanisms associated with water freezing below c. -20°C.

Figure 1. Sakurajima volcano, Japan, with lightning (10th February 2010). The ballistic trajectories represent hot magma clasts, with the electric discharge following an apparently more random path. The close proximity of the electric discharge to the crater rim suggests that the charge separation mechanisms operating are different from those leading to larger-scale meteoric lightning (reproduced with the kind permission of Martin Rietze http://apod.nasa.gov/apod/ap100210.html).

Figure 3. Lightning propagates within the eruption plume of Kirishima volcano, Japan. This suggests that the potential gradient required to cause electrical breakdown in the plume is less than that required for free air. This hypothesis would appear consistent with the presence of many charged particles and ions within the plume. The dominant charging mechanisms here are likely to be sourced in the fragmentation process, specifically fractoemission mechanisms during brittle fracture of magma.

The reader can find many spectacular images of volcanic lightning online. Figure 3 shows lightning occurring during the Kirishima eruption in late January 2011. This image suggests that the lightning is here confined within the plume, implying that the electrical properties of the plume are more conducive to electrical breakdown than the surrounding atmosphere; the plume is acting as a lightning conductor. Large lightning strikes, very similar to those seen from thunderstorms, are also observed from large volcanic clouds some distance from the eruption source, and these lightning strikes pose a fatal hazard in their own right.

Volcanic lightning has been identified as a potential fixation mechanism for nitrogen [13], making this element available for biological processes. The fixation rates are estimated at 10^{13} g NO per year about 4 Ga ago and are comparable with other nitrogen fixation mechanisms, such as meteoric lightning, hot volcanic vents and bolide impacts, during the early history of Earth and other planets.

Figure 2a. The presence of liquids promotes the formation of spherical aggregates of volcanic particles known as accretionary lapilli.

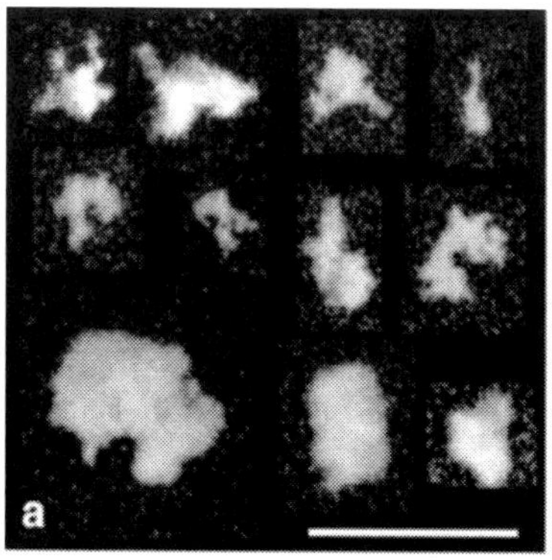

Figure 2b. The absence of liquids creates aggregates with fractal geometries, here imaged in flight. The scale bay is 0.5 mm long.

6. Conclusions

Electrostatic phenomena have a significant impact on volcanic plumes by causing a profound change in the sedimentation behaviour of small ash particles. The presence of lightning is spectacular and can give a detectable signal for the remote monitoring of volcanic activity. Changes in atmospheric potential gradient provide data on the magnitude and sign of charge distribution in volcanic plumes, but much research remains to be done into all these phenomena to improve our fundamental understanding and the implications for eruption monitoring.

References

[1] Dickinson J T, Langford S C, Jensen L C, McVay G L, Kelso J F, and Pantano C G 1988 Fractoemission from fused silica and sodium silicate glasses *J. Vac. Sci. Technol.* A**6** 1084

[2] James M R, Lane S J, and Gilbert J S 2003 Density, construction, and drag coefficient of electrostatic volcanic ash aggregates *J. Geophys. Res.*, **108(B9)** 2435

[3] Gilbert J S, Lane S J 1994 The origin of accretionary lapilli *Bull Volcanol*, **56** 398-411

[4] Mather T A, Harrison R G 2006 *Surveys in Geophysics* **27(4)** 387-432

[5] Sahagian D 1999 Magma fragmentation in eruptions *Nature* **402** 589 - 591

[6] James M R, Gilbert J S and Lane S J 2002 Experimental investigation of volcanic particle aggregation in the absence of a liquid phase *J. Geophys. Res.* **107**(B9)

[7] James M R, Lane S J and Gilbert J S 1998 Volcanic plume monitoring using atmospheric potential gradients *Journal of the Geological Society of London* **155** 587-590

[8] James M R, Lane S J & Gilbert J S 2000 Volcanic plume electrification: Experimental investigation of a fracture-charging mechanism *J. Geophys. Res.* **105** (B7), 16641-16649

[9] Lane S J, Gilbert J S and Hilton M 1993 *Bull. Volc.* 55 481-188

[10] James M R, Wilson L, Lane S J, Gilbert J S, Mather T A, Harrison R G and Martin R S 2008 Electrical charging of volcanic plumes *Space Sci. Rev.* **137** 399 - 418

[11] McNutt S R and Williams E R 2010 *Bull. Volcanol.* **72** 1153–1167

[12] Arason P and Bennett A J 2010 *AGU Fall Meeting 2010*, abstract #AE33B-0278

[13] Navarro-González R, Molina M J and Molina L T 1998 *Geophys. Res. Lett.* **25** 3123-3126

Fixed site monitoring of potential gradient fluctuations near to AC high voltage power lines

J C Matthews

University of Bristol, UK

E-mail: j.c.matthews@bristol.ac.uk

Abstract. A fixed site monitoring station recorded the potential gradient disturbances near to two high voltage power lines during 2008. The full year's results show that the electrical environment downwind of power lines is modified compared to that upwind. Potential gradient disturbance was greater on days when there was rainfall. Humidity was inversely correlated with mean potential gradient when the station was both downwind and upwind of both power lines. Wind speed is weakly correlated with the standard deviation of a 10 minute sample of potential gradient downwind of both power lines, but not upwind. The distributions of mean and standard deviation of potential gradient in 10 minute samples showed that the field was more negative overnight and on days where there was rain, but less variable at night and on dry days. Upwind of the power lines, the average 24 hour trace exhibits the natural background Carnegie curve, with peaks corresponding to increased global thunderstorm activity, while local effects mask this trace when the FSMS is downwind of the power lines. The results show that corona ions can cause potential gradient disturbances downwind of high voltage power lines, most particularly during rain and high humidity, and overnight.

1. Introduction

The Earth has a natural potential gradient (PG) between the ground and the ionosphere. High voltage power lines can produce corona ions when the field gradient near the cables is sufficient to ionise the air. If these ions escape the line, and there is a charge separation between positive and negative ions, then the space charge can be measured as fluctuations in PG using electric field mills [1].

It is useful to know the times and conditions when corona ion emission is most likely to occur. A fixed site monitoring station (FSMS) was constructed in a field in Littleton-Upon-Severn, South Gloucestershire, UK to investigate the fluctuations in PG downwind of a 275 and a 400 kV power line in different meteorological conditions [2].

2. The fixed site monitoring station

The FSMS contains a JCI 131 all weather electric field mill meter recording the vertical P G at 1 s intervals, and a Davis Instruments weather station recording the wind speed and direction, temperature, pressure, humidity and solar radiation at 10 minute intervals. Zero levels were checked approximately every 2 weeks by recording the response when the field mill aperture was placed into a grounded chamber. The field mill output was compared to the recorded PG from a factory calibrated JCI 140 field mill placed 10 m away approximately monthly [2]. After two years of measurements at 1 s intervals, an extensive database has been created of the PG and weather conditions at the FSMS.

Published under licence by IOP Publishing Ltd

The mean and standard deviation of the PG were calculated for every 10 minutes and put into a database with the weather conditions. Data was discarded where there was less than 600 seconds of PG data recorded and where the field meter generated an alert signal from its internal operational health monitoring. Data was separated by wind direction into three zones as shown in figure 1.

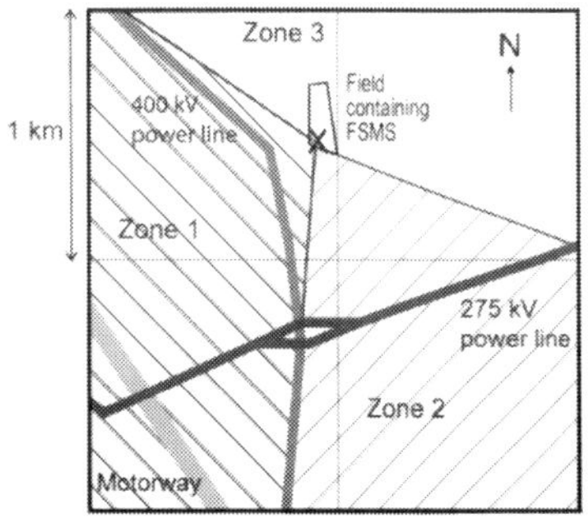

Figure 1. Map showing the position of the FSMS (marked by an X) with respect to a 400 kV power line and a 275 kV power line and the M48 motorway. The FSMS is 175 m from the 400 kV line at its closest point and 750 m from the 275 kV line. The map is split into 3 zones dependent on wind direction. In zone 1, the FSMS is downwind of the 400 kV power line, zone 2, downwind of the 275 kV line and zone 3 upwind of both.

3. Results

3.1. The effect of weather

3.1.1. Rainfall. Rainfall can affect the production of corona ions from HV power lines [3], therefore the PG on dry and rainy days is likely to be different downwind. Days in which there was at least 0.2 mm of rain were separated from days where there was less than 0.2 mm or no rain at all in 2008. There were 150 days with no rainfall, whilst the other 215 days had some rainfall.

The mean and standard deviation for each 10 minute sample were found for each of the three zones. Zone 3 had the smallest difference between rainy and dry days, due to the presence of charged clouds some difference would be expected in all zones. The increase of SD in zones 1 and 2 are expected as rainfall will create space charge. The histograms for zone 1 are shown in figure 2 and using Kolmogorov-Smirnov (KS) tests were shown to be significantly different to each other (at the $p < 0.01$ significance level).

3.1.2. Correlations between weather conditions. The Spearman rank correlation coefficient ρ, was found for the 10 minute mean and SD of PG and the measured weather conditions, all correlations presented here have a probability, p, less than 1% of being caused by chance. No strong correlations are present, but weak correlations are found which while not proof of a power line effect due to the large number of confounders, do give an indication of areas worthy of further investigation.

Temperature does not correlate with any of the PG metrics, other than a very weak correlation between PG SD and temperature in zone 3 ($\rho = -0.21$). Pressure is weakly inversely correlated with SD in zone 1 and (both $\rho = -0.24$), but not zone 3 ($\rho = -0.09$). Low pressure is associated with poor weather, such as increased winds and rainfall, and also storm clouds which have their own charge, so it is likely that these correlations are not due to the power lines. Humidity is inversely correlated with PG mean in all three zones. The correlation is strongest in zone 1 than zone 3 then weakest in zone 2 (ρ = -0.40, -0.31 and -0.27 respectively). Wind speed is weakly correlated with PG SD in zone 1 and (ρ = 0.38 and 0.36) in zone 3 there is near to no correlation with SD. It is important to note that wind speeds can be higher in zone 1.

3.2. The effect of time of day

3.2.1. Distributions at day and night. Mean and SD data from the FSMS were separated into night and day, according to solar radiation data (daytime is when solar radiation > 0), and plotted as histograms (shown for zone 1 in figure 3). The difference between the mean PG at night and day is small in zone

3 yet for zones 1 and 2, the distribution is more negative overnight. For SD, the difference between the distributions in night and day in zone 3 is very small. The variability of PG is consistently greater downwind of a power line during the day than it is during the night, but the difference is small. In zone 2 there is greater difference than in zone 3, where 32.3% of measurements in the day and 38.7% of measurements in the night are found under 20 V m^{-1}. In zone 1 there is greater difference again where 32.3% of measurements in the day and 38.7% of measurements in the night are found under 20 V m^{-1}. All histograms were compared using the KS test and differences were significant to the $p < 0.01$ level.

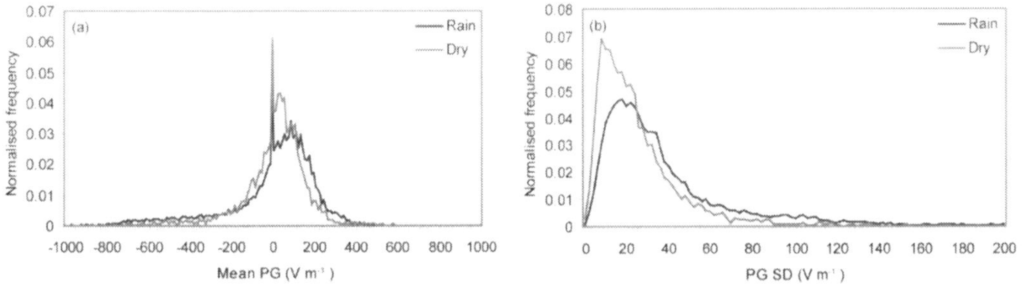

Figure 2. Histograms showing the distributions of mean (a) and SD (b) of PG downwind of a 400 kV power line, separated by rainy or dry days.

3.2.2. Diurnal cycles and the Carnegie curve. Global thunderstorm activity peaks after midday and is at a minimum overnight, GMT. The average global background PG has been shown to correlate well with global thunderstorm activity, and this average curve is known as the Carnegie curve [4].

The mean PG for each hour was averaged throughout 2008 from all three zones. The resulting curves were compared to the Carnegie curve. Only the control data from zone 3 shows a strong correlation, with a Pearson correlation coefficient of $\rho = 0.83$. Zone 2 is weakly correlating ($\rho = 0.52$) while data from zone 1 does not correlate well ($\rho = 0.29$, $p = 0.18$). The average curves of zones 1 and 2 are weakly correlating to zone 3 ($\rho = 0.56$ and 0.50 respectively, $p \sim 0.01$) but the correlation between 1 and 2 is weak and insignificant ($\rho = 0.37$, $p = 0.07$).

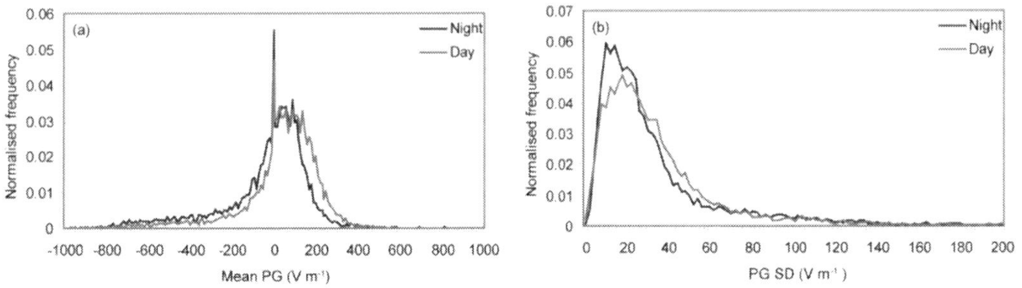

Figure 3. Histograms showing the distributions of mean (a) and SD (b) of PG downwind of a 400 kV power line, separated by day and night.

4. Discussion and conclusions

4.1. Natural sources of PG fluctuation

There are several natural sources of space charge that need to be considered when interpreting the PG measurements downwind of HV power lines. Clouds can often carry charge and can cause changes of PG over several seconds or minutes of up to a few thousand V m^{-1} in extreme cases. Without measurements at the opposite side of the power line, it is impossible to separate the contribution to PG from clouds and from corona ions (though times during the day when there is a lot of cloud cover can be ascertained using solar radiation measurements). The assumption is made in this work that there is not a significantly different amount of PG disruption from charged clouds from south westerly winds, as the distribution of PG measured for all wind directions at the Reading University Atmospheric Observatory [5] do not show the extreme values of PG shown in the distributions.

The FSMS is positioned to the east of the mouth of the River Severn, and is north of the motorway. Both traffic and water can produce ions, but measurements of space charge from water and traffic show that PG modifications are greatly reduced after 2 km [6] making the effect at the FSMS small.

4.2. Weather conditions for the highest corona ion emission

These findings show that the highest levels of corona emission occur in damp atmospheres, whether it is raining or there is very high humidity, or when the wind speed is high. The PG readings at these points would tend to be more negative than average PG values.

Previous results from real DC and AC lines [7] have shown an increase in negative corona ion emission with increasing humidity. However, humidity is often higher overnight when the atmosphere is more stable, and space charge could be trapped in a layer near the ground, increasing the space charge density. The other confounder is that very high humidities are associated with disturbed weather, which can affect atmospheric PG.

As wind speed increases, lighter negative ions produced after corona avalanching are more likely to be 'swept away' by the wind, suppressing the creation of positive corona streamers [8]. This would result in more negative corona loss. Higher wind speeds downwind of the 400 kV line at the FSMS showed median values reducing with increasing wind speed [2], so as predicted, more negative corona ions are escaping the line. The turbulence in the atmosphere increases with increasing wind speed, so the result that higher wind speeds are correlated, albeit weakly, with increased standard deviation of 10 minute samples of PG downwind of both power lines at the FSMS is unsurprising. There is no correlation in zone 3, as there is less space charge in the mixing layer of the atmosphere, the more turbulent atmosphere will not affect PG readings.

4.3. Time of day for the highest corona ion emission

For measurements in zone 3, the results correlated very well with the natural Carnegie curve that is expected during fair weather conditions, measurements from zone 1 and zone 2 did not correlate with the Carnegie curve. The space charge produced from the power lines is high enough to mask the natural Carnegie curve. Overnight, the mean value of PG in zones 1 and 2 is negative for some hours. The most likely explanation for the increase in negative space charge overnight is the increase in humidity, which has been shown to be associated with negative fields.

Acknowledgements

This work was funded by Children with Leukaemia, registered charity number 298405.

References

[1] Fews A P, Wilding R J, Keitch P A, Holden N K and Henshaw D L 2001 *Atmos. Res.***63**(3-4) 271-89
[2] Matthews J C, Ward J P, Keitch P A and Henshaw D L 2010 *Atmos. Environ.* **44** 5093 - 100
[3] Kirkham H 1980 *J. Appl. Meteorol.* **19** 35-40
[4] Harrison R G 2005 *Surv. Geophys.* **25** 441-84
[5] Bennett A J and Harrison R G 2008 *J. Phys. Conf. Ser.* **142** 012046
[6] Muir M S 1977 *J. Atmos. Terr. Phys.***39** 1341-6
Israelsson S amd Lelwala R 1999 *Atmos. Res.* **51**(3-4) 301-7
[7] Hendrickson R C 1986 *Bioelectramagnetics* **7**(4) 369-79
Groom K N and Chalmers J A 1967 *J. Atmos. Terr. Phys.* **29** 613-5
[8] Deno D W and Comber M G 1982 Corona phenomena on AC transmission lines *Transmission Line Reference Book, 345 kV and Above* (Palo Alto: EPRI) chapter 4, pp 122-48

Journal of Physics: Conference Series **301** (2011) 012006 doi:10.1088/1742-6596/301/1/012006

The electrostatic environments of Mars and the Moon

C I Calle

Electrostatics and Surface Physics Laboratory
NASA, Kennedy Space Center, FL 32899

e-mail: carlos.i.calle@nasa.gov

Abstract. The electrical activity present in the environment near the surfaces of Mars and the moon has very different origins and presents a challenge to manned and robotic planetary exploration missions. Mars is covered with a layer of dust that has been redistributed throughout the entire planet by global dust storms. Dust, levitated by these storms as well as by the frequent dust devils, is expected to be electrostatically charged due to the multiple grain collisions in the dust-laden atmosphere. Dust covering the surface of the moon is expected to be electrostatically charged due to the solar wind, cosmic rays, and the solar radiation itself through the photoelectric effect. Electrostatically charged dust has a large tendency to adhere to surfaces. NASA's Mars exploration rovers have shown that atmospheric dust falling on solar panels can decrease their efficiency to the point of rendering the rover unusable. And as the Apollo missions to the moon showed, lunar dust adhesion can hinder manned and unmanned lunar exploration activities. Taking advantage of the electrical activity on both planetary system bodies, dust removal technologies are now being developed that use electrostatic and dielectrophoretic forces to produce controlled dust motion. This paper presents a short review of the theoretical and semiempirical models that have been developed for the lunar and Martian electrical environments.

1. Introduction

Mars and the Earth's moon have very different surface and atmospheric environments. They, however, have one thing in common: their surfaces are covered with a relatively homogeneous layer of dust. This surface dust dominates the electrostatic environment of Mars and plays an important role in that of the moon.

These electrostatic environments are far from being fully understood. The desire for their study goes beyond scientific inquiry. Human and robotic exploration missions to both bodies require a good understanding of this environment to avoid failures and mishaps.

2. Lunar and Martian Electrostatic Environments

2.1. The Moon

At 10^{-12} mbar, the lunar atmosphere is extremely rarefied and its influence in the electrostatic environment near the surface of the moon is negligible. This environment is instead controlled by the flux of photons and ions from the solar wind, cosmic rays, and galactic cosmic rays. The lunar surface is composed of rocks and granular material with grains and dust particles ranging in size from a few centimeters down to a few nanometers. This granular material is expected to be electrostatically charged due to the incident plasma and to the UV from the sun, which releases photoelectrons from the surface of the material.

The plasma and photon fluxes bathing the moon produce a complex electrostatic environment. On the sunlit side, photoelectric charging by solar UV photons dominates. The emission of photoelectrons leaves the surface positively charged to a potential of about 5 to 10 V [1]. On the dark side, plasma electrons dominate and the surface becomes negatively charged to a negative potential or the order of the electron temperature, about -50 to -100 V [2]. The moon is essentially a charged body in a plasma, creating a screening effect in the plasma with a characteristic distance or Debye length of the order of meters on the lunar dayside and of the order of kilometers on the dark side. A Debye sheath is then formed around the moon which reaches a meter or so on the dayside and several kilometers on the nightside [3,4]. The dayside Debye sheath, dominated by photoelectrons, is usually called the photoelectron sheath.

Lunar surface potentials depend greatly on changes in the solar wind and cosmic rays, which affect the plasma environment. When the moon crosses the Earth's plasma, negative surface potentials can reach several kV. Surface potentials reaching -5 kV were observed by Lunar Prospector during periods of intense solar activity [5].

In our airless moon, it would seem likely that the dust on its surface be relatively static. However, the complex electrostatic environment just described affects dust behavior in unexpected ways. There appears to be evidence of dust levitation and dust transport on the moon. This issue has been controversial since it was first raised during the Apollo missions [6]. A horizon glow was reported by Apollo astronauts [7] (Fig. 1). This glow has been interpreted as evidence of transient dust clouds above the lunar surface that could reach several kilometers above the surface. Observations with the lunar Surveyor spacecraft [8] and the Lunar Ejecta and Meteorites Experiment (LEAM) on Apollo 17 indicated the presence of dust clouds [9]. Although no theoretical model satisfactorily explains the phenomenon, it has been suggested that electrostatic charging of the lunar surface due to exposure to charged particles from the solar wind as well as UV radiation could result in the levitation and transport of dust particles [10].

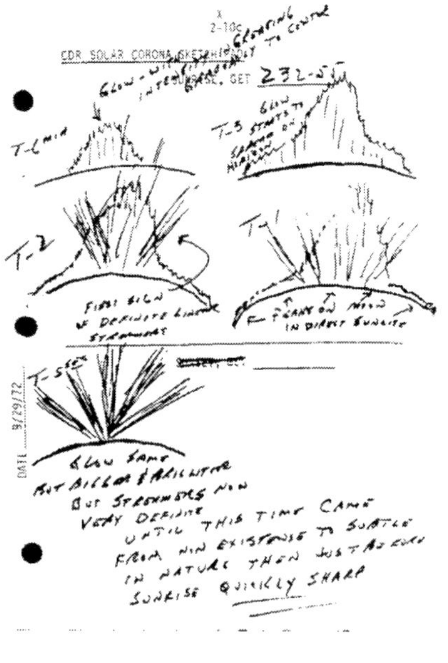

Figure 1. Sketches done from lunar orbit by Apollo astronauts showing streamers reaching up to 100 km from the surface. [Courtesy NASA]

2.2. Mars

2.2.1 *Particle Size Distribution in the Martian Atmosphere*

The surface of Mars is covered with a layer of dust that has been somewhat homogenized by planet-wide dust storms. These dust storms, which have wind velocities up to 32 m/s, have been observed to last for several months (Fig. 2) [11]. Dust devils up to 10 km high were detected during the 1997 Mars Pathfinder mission with daily frequency. NASA's Mars rover Spirit has photographed dozens of dust devils from its location inside the red planet's Gusev Crater (Fig. 3). Opportunity, the mission's second rover, had not been able to capture an image of a dust devil until this past July, six years after its arrival on Mars. Its location, on Meridiani Planum, is not as dusty. NASA's Phoenix Mars Lander also photographed dust devils in the Martian arctic during its 2008 mission.

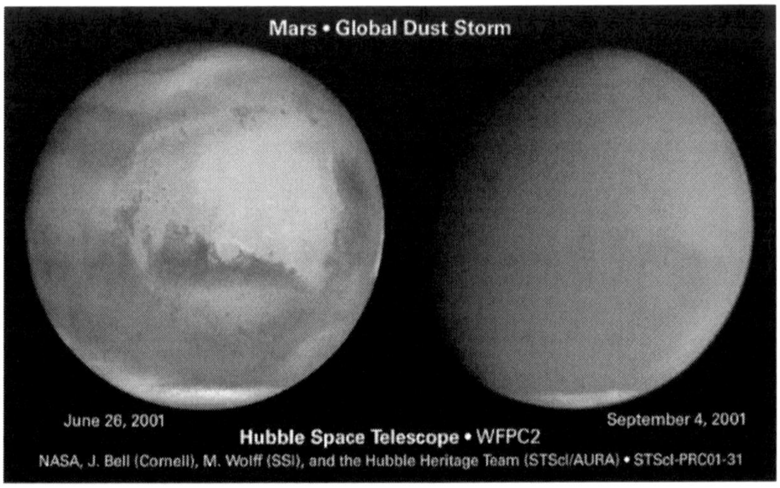

Figure 2. A Martian global dust storm [Courtesy NASA].

These storms, along with the frequent dust devils, carry dust particles that are uplifted into the thin atmosphere of the planet. The actual mechanism that uplifts dust into the atmosphere is still nor clearly understood. From the knowledge of the terrestrial mechanism, it has been proposed that saltation aided by dust devils and the presence of electric fields may uplift dust [12, 13]. From optical data using instruments on orbiting spacecraft and on landers, estimates of the size distribution of dust particles in the atmosphere have been made. These measurements yield a value for particle diameter of 1.5 ± 0.2 µm. However, the size distribution of atmospheric dust changes with the presence of dust devils and dust storms, as one would expect. At the peak of the 2001 global dust storm depicted in Figure 2, particle diameters derived from data from the Thermal Emission Spectrometer on board the Mars Global Surveyor orbiting spacecraft ranged from 2 to 5 µm [14].

Figure 3. A dust devil photographed by the Spirit rover the 488 day of the mission [Courtesy NASA]

Using measurements by the Microscopic Imager (MI) on the Mars Exploration Rovers, Landis *et al* seem to find evidence for a three-component particle distribution for atmospheric dust: *Atmospheric dust* suspended for long periods of time, with diameters in the 2 to 4 µm; *settled dust* raised into the atmosphere by wind and dust devils, with diameters in the 10µm and under range; and *saltating particles*, with diameters greater than 80 µm, that move due to saltation [15].

Particle size distribution data for the Martian soil has been obtained by the Mars Exploration Rovers Spirit and Opportunity at the locations that the two craft have visited. These values are however somewhat skewed because the instrument used to measure them, the Microscopic Imager, has a resolution of 31 μm per pixel, which limits the detection of very small particle sizes. The MI on Spirit has recently completed measurements of the soil on Scamander Crater, obtaining average particle sizes of 220 μm [16].

2.2.2 *Electrostatic Charge of Martian Dust Particles*

Stationary surface sand and dust on Mars may be electrostatically charged due to incident UV radiation reaching the surface. Although the total integrated UV flux over 200-400 nm is comparable to Earth's, shorter wavelengths contribute a larger proportion of this flux [17]. Contact charging may also occur due to collisions between wind-blown dust particles and stationary surface particulate matter.

There is experimental evidence that collisions due to dust motion can result in electrostatic charging of dust particles. In the early 1970s, Eden and Vonnegut observed a glow as well as filamentary discharges as a container with sand was shaken under Martian atmospheric conditions [18]. Shortly after, Mills suggested that triboelectrically produced glow discharges in the Martian atmosphere might explain the apparent absence of carbonaceous matter on the planet [19]. More recently, we were able to observe glow discharges caused by collisions of Mars simulant dust particles under atmospheric conditions mimicking those of Mars and showed that these discharges altered known organics on Mars simulant dust [20]. Fabian et al and Krauss *et al* designed experiments to demonstrate that the vertical motion of dust is sufficient for this charging mechanism to produce strong electric fields capable of creating electrical discharges in a low-pressure CO_2 atmosphere [21, 22]. They were able to detect discharges both visually and electronically and measured the discharge frequencies and intensities. The range of pressures required to efficiently produce these discharges was also examined, demonstrating that electrical discharges can occur under conditions expected on the Martian surface. Additionally, a simple theoretical model has been created to constrain the parameters involved in the dust charging experiment. Their model supports the ideas developed in their experiment. Similar theoretical calculations had been performed earlier by Melnik and Parrot [23].

Accurate determinations of the charge and particle size distribution for particles in the Martian atmosphere could be obtained when dedicated instrumentation is flown on a future mission. With that in mind, our laboratory, in collaboration with the University of Arkansas, developed a Dust Particle Analyzer instrument that is capable of determining the electrostatic charge and the aerodynamic diameter of dust particles in the Martian atmosphere [24]. Aerodynamic diameters and electrostatic charge values of Martian simulant dust particles in a vacuum chamber at simulated Martian atmospheric conditions are shown in Fig. 4.

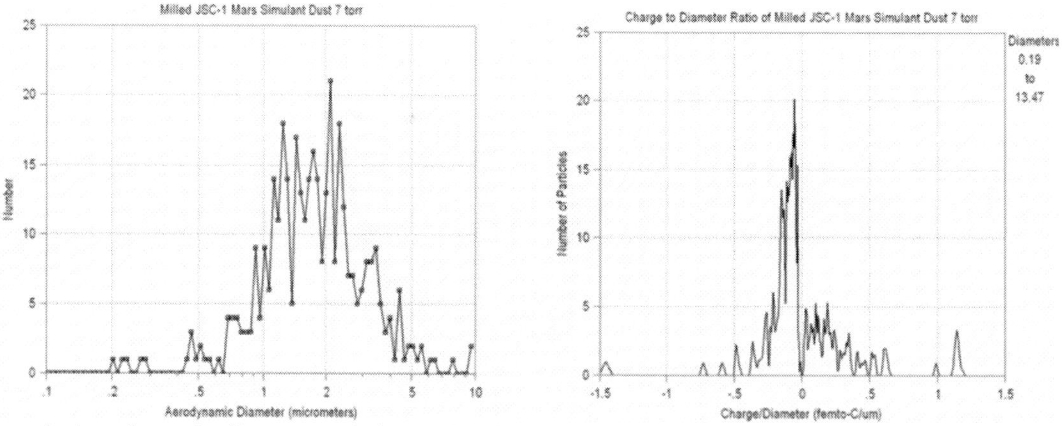

Figure 4. Aerodynamic diameter and charge density of Martian simulant dust particles measured with NASA's Dust Particle Analyzer under simulated Martian atmospheric conditions.

3. Challenges and Possibilities for Exploration Missions

The complex electrostatic environments of Mars and the moon present challenges for human and robotic exploration. On both surfaces, dust can deposit on the surfaces of solar cells, equipment, thermal radiators, and other hardware likely to be used in exploration missions. Lunar dust is expected to be electrostatically charged by the incident plasma and by the solar UV. Dust in the Martian atmosphere is also expected to be charged by dust collisions in dust devils and dust storms.

The presence of electrostatically charged dust on Mars and the moon also presents opportunities to enable exploration missions. Our laboratory has been developing a dust mitigation technology that takes advantage of these electrostatic environments [25-27]. The Electrodynamic Dust Shield (EDS), a system based on the generation of changing non-uniform electric fields able to accelerate charged dust particles with the dielectrophoretic force, has proven to be extremely efficient at clearing dust at lunar and Martian environmental conditions. In the EDS, sets of independent electrodes sandwiched between two dielectric layers to form a multi-layer coating, generate non-uniform electric fields that move across the surface when out of phase signals are applied to the electrodes (Figure 5). Charged dust particles on the surface are carried along with the traveling field. The EDS coating can be applied to metallic and electrically insulating surfaces.

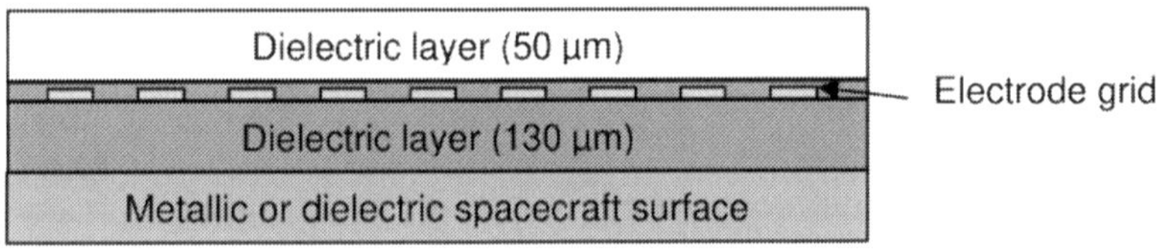

Figure 5. Schematic diagram of the multi-layer Electrodynamic Dust Shield coating for metallic or dielectric surfaces.

The EDS has been implemented on solar panels, optical instruments, viewports, visors, thermal radiators, as well as on flexible materials and fabrics. For applications requiring transparent EDS coatings (solar panels, optical instruments, habitat viewports, and astronaut visors) indium tin oxide transparent electrodes with transparent dielectrics have been successfully developed and tested under Martian and lunar simulated environmental conditions (Figure 6). Inclusion of this technology along with other instrumentation for the accurate measurements needed will enhance and enable future missions.

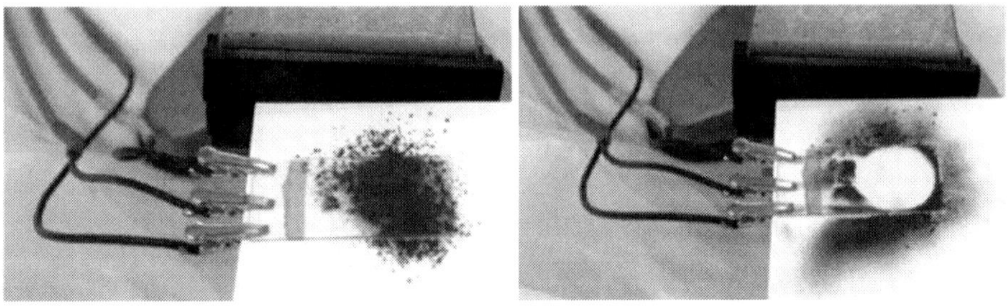

Figure 6. Photographs of a transparent EDS on glass before (left) and after dust removal (right).

References

[1] Manka R H 1973 Plasma and potential at the lunar surface *Photon and particle interactions with surfaces in space* ed R J L (NY: Grard Springer) pp 347-361

[2] Halekas J S, Mitchell R P, Lin L L, Hood M, Acuña H and Binder A 2002 Evidence for negative charging of the lunar surface in shadow *Geophys Res. Lett.* **29** 1436

[3] Singer S F and Waker E H 1962 Photoelectic screenign of bodies in interplanetary space, *Icarus* **1** 7-12

[4] Freeman J W and Ibrahim M 1975 Lunar electric fields, surface potential and associated plasma sheaths *Moon* **14** 103-114

[5] Halekas J S, Bale S D, Mitchell D L, and Lin R P 2005 Electrons and magnetic fields in the lunar plasma wake *J. Geophys. Res.* **110** A07222
[6] Gold T and Williams G J 1973 Electrostatic Transportation of Dust on the Moon *Photon and Particle Interactions with Surfaces in Space* ed R J L Grard (Reidel: Dordrecht) pp 557-560
[7] Zook H A and McCoy J E 1991 Large Scale Lunar Horizon Glow and a High Altitude Lunar Dust Exosphere *Goephys. Res. Lett.* **18**, 2117
[8] Rennison J J and Criswell D R 1973 Surveyor Observations of Lunar Horizon Glow *The Moon* **10**, 121
[9] Berg O E Wolf H and Rhee J 1975 Lunar Soil Movement Registered by the Apollo 17 Cosmic Dust Experiment *Interplanetary and Zodiacal Light* 233, Springer-Verlag, Heidelberg
[10] Horanyi M 1998 Electrostatic Charging of Lunar Dust *LPS* **29**, 1527
[11] Gaier J R, Perez-Davis, M E, Marabito, M 1990 Aeolian Removal of Dust From Photovoltaic Surfaces on Mars *NASA Technical Memorandum* 102507
[12] Reno N O and Kok J F 2008 Electrical activity and dust lifting on Earth, Mars and beyond, *Space Sci. Rev.* **137** 19-134
[13] Greely R *et al* 2003 Martian dust devils *J. Geophys. Res.* **108** 5041
[14] Wolf and Clancy 2003 Constraints on the size of Martian aerosols from Thermal Emission Spectrometer observations, *J. Geophys. Res. (Planets)* **108** 5097
[15] Landis G A, Herkenhoff K, Greeley R, Thompon S, Whelley P and the MER Athena Science Team 2006 Dust and sand deposition on the MER solar arrays as viewed by the Microscopic Imager *LPS* **37** 1937
[16] Siebach, Ardidson, Cabrol, and the Athena Science Team 2010 Recent Spirit results: Microscopic Imager analysis of particle properties in Scamander Crater, west of home *LPS* **41** 2548
[17] Cockell CS, Catling D, Davis WL, Kepner RN, Lee PC, Snook K, McKay CP 2000 The ultraviolet environment of Mars : biological implications past, present and future *Icarus* **146** 343-359
[18] Eden H F and Vonnegut B. 1973 Electrical breakdown caused by dust motion in low-pressure atmospheres: Considerations for Mars *Nature* **280** 962
[19] Mills A, 1977 Dust clouds and frictional generation of glow discharges on Mars *Nature* **268** 614
[20] Hintze P E, Buhler C R, Schuerger A C, Calle LM, and Calle C I 2010 Alteration of five organic compounds by glow discharge plasma and UV light under simulated Mars conditions *Icarus* **208** 749-757
[21] Fabian A, Krauss C, Sickafoose, A, Horanyi M, Robertson S 2001 Measurementsof electrical discharges in Martian regolith simulant *IEEE Trans. Plasma Sci.* **29** 288–291
[22] Krauss C E, Horanyi M, Robertson S 2006 Modeling the formation of electrostatic discharges on Mars *Journal of Geophysical Research-Planets* **111** E02001
[23] Melnik O and Parrot M Electrostatic discharge in martian dust storms 1998 *J. Geophys. Res.* **29** 117
[24] Buhler C R, Mazumder M K, Calle C I, Clements J S, Srirama P K and Chen A 2008 Development of the dust particle analyzer, *Proceedings of the Pan Pacific Imaging Conference* June 2008 Tokyo (Japan)
[25] Calle C I *et al*, 2010 Dust removal technology demonstration for a lunar habitat, *AIAA Space 2010* August 2010 Anaheim, CA (USA)
[26] Calle C I, Buhler C R, McFall J L and Snyder S J 2009 Particle removal by electrostatic and dielectrophoretic forces for dust control during lunar exploration missions, *J. Electrostatics* **67** 89–9
[27] Calle C I, Arens E E, McFall J M, Buhler C R, Snyder S J, Geiger J K, Hafley R A, and Taminger K M 2009 Reduced gravity flight demonstration of the dust shield technology for optical systems, *IEEE Aerospace Conference* March 2009 Big Sky MT (USA)

Measuring Martian lightning

K L Aplin[1,2], C J Davis[2,3], W J Bradford[2] and K L Herpoldt[1,2]

1. Department of Physics, University of Oxford, Denys Wilkinson Building, Keble Road, Oxford OX1 3RH UK

2. RAL Space, R25, Harwell Oxford, Didcot OX11 0QX UK

3. Department of Meteorology, University of Reading, PO Box 243, Reading RG6 6BB UK

k.aplin1@physics.ox.ac.uk

Abstract. Lightning is expected to occur on Mars within dust devils and dust storms, which are likely to discharge in the low pressure carbon dioxide environment. Despite this, radio emissions from Martian lightning have not yet been conclusively observed by remote sensing, nor have there been any in situ measurements of Martian atmospheric electricity. We report laboratory experiments to simulate Martian electrical discharges and measure the radio emissions from them, to facilitate searches of Martian lightning from spacecraft data. Voltage transients were observed in a tank in which Martian analogue simulant was allowed to become triboelectrically charged, and then fall to a sensing electrode. A plausible explanation for our results is that electrical discharges have taken place, caused by charge separation on differently sized particles as predicted by theory. Unlike in previous reports, we have been able to detect discharges without adding glass microballoons to the Martian analogue dust to facilitate the charging, although the addition of glass microballoons did enhance the transient rate.

1. Introduction

Lightning has long been studied on Earth and more recently has been observed on other planets throughout the Solar System. Lightning is expected to occur on Mars because of triboelectrification within regional-scale dust storms and localised dust devils [1]. Terrestrial dust devils become highly electrically charged, up to ~100kVm^{-1} [2], and based on the Paschen curve for 10mbar CO_2, discharges are expected from Martian dust devils at ~20kVm^{-1}. Martian lightning is expected to play a role in the production of trace organic molecules that may be important for biological processes [3]. Despite this clear scientific significance, there have never been any electrical measurements in the Martian atmosphere, and neither do we know of any forthcoming missions carrying relevant instrumentation. We are therefore investigating techniques for remote sensing of radio emissions generated by Martian electrical discharges. This has been attempted previously, but with no conclusive results [2]-[5]. Our work is in two parts, laboratory experiments to characterise the optical and electrical signals expected, and analysis of spacecraft data to search for electromagnetic signals emitted from the Martian atmosphere.

2. Laboratory simulations of the Martian atmosphere

2.1. Experiment

We have set up a laboratory experiment intended to generate and detect electrical discharges in a Martian analogue atmosphere, based on [6]. The experiment uses a cylindrical Perspex tank approximately 1m long of diameter 0.25m. In the top of the tank there is a funnel sealed by a plunger, which is accessible from outside the tank. The funnel is filled with 500g JSC Mars-1 Martian regolith simulant [7], (baked to remove adsorbed water) and the tank is evacuated and filled with 7-9 mbar CO_2. The plunger is then released and the dust falls to the bottom of the tank through another inverted funnel. The top metal plate is electrically grounded, and the voltage on the bottom metal plate is measured with a Keithley 2000 DMM, with the voltage logged to a PC via an IEEE-488 interface. The hole between the funnels, through which the dust falls, is 2.5cm diameter, wide enough for most of the falling dust to pass through without touching it. This ensures the majority of the dust charging occurs purely from dust-to-dust contact, although there will be some charging from the dust touching the funnel walls. The tank is mounted on a support frame so that after each dust drop it can be rotated to reload the funnel, Figure 1.

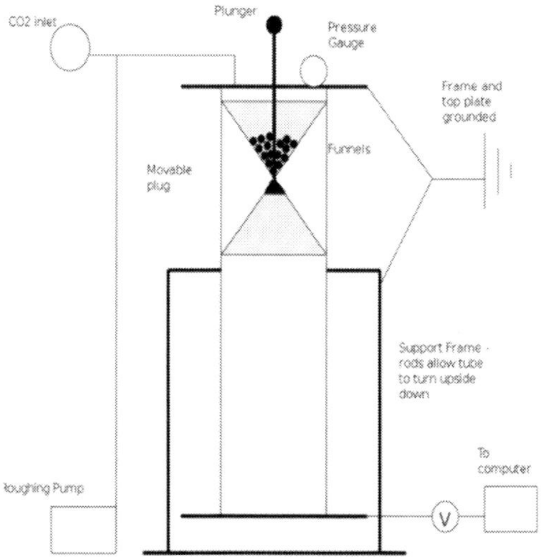

Figure 1 Martian analogue experiment, showing dust loaded before a drop. The CO_2 line was disconnected before a dust drop and subsequent rotation of the tank to reload the dust.

In [6] discharges in the dust were detected, by spikes in the voltage recorded at the sensing plate, and also visually when in a dark room. However, discharges only occurred when the JSC Mars-1 simulant was mixed with glass microballoons, implying that the potentials generated by dropping Mars simulant dust alone were not adequate for discharges.

2.2. Results

In total 22 dust drops were carried out for which data were recorded, 13 drops with dust only, and 9 drops with dust plus microballoons. The results were repeatable, with examples shown in Figure 2. All dust drops, activated by pressing down the plunger to release the dust, took place after 100s logging to record the background noise level of the system, ±1V. The dust drops (Figure 2(a)) show a sharp positive peak of ~12V, followed by a few ~2-4V negative spikes, then a slow smooth decay of the voltage. The results with microballoons (Figure 2(b)) are different, always commencing with a

Journal of Physics: Conference Series **301** (2011) 012007 doi:10.1088/1742-6596/301/1/012007

negative transient of ~10V which then appears to be effectively neutralised by numerous bipolar spikes.

The positive transient on releasing the Martian analogue dust is consistent with the larger size fraction, which is expected to become positively charged with respect to the smaller particles [8], arriving at the bottom of the tank first. The smaller dust particles are expected to be negatively charged and settle slowly, which explains the gentle decay of the positive voltage in Figure 2(a). The negative spikes in Figure 2(a) are too large to be caused by any single particle (which would have to be very large, and therefore would be expected to be positive in any case), and are well above the system noise level. Figure 2(b) shows many more transients, but with no overall decay, as in 2(a). The differences between Figure 2(a) and (b) and the previous observations of discharges when microballoons were added suggests that the spikes are discharges caused by charge separation between the falling particles.

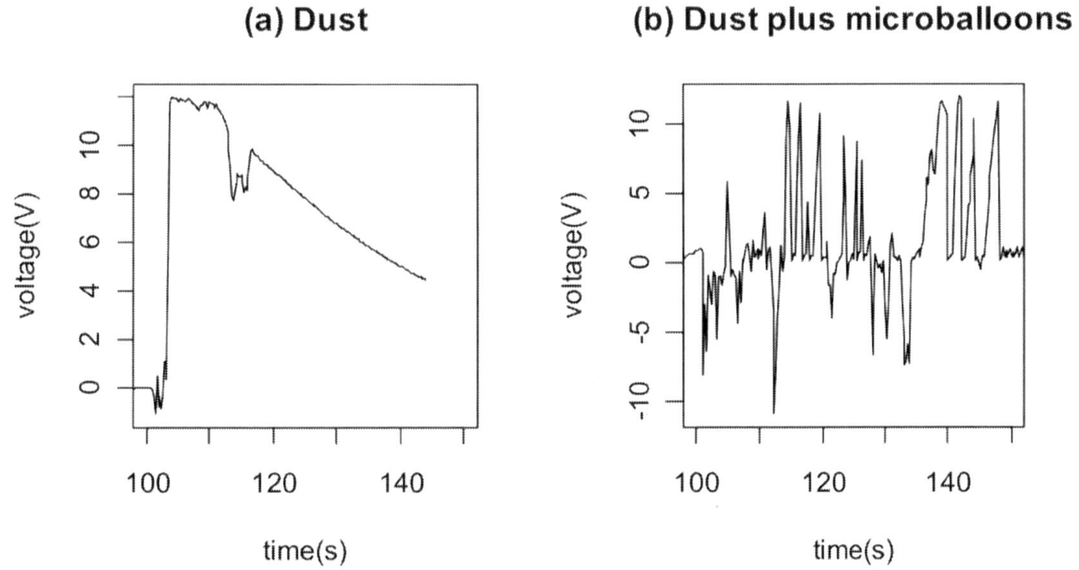

Figure 2 Typical dust drop results at 9mbar (a) ~490g dust only, representative of 12 out of 13 drops (b) ~250g dust with ~230g microballoons, representative of 8 out of 9 drops.

The differences between Figure 2(a) and (b) can be explained by considering that the glass microballoons are higher in the triboelectric series than the JSC Mars-1 simulant (which is located between stainless steel and Teflon [9]), and therefore likely to charge positively with respect to the dust. Firstly, this implies that the Martian analogue dust is likely to become negatively charged when the microballoons are present, causing negative initial transients, as the larger dust particles fall to the sensing electrode first, Figure 2(b). Secondly, the minimum electric field for a gaseous electrical discharge V_{crit} for a material of ionization energy E_i and with mean free path λ is given in [10] as

$$V_{crit} \equiv 2\left(\frac{E_i}{e\lambda}\right) \qquad (1).$$

Since the glass microballoons reduce the mean radius of the particle distribution, this will increase the mean free path and reduce V_{crit}, increasing the likelihood of discharges in the glass microballoon mix compared to the dust only case. Although quantitative estimates are difficult without accurate measurements of the particle size distribution, this is consistent with the observations.

These results are broadly consistent with [6], although we observe voltage spikes in our dust-only drops as well as the glass microballoon-dust mixture. Unlike [6] however, a faint purple glow was only seen in one of the dust only cases, but this could have been because the experiments were carried out in a well-lit room. Electrical discharges still remain the most plausible explanation for the voltage spikes seen.

3. Detecting radio emissions from Martian electrostatic discharges

The presence of electrical discharge on Mars has been inferred from an observed increase in the non-thermal microwave emission during a Martian dust storm [3], but no evidence of electrical discharge associated with dust storms was observed when examining the background radio noise between 4.0 and 5.5 MHz in data from the MARSIS instrument onboard the Mars-Express spacecraft [4]. Although the MARSIS instrument would easily detect lightning discharges of similar intensity to those found on Earth [4] this does not rule out the possibility of discharges that emit a signature at lower frequencies (such as sprites) [11] or much higher frequencies (such as coronal discharge) [5], which we may have observed in our laboratory experiments.

4. Conclusions

Our results suggest that electrical discharges have been observed in Martian analogue dust in a simulated Martian atmosphere of 9mbar CO_2. The effects observed are consistent with the theory of triboelectric particle charging, for example, the number of discharges observed increases when glass microballoons are added to the mixture. Future experiments will be carried out in a dark room, to facilitate optical detection, and with a radio spectrum analyzer to characterize the discharges. Radio observation of corona discharges will enable targeted spacecraft observations to be made.

References

[1] Aplin K L 2006 Atmospheric electrification in the Solar System *Surv. Geophys.* **27**, 1, 63-108
[2] Jackson T L and Farrell W M 2006, Electrostatic fields in dust devils: an analog to Mars, *IEEE Trans. Geoscience Remote Sensing*, **44**, 10, 2942-2949
[3] Ruf C, Renno N, Kok J, Bandelier E, Sander M, Gross S, Skjerve L, and Cantor B 2009 Emission of non-thermal microwave radiation by a Martian dust storm *Geophys. Res. Lett*, **36**, L13202
[4] Gurnett D, Morgan DD, Granroth L J, Cantor B A, Farrell W M and Espley J R 2010, Non-detection of impulsive radio signals from lightning in Martian dust storms using the radar receiver on the Mars Express spacecraft, *Geophys. Res. Letts.* **37**, L17802
[5] Zarka P, Farrell W M, Fischer G and Konovalenko A, 2008 Ground- and space-based radio observations of planetary lightning *Space Sci Revs.* **137**, 1-4, 257-269
[6] Krauss C E, Horanyi M and Robertson S 2003 Experimental evidence for electrostatic discharging of dust near the surface of Mars *New. J. Phys* **5**, 70.1-70.9
[7] Allen C, Morris R, Lindstrom D, Lindstrom M, and Lockwood J 1997 *Lunar and Planetary Institute Sci. Conf. Abs.*, **28**, 27.
[8] Lacks D and Levandovsky A 2007 Effect of particle size distribution on the polarity of triboelectric charging in granular insulator systems *J. Electrostatics* **65**, 2, 117-112
[9] Sharma M, Clark D W, Srirama P K and Mazumder M K 2008, Tribocharging characteristics of the Mars dust simulant (JSC Mars-1) *IEEE Trans. Industry Apps.*, **44**, 1, 32-38
[10] Treumann R A, Klos Z and Parrot M, 2008 Physics of electric discharges in atmospheric gases, *Space Sci. Revs.* **137**, 1-4, 133-148
[11] Farrell W M, Kaiser M, Desch M, Houser J, Cummer S, Wilt D and Landis G 1999 Detecting electrical activity from Martian dust storms *J. Geophys. Res.*, **104**, 3795–3801

Journal of Physics: Conference Series **301** (2011) 012008 doi:10.1088/1742-6596/301/1/012008

Asteroid electrostatic instrumentation and modelling

K L Aplin[1,2], N E Bowles[1], E Urbak[1], D Keane[2,3] and E C Sawyer[2]

1. Department of Physics, University of Oxford, Denys Wilkinson Building, Keble Road, Oxford OX1 3RH UK

2. RAL Space, R25, Harwell Oxford, Didcot OX11 0QX UK

3. Murray Edwards College, New Hall, University of Cambridge, Cambridge CB3 0DF UK

k.aplin1@physics.ox.ac.uk

Abstract. Asteroid surface material is expected to become photoelectrically charged, and is likely to be transported through electrostatic levitation. Understanding any movement of the surface material is relevant to proposed space missions to return samples to Earth for detailed isotopic analysis. Motivated by preparations for the Marco Polo sample return mission, we present electrostatic modelling for a real asteroid, Itokawa, for which detailed shape information is available, and verify that charging effects are likely to be significant at the terminator and at the edges of shadow regions for the Marco Polo baseline asteroid, 1999JU3. We also describe the Asteroid Charge Experiment electric field instrumentation intended for Marco Polo. Finally, we find that the differing asteroid and spacecraft potentials on landing could perturb sample collection for the short landing time of 20min that is currently planned.

1. Introduction

The work presented here was motivated by the study of an Asteroid Charge Experiment (ACE) for the Marco Polo mission. Marco Polo is a sample return mission currently under consideration by the European Space Agency, and, as the sampling process and surface material are both likely to be influenced by electrostatics, the ACE instrument was proposed to measure electric fields at the asteroid visited by Marco Polo.

By analogy with the Moon, there is likely to be considerable photoelectric charging of asteroid surfaces. Electrostatic dust levitation has been proposed as a method to redistribute particles [1,2], and also as a loss mechanism for smaller particles, not bound by the gravitational field of the asteroid [1]. Asteroid electric charge has never been measured, but simple estimates predict that an electric potential (~1 kV) can be attained on the dark side compared to the sunlit side, which becomes slightly positively charged by photoelectron emission. These differences are enhanced further at the terminator (the day/night boundary), when fields could reach ~100-300 kV/m [1].

Section 2 describes electric field modelling for a real example, the Itokawa asteroid, visited by the Japanese Hayabusa mission in 2005, and for which a full shape model was obtained. The ACE instrument specification is described in section 3.

2. Modelling the electrical environment of an asteroid

The most electrically significant regions of an asteroid are expected to be solar irradiation boundaries, i.e. the terminator, and shading caused by local topography. Electrostatic modeling was used to confirm the ACE electric field instrumentation requirements. The Marco Polo baseline asteroid (1999JU3) is not characterised in detail, however a full shape model for another asteroid, Itokawa, is available [4]. 1999JU3 and Itokawa are similar (Table 1), but asteroids like Itokawa are expected to contain iron, unrepresentative of the electrical conditions on 1999JU3, a carbonaceous asteroid. The Itokawa shape model and orbital parameters have therefore been assumed, with the electrical properties of a typical carbonaceous asteroid.

Three finite element analysis electrostatic modelling packages were used, FEMM and Ansoft Maxwell SV for 2D calculations, and ALGOR FemPro for 3D. The results were insensitive to the range of electrical conductivities and dielectric permittivities expected for a carbonaceous asteroid ($0.113<\varepsilon_r<112.9$ and $10^{-9}<\sigma<10^{-3}$ S/m), therefore values in the middle of these range were selected. The dayside potential was set at +5V and the nightside at -1000V [1] (these values are related to the orbital distance, and insensitive to the composition of the asteroid). Firstly, horizontal electric fields across the asteroid surface at the terminator for a cross-section through a spherical asteroid were determined using two different packages, for verification. For a spherical asteroid, the maximum terminator field ~50kV/m and electric fields at the sub-solar point ~50V/m, consistent with analytical results [1]. Following this, a 3D model was established based on the Itokawa shape model. This proved useful for visualization of the potential (Figure 1), but the mesh resolution limited electric field calculations.

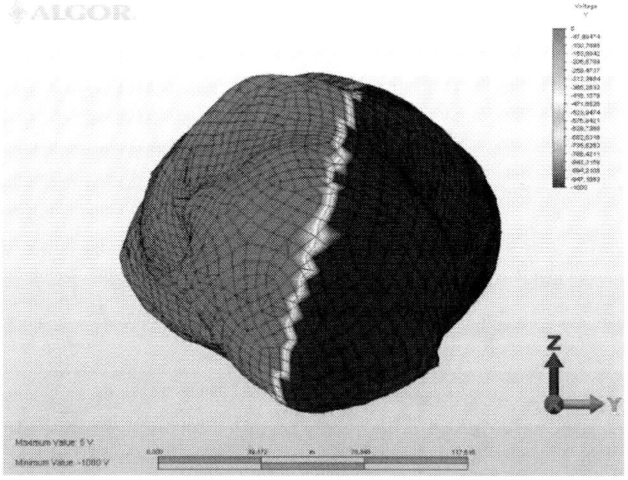

Figure 1 3D model results for Itokawa, showing local surface potential and including the terminator region (Red (on left): +5V (dayside), blue (on right): -1000 V(nightside)

Table 1. Comparison of 1999JU3, asteroid proposed for the Marco Polo mission, and Itokawa, whose shape model is used here. 1 Astronomical Unit (AU) is the distance from the Sun to the Earth.

Name [Reference]	**1999 JU3 [3]**	**Itokawa [4]**
Period (hours)	7.6	12.1
Type	Carbonaceous chondrite, representative of primitive solar nebula	Stony, made of iron and magnesium silicates
Diameter	900m	500 x 300 x 200 m
Orbit (AU)	0.96-1.41	0.95-1.70

The 3D model was used to investigate the effect of shadowing for a segment of Itokawa. It suggested that localised regions of high field (~1kV/m) occur at dayside shadow boundaries (Figure 2). Following [1] it is possible to estimate the typical sizes of particles that can be electrostatically levitated: ~300μm at the terminator and ~100μm in the shadow region, representing a substantial fraction of a Moon-like surface particle distribution [5]. Photographs of the Eros asteroid show that the "ponds", craters filled with smooth deposits, are thought to have been formed by the build up of dust, whereas regions exposed to solar radiation are stripped of smaller particles [4,6]. Electrostatically transported material could therefore form a significant proportion of the sample. A detailed understanding of the size fraction most likely to be have been electrostatically moved or even removed at the sampling site can only be achieved by electrical measurements, for which instrumentation is described in the next section.

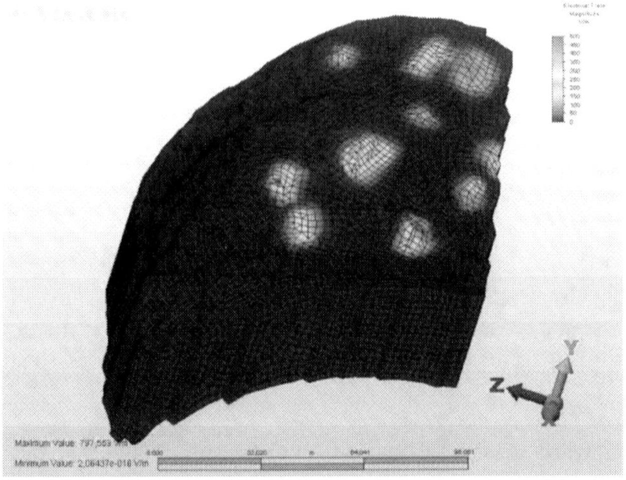

Figure 2 3D model results showing electric fields generated by local shadowing. The sun is illuminating the asteroid from the left, causing shadowed regions (lighter colours. It is assumed that these regions acquire the nightside potential of -1kV.

3. Asteroid electric field measurements

The Asteroid Charge Experiment (ACE) comprises electric field detectors for both remote and in situ sensing of the asteroid potential (it also includes an electron spectrometer and a radiation detector, which are not discussed here). Its scientific aims are to characterise the electrical environment of the asteroid and the dosage the sample receives on its return to Earth.

3.1. Instrument description and expected signal levels

The electric field instruments are essentially two sets of isolated electrodes, one set for in situ sensing and one set for remote sensing. The in situ sensing electrodes are three physically-separated conducting electrodes for horizontal and vertical electric field sensing. The electrodes acquire the local potential, and differencing the potentials across them gives the electric field. It is intended that they will be mounted on the spacecraft legs to give an indication of both horizontal and vertical electric fields, Figure 3.

The remote sensing electrodes rely on the principle that displacement and induced currents in the sensing electrodes are proportional to electric field changes from the asteroid during orbit and approach. This can be used to detect the whole body potential on the asteroid at a distance. Two adjacent electrodes of differing geometry are used to separate out the currents caused by the asteroid electric field and currents caused by local charged particles impacting the electrodes, similar to [7]. Estimates suggest that the change in induced potential on the plate is expected to be in the range 1mV-1V (0.1m edge plate, with 10pF capacitance) which is readily detectable with a straightforward amplifier circuit, whereas the displacement current is expected to be small, ~1fA for this size of plate.

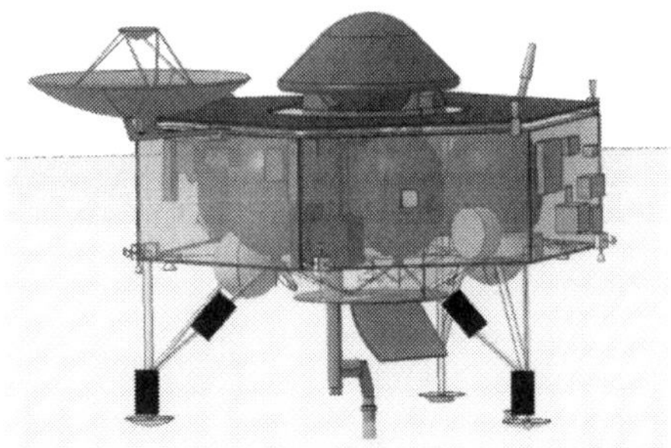

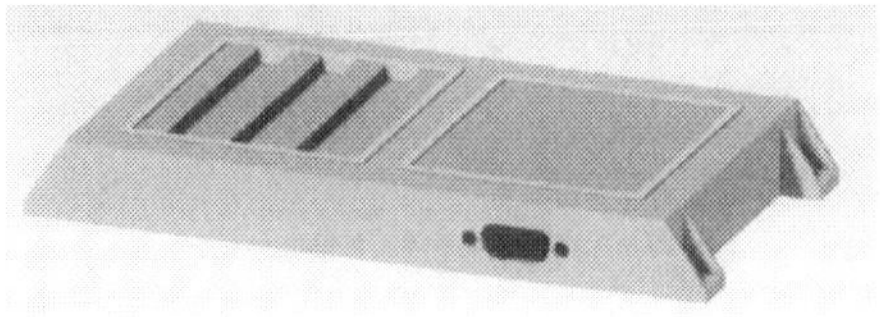

Figure 3 Electric field sensing electrodes (left) Spacecraft showing suggested locations for in situ electrodes, marked in black (not to scale). (right) Remote electric field sensor concept, showing isolated metallic plates of differing geometry.

3.2. Electrostatic effects on sampling
The spacecraft will be positively charged on arrival at the asteroid [8], which is expected to become positively charged on the dayside from photoemission, and negatively charged on the nightside from the solar wind. If the spacecraft capacitance is 200pF, then current flow is ~10pA if the spacecraft lands on the dayside as planned, suggesting that the spacecraft-asteroid potential difference may change by only ~1V during the proposed 20min landing time. One way to avoid electrostatic effects on sampling would be to use the remote measurement of asteroid potential, made before landing, and drive the (isolated) sampling apparatus at this voltage. Further work is needed to study this in detail.

4. Discussion
Theoretical and modelling work presented here and elsewhere [1] suggests that substantial electric fields can develop at asteroid terminators and between shadowed and sunny regions. The asteroid-spacecraft potential difference implies that a current will flow when the spacecraft lands, and that this may not equilibriate during the 20min scheduled for sampling collection. Under these circumstances the spacecraft itself may modify the sample collection, most likely by electrostatic repulsion. Remote sensing of the asteroid potential and driving the sample mechanism at this potential could provide a solution to this problem. It is also essential to measure the electrostatic environment at asteroid surfaces to understand the context of the sample taken, before the charge state of the sample is modified. The ACE instrument contains separated electrodes for in situ measurement of electric fields and also electrodes for remote sensing of the asteroid charge through induction.

References
[1] Lee P 1996 Dust levitation on asteroids *Icarus* **124,** 181–194
[2] Colwell J E, Culbis A A S, Horányi M and Robertson S 2005 Dust transport in photoelectron layers and the formation of dust ponds on Eros *Icarus* **175**, 159-169
[3] Michel P and Delbo M 2010 Orbital and thermal evolutions of four potential targets for a sample return space mission to a primitive near-Earth asteroid *Icarus* **209**, 520-534
[4] Fujiwara A et al 2006 The rubble-pile asteroid Itokawa as observed by Hayabusa *Science* **312**, 1330-1334
[5] Heiken G H, Vaniman D T and French B M 1991, *Lunar source book*, CUP
[6] Hughes A L H, Colwell J E and deWolfe A W 2008 Electrostatic dust transport on Eros: 3-D simulations of pond formation *Icarus* **195**, 630-648
Miyamoto H et al 2007 Regolith migration and sorting on asteroid Itokawa *Science* **316**, 1011-1014
[7] Bennett A and Harrison R G 2008 Surface measurement system for the atmospheric electrical vertical conduction current density, with displacement current correction *J. Sol.-Terr. Phys.* **70**, 1373-1381
[8] Whipple E C 1981 Potentials of surfaces in space *Rep. Prog. Phys.*, **44**, 1197-1250

Journal of Physics: Conference Series **301** (2011) 012009 doi:10.1088/1742-6596/301/1/012009

The influence of unburned carbon particles on electrostatic precipitator collection efficiency

M Jędrusik[1,2] and A Świerczok [2]

[2] Wrocław University of Technology, Institute of Heat Engineering and Fluid Mechanics, Wyb. Wyspiańskiego 27, 50-370 Wrocław, Poland

Abstract. Laboratory tests have shown that the chemical composition of fly ash (in that unburned coal) as well as its size distribution has significant influence on the dust cleaning process. Likewise the design of discharge electrodes has shown a strong influence on the dust cleaning. Tests of precipitation efficiency were carried out on a laboratory electrostatic precipitator (ESP) model using fly ash samples of diverse size distribution and unburned coal content collected from several grate boilers. Test results show explicit dependency of the ESP precipitation efficiency on physical and chemical characteristics of the fly ash, design of discharge electrodes and amount of electrical energy delivered to the ESP. Mercury concentration measurements show higher levels in the fly ash than in the fired coal indicating high sorption capacity of the fly ash. Prior observation suggests good mercury adsorption on fine fly ash particles in the presence of elemental coal. Hence the improvement of ESP collection efficiency of fine particles containing unburned coal may help decrease the emission of mercury.

1. Introduction

Studies of the dust separation process in electrostatic precipitators [1, 2, 3] have shown that chemical composition of fly ash – mainly containing unburned coal – has significant influence on precipitation efficiency. New dust emission standards for small power plants using grate boilers as well as introduction of deNOx systems in big power plants – especially with low NOx burners – will require even more efficient precipitation of fly ash because of its high content of unburned coal.

Although theoretic principles of dust separation process in electrostatic precipitators are well known [4, 5], further studies of the influence of construction and electric parameters of discharge electrodes on ESP performance efficiency are essential. Particularly studies on optimal selection of ESP discharge electrode construction seem very important.

2. Experimental

The test bench was comprised of one stage electrostatic precipitator (ESP) model chamber with a set of discharge and collecting electrodes, ESP inlet and outlet air ducts, dust particle feeder, high voltage supply unit (negative polarity) and exhaust fan [3].

[1] maria.jedrusik@pwr.wroc.pl

2.1. Fly-ash analysis

Physic-chemical analyses of the fly-ash used in tests are given in table 1. Figure 1 shows results of fly-ash particle size distribution.

Table 1. Properties of the fly ash

Parameter	Units	Fly ash type		
		G	A	K
Chemical composition of ash:	%			
SiO_2		28.99	25.74	40.39
Fe_2O_3		3.67	5.91	10.01
Al_2O_3		17.14	24.31	18.09
TiO_2		0.86	1.24	0.96
CaO		2.82	2.18	5.44
MgO		1.01	0.50	1.47
SO_3		2.26	5.76	1.27
K_2O		2.68	3.58	2.08
P_2O_5		0.01	2.10	-
Na_2O		1.14	2.82	1.47
Unburned carbon in ash:	%	28.60	23.31	17.92
Density	kg/m^3	2090	2320	1760
Resistivity	ohm cm	5×10^7	1×10^8	2×10^7

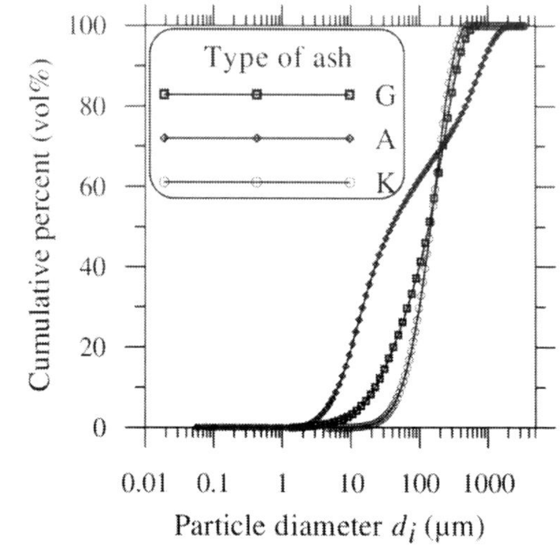

Figure 1. Particle size distribution of tested fly ashes.

The tested fly-ash samples G, A and K characterize high unburned coal contents in a range of 18-28%. The measured fly-ash resistivity in laboratory conditions was 10^7-$10^8\Omega\times$cm and remained almost steady with temperature changes from 20°C to 215°C.

The particle size distribution of tested fly ashes was measured using a Mastersizers 2000 analyzer of Malvern Instruments, and the resultant distribution curves (figure 1) show characteristic shape of fly ashes from grate boilers.

2.2 Discharge electrodes configurations

The discharge electrode (DE) shape and basic dimensions are depicted by figure 2 and voltage-current characteristics of tested electrode configuration by figure 3.

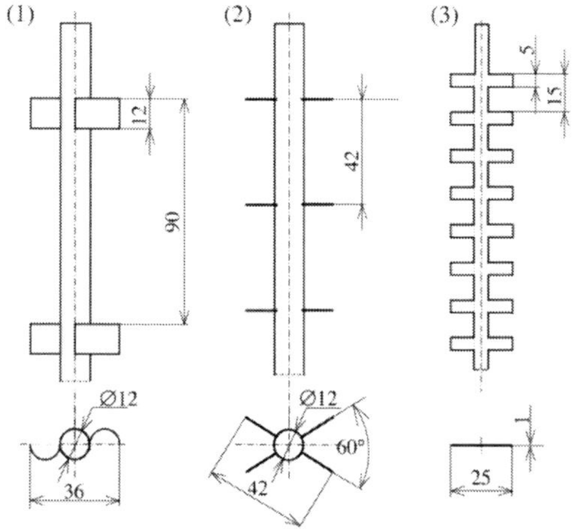

Figure 2. Design of the discharge electrodes, dimension in mm.

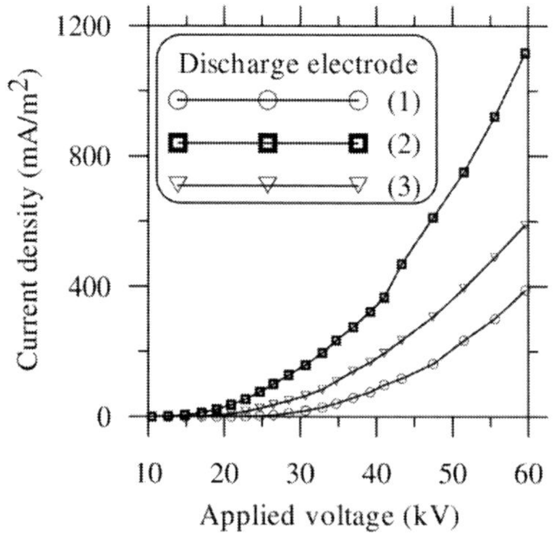

Figure 3. Current-voltage curves of discharge electrodes tested under clean air conditions.

The voltage-current characteristics of tested discharge electrodes clearly differentiate construction of electrode for: (1) low-current electrodes and (2) high-current electrodes, so called 'aggressive'.

3. Results and discussion

Performance tests were conducted on a bench scale laboratory ESP with fly ash samples described in table 1 and figure 1. The test results are presented as a function of ESP applied voltage for different ash samples on figure 4 and figure 5 for discharge electrode type (2) and (3) respectively. For all tests the air flow velocity in the ESP was held constant at 0.8 m/s and the dust concentration at 0,2 g/m^3.
For each level of the supply voltage three collection efficiency tests were carried out.

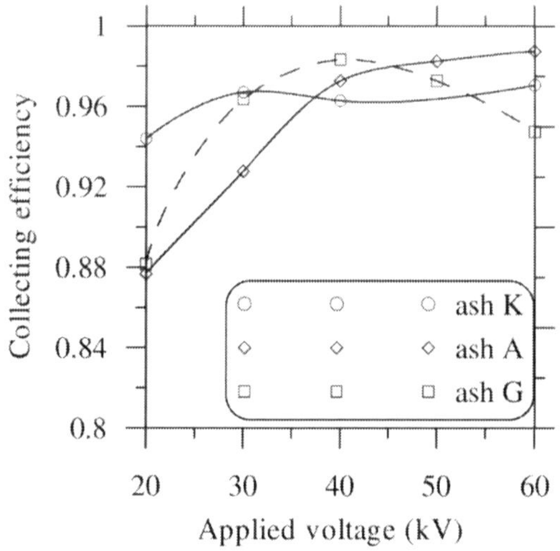

Figure 4. Collection efficiency as a function of applied voltage for discharge electrode type (2).

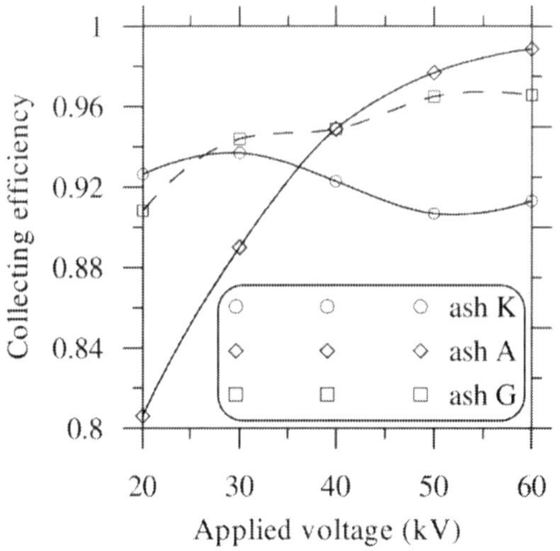

Figure 5. Collection efficiency as a function of applied voltage for discharge electrode type (3).

Collection efficiency of fly ash type A for both tested discharge electrodes shows a typical shape (the increasing tendency with increase of applied supply voltage). In a case of fly ash type K the influence of applied voltage on collection efficiency is negligible although with type (2) discharge electrode the results were better. Collection efficiency of fly ash type G and discharge electrode (2) decreases when applied voltage over crosses a level of 40 kV, but for discharge electrode (3) presents slight increase only but at lower values. It should be mentioned that this fly ash contains more than 28% of unburned coal. It seems that there is a certain applied voltage level above which further collection efficiency will not be possible.

The above observations show that besides the unburned coal contents of fly ash and its resistivity other factors also affect the precipitator collection efficiency. It calls on further investigation on such fly ashes with special consideration to its chemical composition and unburned coal content for respective particle size distribution ranges.

4. Studies on Mercury contents in fly ashes

The study of mercury content in fly ash at the ESP inlet has shown that it is several times higher than the mercury contents of the fired coal (for hard coal it is in a range of 50 to 150 ppb), suggesting a high sorption capacity of the fly ash [6]. Typical samples of fly ash G taken at power plant conditions have shown the following mercury content values:

- from ESP inlet: 1.0925 mg/kg (ppm)
- inside ESP: 0.6615 mg/kg (ppm)
- from ESP outlet: 7.5303 mg/kg (ppm)

Additionally the mercury content for respective particle size distribution ranges of fly ash sample G collected from typical ESP has been measured and the results presented in table 2.

Tabel 2. Mercury content (ppm) of fine fly ash particles

Fly ash sample	Dust diameter (μm)			
	0-3	3-10	10-24	24-45
(G)	9.0827	6.2917	3.6420	1.0657

The above experimental results confirm professional literature suggestion about mercury accumulation in fine particles emitted from coal combustion processes [7]. Therefore a serious improvement of fine particles collection efficiency becomes critically necessary for also reducing the mercury emission after combustion processes.

5. Summary

Analysis of experiment results showed some problems in collection of fly-ash from stocker-fired boilers:

- decrease of collection efficiency for high content of unburned carbon in the fly ash,
- no clear relationship between the supply voltage and collection efficiency of fly ash,
- the improvement of bulk fine particle collection efficiency which may significantly reduce the mercury emission into atmosphere.

Acknowledgments

These studies were carried out as a part of developing project no. N R06 0014 06/2009 funded by the Polish National Centre for Research and Development (NCBiR) (chapters 2 & 3).

The studies were financed by the Polish Ministry of Science and Higher Education (MNiSzW) (chapter 4).

References

[1] Barranco R, Gong M, Thompson A, Cloke M, Hanson S, Gibb W, Lester E 2007 The impact of fly ash resistivity and carbon content on electrostatic precipitator performance, *Fuel* **86** 2521-2577

[2] Senneca O 2008 Burning and physico-chemical characteristics of carbon in ash from a coal fired power plant *Fuel* **87** 1207-1216

[3] Jędrusik M and Świerczok A 2009 The influence of fly ash physical & chemical properties on electrostatic precipitation process *Journal of Electrostatics* **67** 105-109

[4] White H J 1990 *Industrial Electrostatic Precipitation* (prep.), (International Society for Electrostatic Precipitation, Library of Congress Catalog Card No 62-18240)

[5] Parker K R 1997 *Applied Electrostatic Precipitation* (London: Blackie Academic & Prof.)

[6] Hower J C, Senior C L, Suuberg E M, Hurt R H, Wilcox J L 2010 Mercury capture by native fly ash carbons in coal-fired power plants *Progress in Energy and Combustion Science* **36** 510-529

[7] Sung-Jun L, Yong-Chil S, Ha-Na, J, Kyu-Shik P, Jeom-In B, Hi-Soo A, Kwang-Chul S 2006 Speciation and mass distribution of mercury in a bituminous coal-fired power plant, *Atmosph. Envir.* **40** 2215-2224

PEF and UV combined system for pathogen microorganisms inactivation in liquid food products

R. Cramariuc[1], M. Popa[2], A. Tudorache[3], E. Brînduşe[3], A. Kontek[3], A. Mitelut[2], L. Fotescu[3], B. Cramariuc[4], M. Geicu[2], L. Nisiparu[5]

[1]Competence Center in Electrostatics and Electrotechnologies, Bucharest, Romania, [2]University of Agronomic Science and Veterinary Medicine, Bucharest, Romania, [3]Research and Development Institute in Viticulture and Vinification Valea Călugărească, Romania, [4]IT Center for Science and Technology, Bucharest, Romania, [5]Carol Davila University of Medicine and Pharmacy, Bucharest, Romania

Email: raducramariuc@yahoo.com

Abstract. Pulsed electrical field (PEF) treatment is a non-thermal food preservation technology based on the use of the electrical field in impulses applied in order to inactivate and control pathogen microorganisms in foods. This technology is highly appreciated for its ability to prolong the shelf life of the treated product without the use of heat and also for its ability to preserve the product's sensory qualities and nutritional value as well as for the microbiological control of the treated products. This paper presents the PEF and UV treatment methods, or a combination between the two, for microbe inactivation in liquid products. The experiments were carried out using yeasts, lactic bacteria and acetic bacteria in the following systems: stand-alone treatments (PEF or UV) or in combination (UV+PEF or PEF+UV). The results of these experiments showed that one can obtain total inactivation of microorganisms using the combined UV+PEF system, thus leading to the possibility of increasing liquid food products quality as compared to the quality obtained using thermal pasteurization.

1. Introduction

Pulsed electrical field (PEF) treatment is a non-thermal technology for food preservation based on the use of a pulsed electrical field to inactivate pathogen microorganisms in food products and to control their content in products under treatment. This technology is highly appreciated for its ability to prolong shelf life without the use of heat, and also to preserve the sensory qualities and nutritional value of the product, as well as to microbiologically control the products under treatment. Literature in this field mentions that microorganism inactivation using PEF technology alone with naturally contaminated liquid food products as being difficult. Thus, the possibility of obtaining higher microbe inactivation rates was reported only in the case of combining PEF treatment with other effects, such as: heat, pressure or antimicrobial agents [1-2]. Paper [2] mentions the use of a combined PEF + UV treatment for the pasteurization of fresh apple juice. The results in the UV + PEF combined system were better from the microbial inactivation point of view as compared to the application of the PEF technology alone.

In the present work we present the PEF and UV treatment methods, or a combination between the two, for microbe inactivation in liquid products. The experiments were carried out using yeasts, lactic

Published under licence by IOP Publishing Ltd

bacteria and acetic bacteria in the following systems: stand-alone treatments (PEF or UV) or in combination (UV+PEF or PEF+UV). The aim is to develop a technology based on the combined treatment for the pasteurization of grape must and for the microbiological stabilization of wines.

2. Materials and methods

2.1. *Materials under treatment*

The experiments have been carried out on microorganism suspensions in distilled water sterilized by autoclaving for 30 minutes at 1 bar. The microorganisms to be tested were taken from the microorganism collection of Valea Calugareasca IC-DVV, from the category of microbes present in wines: for yeasts *Saccharomyces cerevisiae* was used, strain PN III-11, for lactic bacteria, *Oenococcus oenos* strain 149 and for acetic bacteria strain 11 of the *Acetobacter aceti* species. A series of consecutive replications on liquid mediums led to suspensions of microorganisms which, after several dilutions, were used to obtain certain inoculation titres.

Active microorganisms were highlighted using the microbiologic control method involving filtering through 0.45 μm pore membranes, followed by incubation on mediums specific to each type of microorganism: (a) for yeasts, agar based on yeast extract, peptone and glucose to which chloramphenicol was added; (b) For lactic bacteria, agar based on yeast extract, meat extract, peptone, sodium acetate, magnesium sulfate, glucose, to which acitidione was added; (c) For acetic bacteria, agar based on yeast extract to which acitidione and bromophenol blue were added. The experiments were carried out such that insight into following aspects is gained: (1) The influence of the duration of application of ultraviolet (UV) rays on microorganism inactivation; (2) The influence of the duration of application of pulsed electrical field (PEF) on microorganism inactivation; (3) The effect of the UV and PEF treatment, each alone and combined, on microorganism inactivation.

2.2. *UV radiation processing*

UV radiation exposure has been carried out in an UV rays water sterilizer with the following specifications: Working fluid – filtered water; Water flow = 0.1-0.3 l/min; Maximum power in the radiation chamber = 1 bar; UV wavelength = 254 nm; UV lamp power = 8 W; Power source = 220 V / 50Hz; Dimensions = 260 x 160 x 82 mm. Materials exposure to UV rays was carried out in lot conditions of 750 ml, from the products under treatment that have already been mentioned in section 2.1. The above mentioned apparatus allows continuous flow functioning. The materials under treatment have been exposed at different times which have been mentioned in the results section. After exposure, the materials have been transferred to sterile bottles and either analyzed or treated with high voltage impulses (PEF). The treatments were carried out in the following order: UV + PEF or PEF + UV. The results are presented in section 3.

2.3. *High voltage impulses (PEF) processing*

High voltage processing has been carried out in a concentric cylinder PEF treatment chamber with enhanced field area. The HV electrode was made from stainless steel with the diameter 60mm and the length 450mm. The materials under UV rays treatment were introduced in the PEF treatment chamber of 750 ml capacity. The HV impulses had the following characteristics: maximum impulse voltage: 30 kV; pulse front duration: 100 ns; pulse duration: 1 μs. The variable treatment duration is mentioned in the results section. The PEF system was cleaned and sterilized before and after each attempt.

3. Results

The tables and graphs below show the experimental results. Simple UV treatment led to a 36% - 62% decrease of microorganism count, depending on the type of microorganism and duration of exposure. The inactivation rate was higher for bacteria. In the case of yeasts the lethality rate was 55% for a 5 minutes UV treatment. Under the same conditions, lactic bacteria were inactivated at a rate of 59%

and acetic bacteria at a 62% rate. The increase of UV treatment duration to 10 minutes did not significantly increase microorganism inactivation rate (Table 1; Figure 1a).

Table.1. Influence of UV treatment on microorganism viability.

Type of microorganism		V1 – 2 min.	V2 – 5 min.	V3 – 10 min.
Yeasts	Viability (%)	47	45	47
	Lethality (%)	53	55	53
Lactic bacteria	Viability (%)	64	41	44
	Lethality (%)	36	59	56
Acetic bacteria	Viability (%)	54	38	38
	Lethality (%)	46	62	62

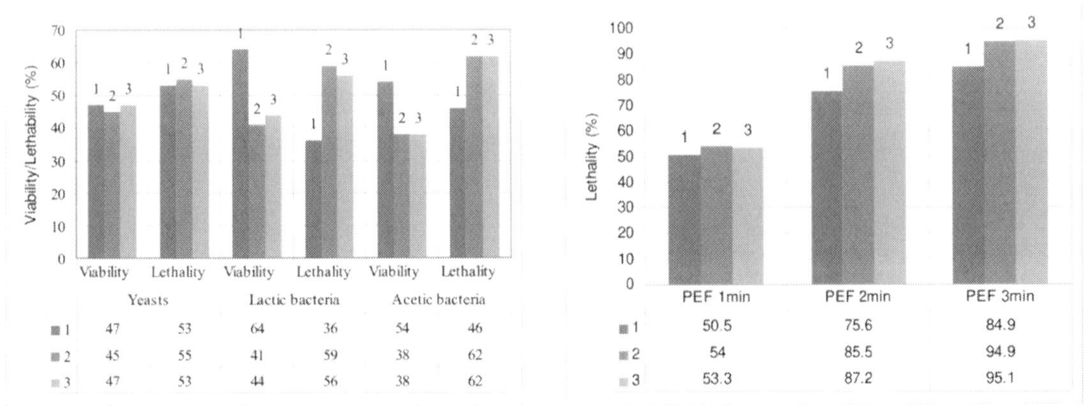

a) b)

Figure 1. (a) Influence of UV treatment on microorganism viability: 1) 2min; 2) 5min; 3) 10min; (b) Influence of the duration of application of pulsed electrical field (PEF) on microorganism inactivation: 1) Yeast; 2) Lactic bacteria; 3) Acetic bacteria;

Simple PEF treatment reduced microbe load in suspension, but the reduction rate was different depending on the type of microorganism and the duration of the treatment (Figure 1b). First degree correlations were established between the microbe load (expressed in CFU/ml) of microorganism groups (yeasts, lactic and acetic bacteria) and PEF treatment duration. Only regression equation parameters differ (Table 2).

Table 2. Correlation between microbial load and PEF treatment duration.

Type of microorganism	Form of regression line	Regression coefficient
Yeasts	$y = 17.181x + 35.951$	0.9338
Lactic bacteria	$y = 20.471x + 37.198$	0.9115
Acetic bacteria	$y = 21.382x + 35.417$	0.8829

Table 3. Influence of UV + PEF treatment on microorganism viability.

Type of microorganism		UV 5 min	PEF 2 min	UV+PEF (5min + 2min)
Yeasts	Viability (%)	48.5	22.0	4.5
	Lethality (%)	51.5	78.0	95.5
Lactic bacteria	Viability (%)	39.0	9.1	2.8
	Lethality (%)	61.0	90.9	97.2
Acetic bacteria	Viability (%)	37.6	6.1	2.1
	Lethality (%)	62.4	93.9	97.9

The combined treatments UV + PEF (Table 3 and Figure 2a) and PEF + UV (Table 4 and Figure 2b) proved to be very efficient leading to an almost complete inactivation of microorganisms. The inactivation rate was higher in the case of PEF + UV treatment, for both lactic and acetic bacteria. Yeasts were more treatment resistant, with a 96% lethality rate under both treatments conditions (Table 4 and Figure 2b).

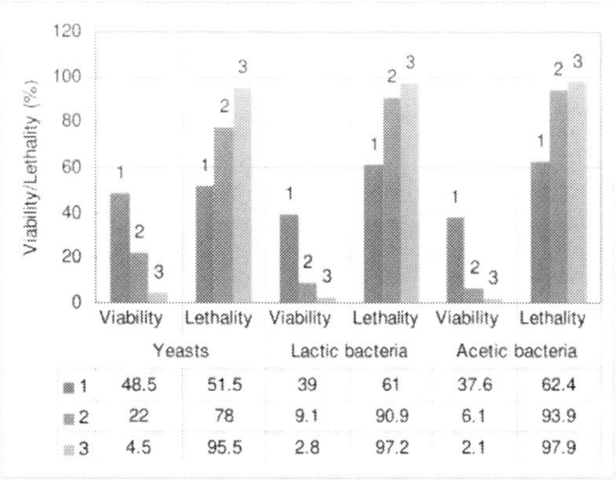

a)

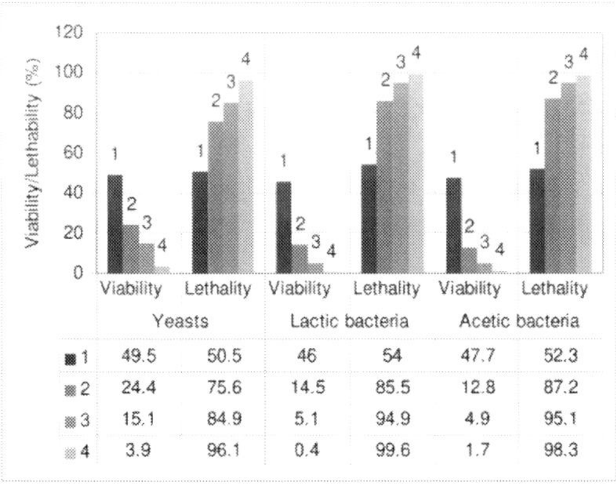

b)

Figure 2. (a) Influence of UV + PEF treatment on microorganism viability: 1) UV 5min; 2) PEF 2min; 3) UV+PEF (5min+2min); (b) Influence of PEF + UV treatment on microorganism viability: 1) PEF 1min; 2) PEF 2min; 3) PEF 3min; 4) PEF+UV (2min+5min);

Table 4. Influence of PEF + UV treatment on microorganism viability.

Type of microorganism		PEF 1 min	PEF 2 min	PEF 3 min	PEF + UV (2min+5min)
Yeasts	Viability (%)	49.5	24.4	15.1	3.9
	Lethality (%)	50.5	75.6	84.9	96.1
Lactic bacteria	Viability (%)	46.0	14.5	5.1	0.4
	Lethality (%)	54.0	85.5	94.9	99.6
Acetic bacteria	Viability (%)	47.7	12.8	4.9	1.7
	Lethality (%)	52.3	87.2	95.1	98.3

4. Conclusions

The paper presents a combined system for microbial inactivation of microorganisms (yeasts, lactic bacteria, acetic bacteria, in different concentrations in distilled water) involved in the processing of grape products. The results confirm PEF + UV system superiority over the PEF alone treatment system. UV radiation treatment is used only in PEF + UV combination when applying the Hurdle strategy. In conclusion, both non-thermal PEF and UV treatments, simple or combined, are promising for the field of enology, as they are inexpensive and have the ability to replace thermal treatments which have negative effects on product quality. However, combined treatments improve system performance and are therefore more efficient.

References

[1] R. Cramariuc, A. Tudorache, M. Popa, E. Brinduse, L. Nisiparu, A. Mitelut, M. O. Turtoi, L. Fotescu, Corona discharge in electroporation of cell membranes, 2008, *J. Phys: Conf. Series* 142, 012062.

[2] F. Noci, M. Reiner, M. Walking-Ribeiro, D. A. Cronin, D. J. Morgan, J. G. Lyng, Ultraviolet irradiation and pulsed electric fields (PEF) in a Hurdle strategy for the preservation of fresh apple juice, 2008, *J. Food Eng.*, 85(1), 141-146.

NOx removal using a wet type plasma reactor based on a three-electrode device

J Jolibois, K Takashima and A Mizuno

Department of Environmental and Life Sciences, Toyohashi University of Technology Tempaku-cho, Toyohashi, Aichi, Japan

E-mail: mizuno@ens.tut.ac.jp

Abstract. In this paper, a wet type plasma reactor based on a three electrode device is investigated experimentally in order to remove NO and NOx at low flow rate. First, a comparison of cleaning performances of gas exhaust has been performed when the surface discharge operates in DBD or SD modes. From these previous results, the second part of study has consisted to improve the electrochemical conversion of the wet type plasma reactor by adding a coil between the AC HV power supply and the surface discharge. The parametric study has been performed with 100 ppm of NO content in gas flow at room temperature and atmospheric pressure for a flow rate of 1 L/min. For each electrical parameter tested, an electric characterization and measurement of NOx content via FT-IR has been conducted. The results highlight a better cleaning of gas exhaust when the surface discharge operates in DBD mode. Moreover, the presence of solution promotes the arc transition when the operating mode is SD, resulting a reliability reduction of plasma device. In addition, the measurements show that the insertion of coil in the electrical circuit improves the NOx removal at a given power consumption for the DBD operating mode.

1. Introduction

Pollution control represents a capital issue for the transport industry. Indeed, the increasing concern over emission of air pollutants on environmental and human health has motivated the research to remove these harmful gases. Therefore, several pollution control technologies, environmentally acceptable and energy-efficient, have emerged in recent years. Among these, the plasma devices (also called plasma reactors) appear as being suitable to reduce pollution and especially to remove NOx [1-5]. NOx removal of gas exhaust via non thermal plasma reactor is divided into dry and wet way. Several studies have shown that the wet type reactor allows to obtain significant results for gas exhaust cleaning compared to the dry reactor. The underlying process is the NOx dissolution into water and its conversion to nitrate (NO_3^-) and nitrite (NO_2^-) ions [1].

The objective of this work is to define an optimal configuration in terms of electrochemical conversion efficiency and reliability, of a wet type plasma reactor.

In this paper, we present a study on NOx removal performed via a wet type plasma reactor based on a three-electrode device. Depending on the applied electrics parameters [6, 7], the surface discharge shape (i.e. three-electrode design) enables to select the operating mode of the discharge. Each discharge regime is stable at room temperature and atmospheric pressure with a standard relative humidity (i.e. 40%). In a first part, the performances of NOx removal obtained by using the DBD and SD operating modes are shown. In a second part, effect of a coil inserted between the AC HV power supply and the plasma device (in DBD mode) on polluted gas is characterized.

2. Experimental setup

Figure 1 shows a schematic illustration of the discharge setup. The plasma device consists of two electrodes flush mounted on each side of a dielectric plate, plus two counter-electrodes placed on the top side of the insulating wall. These counter-electrodes are separated relatively to the electrode #1 by an air gap of 25 mm, named SD gap. Each electrode is made of aluminium strip whose ends are oval in order to reduce edge effects. Upper electrodes are 60 mm-long (in spanwise) for 10 mm in width. The lower one is 30 mm wide for 60 mm-long. The dielectric used is a PMMA flat plate (140 × 100 mm^2) of 3 mm-thick.

The bottom electrode is grounded when the electrode #1 is connected to an adjustable AC HV power supply. The AC power supply is obtained with the help of a transformer supplied by a power amplifier (NF Corporation, model 4510). The transformer can supply a maximum peak voltage of 20 kV at driving frequencies up to 5 kHz. Depending on discharge operating mode, the two other electrodes are either grounded, corresponding to DBD mode, or connected to a negative DC power supply (Matsusada Precision, model HAR-50R6), i.e. SD mode.

The AC voltage applied to electrode #1 is measured by using an HV probe (Tektronix, model P6015A). Electric currents flowing through the system are deduced from voltages across shunt resistors of 100 Ω connected to a fast digital oscilloscope (Tektronix, model DPO 2024). From voltage and current curves, the total electrical power consumption can be calculated.

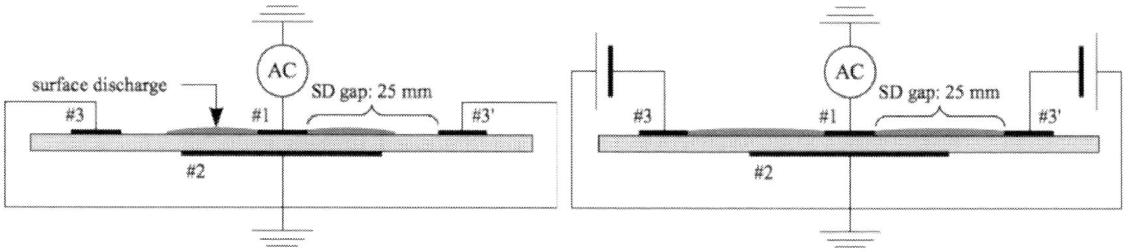

Figure 1. Scheme of plasma device used: DBD (left) and SD (right) modes.

Figure 2 displays a schematic side-view of the wet type plasma reactor used in this study. It is composed of two parts. The first part consists of a wet reactor with two openings allowing the inlet and exhaust of gas flow. This reactor made of polypropylene plate has a rectangular cross-section (100 × 30 mm^2) for 150 mm-long. The liquid inside the reactor is a sodium sulfite solution (150 mL, 1 mol/L) which is used as NOx gas absorbent. Moreover, it allows to re-established the pH level by reduction NO_3^- ions to N_2 or NH_3. The second part of the plasma reactor corresponds to the surface discharge placed on the top plate of the reactor. The whole surface discharge and wet reactor forms the wet type plasma reactor.

The experiments are carried out at room temperature and atmospheric pressure. Total gas flow rate is regulated to 1 L/min by the mass flow controller. The inlet gas composition is NO (100 ppm) and dry air. NO and NOx concentrations (with [NOx] = [NO] + [NO_2]) are measured using FT-IR spectroscope gas analyzer (BEST SOKKI, model SESAM 3-N).

In the cleaning experiment, the discharge is ignited twice 3 min, and this 3 min after the start of recording of gas exhaust concentration. Each ignition is spaced 3 min, bringing the total duration of experiment at 15 min. Moreover, the determination of average concentration is not achieved when gas concentration reaches a steady state.

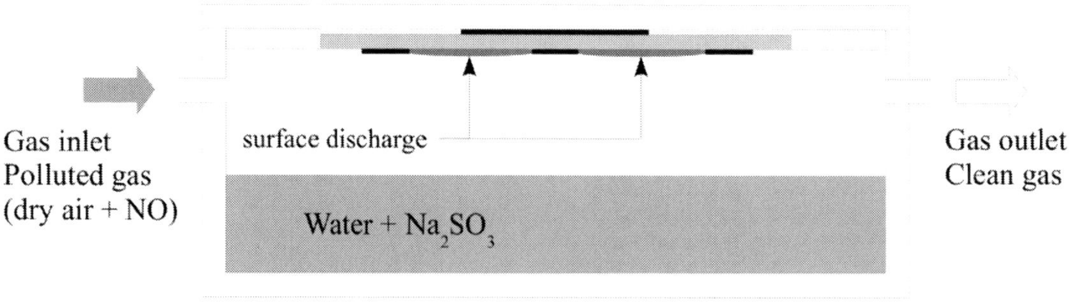

Figure 2. Schematic side-view of the wet type plasma reactor used

3. Results

3.1. NOx removal as function of discharge operating mode

In this section, we focus on the ability of the plasma device to remove NOx depending of the operating mode of the discharge. The initial NO content is fixed at 100 ppm and the polluted air flow rate is equal to 1L/min.

First, we study the cleaning of gas exhaust when the surface discharge operates in DBD mode. In this case, the waveform applied is a sine one, at frequency set to 1 kHz. The peak voltage is between 6 and 13 kV. Figure 3 shows the evolution of NO and NOx concentrations versus the power consumed. For both gases, we can see that the initial content of pollutants is decreased with the injected power. The NO and NOx removal reaches 80% and 40 %, respectively, for a power consumed about 2.25 W, corresponding in our case to an energy density of 135 J/L.

However, the computation of the average concentration is not realized during the steady state, resulting in a bias on the final value of the pollutant content. For example, the minimum concentration of NO in the steady state is about 7 ppm. In the case, the NO removal efficiency reached 93%.

The removal of NO proceeds mainly via oxygen species (O and O_3) in air, leading to the formation of NO_2. In the presence of the solution (wet condition), NO and NOx concentrations decrease due to the absorption of NO_2 into the liquid and / or reaction of NO_2 with OH radicals to forms NO_3^- ions.

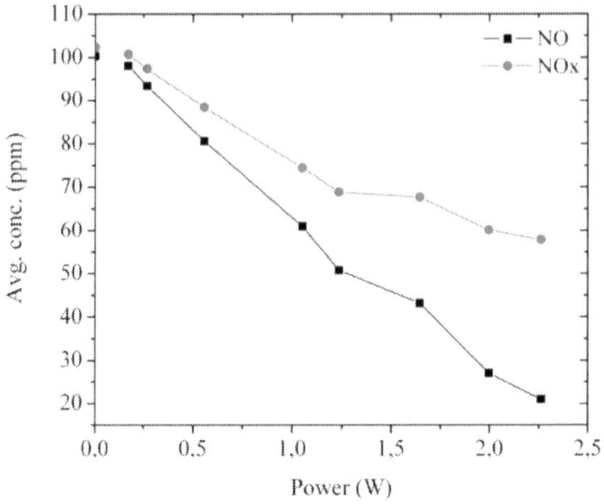

Figure 3. NO and NOx concentrations versus power consumed by plasma device in DBD operating mode.

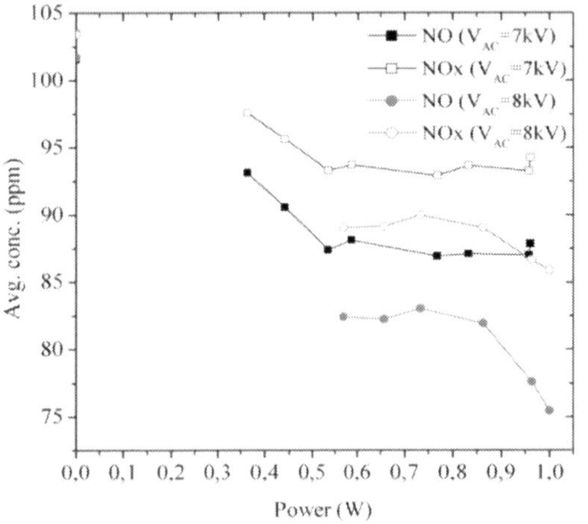

Figure 4. NO and NOx concentrations versus power consumed by plasma device in SD operating mode.

Figure 4 presents the evolution of NO and NOx concentrations versus the power consumed in SD mode. Here, the experimental setup is the same as above, except the AC voltage applied equal to 7 and 8 kV peak. Moreover, the negative DC component is ranging between 0 to -11 kV. From this figure, several results appear. First, one can see that at given electrical power consumption the plasma device seems less effective for cleaning exhaust gas compared to the previous case. For example, at 1 W the minimum concentration of NO is about 75 ppm with the SD mode while it reaches 62.5 ppm with the DBD actuation. Then, we observe that the NO and NOx concentrations as a function of power consumed have an abnormal evolution. Indeed, we see that for an electric power range between 0.5 and 0.85~0.9 W, the injected power into the discharge does not improve the treatment of pollutants. On the contrary, outside this range a decrease of NO and NOx content is obtained.

However, it has not been possible to verify the phenomenon with others electrical parameter because the presence of the solution inside the reactor promotes the transition to arc of the discharge.

3.2. Effect of a coil on NOx removal

In this section, we study the influence of a coil on gas exhaust treatment. To achieve this, a coil is placed between the AC HV power supply and the plasma device. The NO concentration is fixed at 100 ppm and the polluted air flow rate is equal to 1L/min.

As reported in previous section, the SD operating mode is unstable (transition to the arc) due to the presence of Na_2SO_3 solution. Therefore, we study the cleaning of gas exhaust when the surface discharge only operates in DBD mode. In this experiment, the waveform applied is a sine one with a peak voltage ranging from 6 to 12 kV, at frequency set to 1 kHz. Figure 5 highlights the NO and NOx concentrations as a function of power consumed without and with coil. It appears clearly that the coil integrated in the electrical circuit allows to improve the cleaning of gas exhaust, except the first measurement point. For example, at 1 W the NO concentration reaches 35 ppm in presence of the coil, whereas the NO content is only 65 ppm with the standard case (i.e. without coil). If we compare the NOx content in presence or absence of coil, we remark that the concentration difference is less significant, i.e. 65 ppm with coil against 80 ppm. It means the inductance allows a higher radicals production, resulting an effectiveness improvement of the NO treatment process. Similarly, it promotes the absorption of NO_2 in the solution of sodium sulphite, which leads to a decrease in NOx content.

The lower efficiency of gas exhaust cleaning for the first measurement point can be explained by the fact that the coil is resistive, about 4 kOhm/H, so the voltage applied across the plasma device is lower compared to the case without coil.

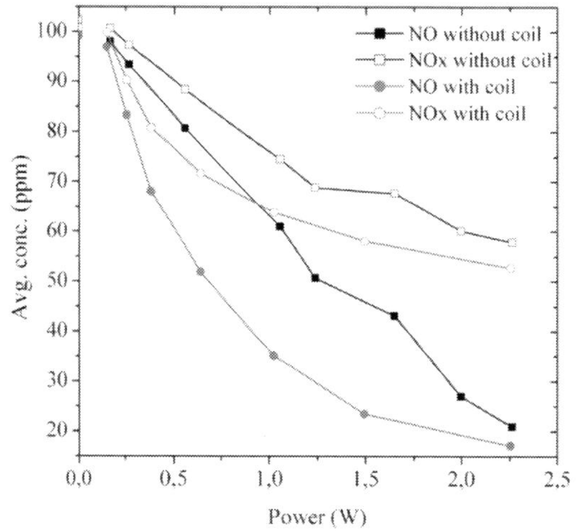

Figure 5. NO and NOx concentrations versus power consumed by plasma device without and with coil (L= 1 H).

4. Conclusion

In this paper, we have shown the capability of a surface discharge to clean gas exhaust with 100 ppm of initial NO content for a flow rate of 1 L/min. Depending on our experimental conditions, it appears that a better cleaning of gas exhaust is obtained when the surface discharge operates in DBD mode. In addition, the evolution of pollutant content presents an abnormal evolution during the SD mode, requiring further investigation. However, the presence of the solution promotes the arc transition of this operating mode, resulting a reliability reduction of plasma device and prevents the use of others electrical parameter.

And finally, adding a coil in the electrical circuit improves the efficiency of gas exhaust treatment by increasing radicals production and promoting NO_2 absorption into Na_2SO_3 solution.

References

[1] A. Mizuno, K. Shimizu, T. Matsuoka and S. Furuta, 1995, *IEEE Trans. Ind. Appl*, **31**, pp. 1463.
[2] T. Kuroki, M. Takahashi, M. Okubo, and T. Yamamoto, 2002, *IEEE Trans. Ind. Appl.*, **3**, pp. 1204.
[3] H.H. Kim, G. Pietro, K. Takashima, S. Katsura and A. Mizuno, 2002, *J. Electrostat.*, **55**, pp. 25.
[4] N. Harada, T. Moriya, T. Matsuyama, H. Yamamoto and S. Hosokawa, 2007, *J. Electrostat.*, **65**, pp. 37.
[5] S.M Thagard, Y. Kinoshita, H. Ikeda, K. Takashima and A. Mizuno, 2010, *IEEE Trans. Ind. Appl*, **46**, Issue 6, pp. 2165.
[6] K. Takashima, N. Zouzou, E. Moreau, A. Mizuno and G. Touchard, 2007, *IJPEST*, **1**, No 1, pp. 14.
[7] E. Moreau, R. Sosa and G. Artana, 2008, *J. Phys. D: Appl. Phys.*, **41**, No 11, pp 5204.

Collection of submicron particles using DBD electrostatic precipitator in wire-to-square tube configuration

R. Gouri[1,2,3], N. Zouzou[2], A. Tilmatine[3], E. Moreau[2] and L. Dascalescu[2]
[1] Department of Electrical Engineering, University of Béchar, 08000, Béchar, Algeria
[2] Institut P' - CNRS - University of Poitiers – ENSMA, 86962, Futuroscope, France
[3] IRCOM laboratory, Djillali Liabes University, 22000, Sidi-Bel-Abbes, Algeria

E-mail : r.gouri@gmail.com

Abstract. The aim of the present work is to investigate the capability of a Wire-to-Square Tube Electrostatic Precipitator (or WST-ESP) to collect submicron particles using a Dielectric Barrier Discharge (DBD). A parametric study is carried out to evaluate the effects of four control variables that might affect the collection efficiency of the WST-ESP: diameter of the corona wire electrode, width of the ground electrode and the area of the tube section. The experimental data show that the wire diameter has a negligible effect on the WST-ESP collection characteristics. However, better overall performances can be obtained by putting in parallel several WST-ESPs with reduced section. The extension of the ground electrode and its discretization increase the collection efficiency.

1. Introduction

Increasing the collection efficiency of electrostatic precipitators with minimum capital and running costs is a big challenge [1]. In previous experimental work [2-3], it has been shown that the DBD can be successfully employed for the collection of submicron particles. In the case of a wire-to-cylinder ESP, the collection efficiency reaches 99 %. The problem is that this geometry is not suitable for implementation in an industrial environment, especially at high flow rates. A WST-ESP, which allows us to put in parallel several ESPs, may be a solution to the problem of industrial scaling.

The goal of the work presented in this paper is to evaluate the collection efficiency of submicron particles in a lab-scale dielectric barrier discharge ESP. A parametric study was carried out, in order to point out the effect of wire diameter, ground electrode size and tube section on ESP performances.

2. Experimental set-up

The experimental setup is illustrated in figure 1. Dry clean air (relative humidity < 5%) is introduced into a custom-designed smoke generator, where burning of incense sticks generates submicron particles with a mean size of about 0.3 µm. The particles are entrained by the airflow through the ESP. A small amount of the exhaust is connected to a diluter with controlled additional clean air. The particle concentration in the diluted sample is measured using an aerosol spectrometer (Pallas Aerosol Technology, Model Wellas-1000, sensor range of 0.18-40 µm, concentration up to 10^5 particles / cm^3). The flow rate inside the measurement cell is fixed at 5 l/min. However, the flow rate inside the ESP is adjusted between 1.6 and 20 l/min and measured using a floating ball flow meter.

The basic configuration of the WST-ESP (Figure 2) consists of a glass square tube provided with two electrodes, one of which is grounded and the other is connected to a High Voltage (HV). The HV electrode consists of a stainless steel wire (0.20 mm in diameter) aligned on the central axis of the dielectric tube. The grounded electrode is made of aluminum tape strips (80 mm width and 80 µm thick) and is placed on the external surface of the tube.

The power supply system consists of a high voltage power amplifier (Trek, 30/20C, ±30 kV, ±20 mA), a function generator (TTI, TG1010, 10 MHz), a current probe (shunt resistor of 100 Ω), a high voltage probe (internal probe of the amplifier, and a digital oscilloscope (Lecroy 424, 200 MHz, 2 GS/s). The performance of the ESPs is studied at different values of HV amplitude (4-26 kV) and frequency (0.3-1000 Hz). All the experiments are carried out at atmospheric pressure and room temperature with controlled air flow rate.

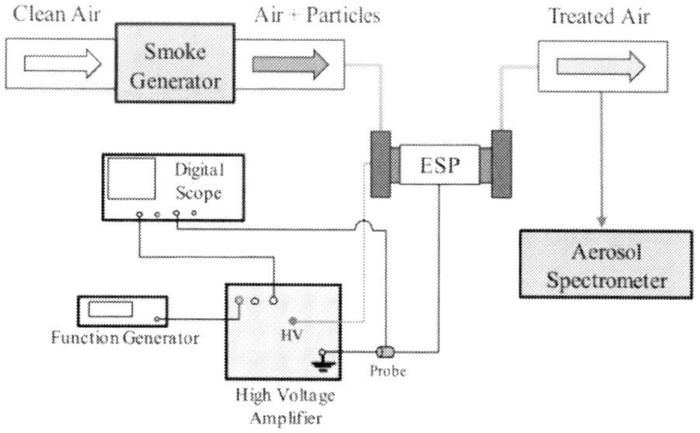

Figure 1. Schematic illustration of the experimental setup.

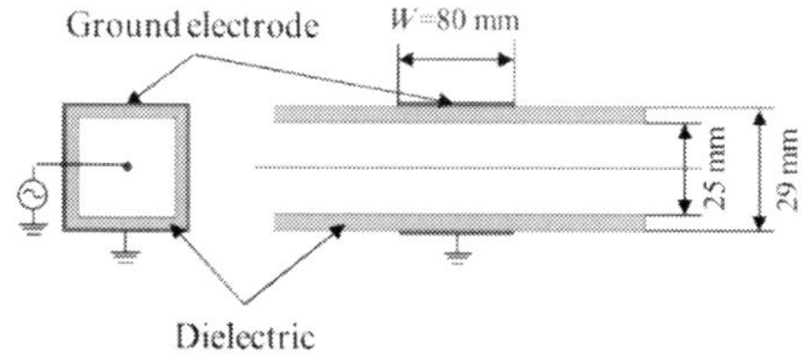

Figure 2. Cross-view of the WST-ESP

3. Results and Discussion

3.1. Current waveform

The current of the ESP includes only a few current pulses during the positive half-cycle, while there are numerous current pulses during the negative one (Figure 3). In the positive voltage half-cycle, the plasma is characterized by a glow-like regime. However, the Trichel pulses dominate the negative voltage half-cycle.

3.2. Collection efficiency

The total number collection efficiency (η) is defined as follows [4]:

$$\eta = 1 - \frac{N_{ON}}{N_{OFF}} \tag{1}$$

where N_{ON} and N_{OFF} are the number of particles per cm^3 for a given size-class with and without plasma, respectively.

To highlight the variations of the collection efficiency, especially when it is between 90 % and 100%, the ESP performance will be characterized by the penetration (P), which is given by:

$$P = 1 - \eta \tag{2}$$

The time-averaged power consumption ($P_{elec.}$) of the discharge is calculated from the measurement of the current (I) and the voltage (V) measured over one cycle:

$$P_{elec.} = \frac{1}{T}\int_{t}^{t+T} I(t)\cdot V(t)\cdot dt \tag{3}$$

In order to improve the reproducibility of the average power calculation, 10 cycles are taking into account using an acquisition memory of 250 kpts/Ch.

3.2.1. Effect of the wire diameter Figure 4 shows the effect of the wire diameter on the particle penetration of the WST-ESP. The experimental data point out that this factor has no impact on the ESP performance. This is an interesting result which shows that it is possible to use bigger wire with better mechanical characteristics without deteriorating the ESP performance.

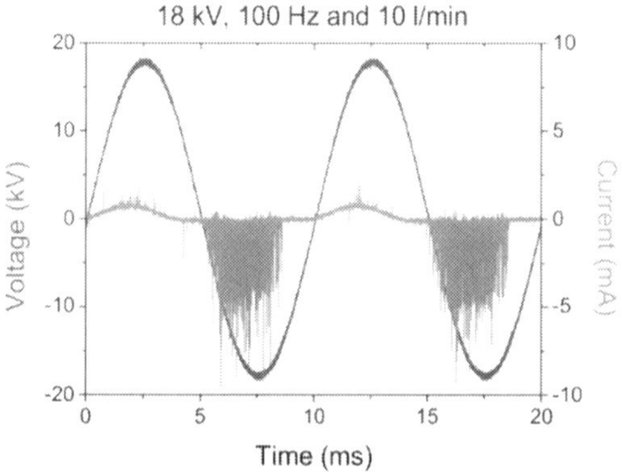

Figure 3. Time evolution of WST-ESP applied voltage and discharge current.

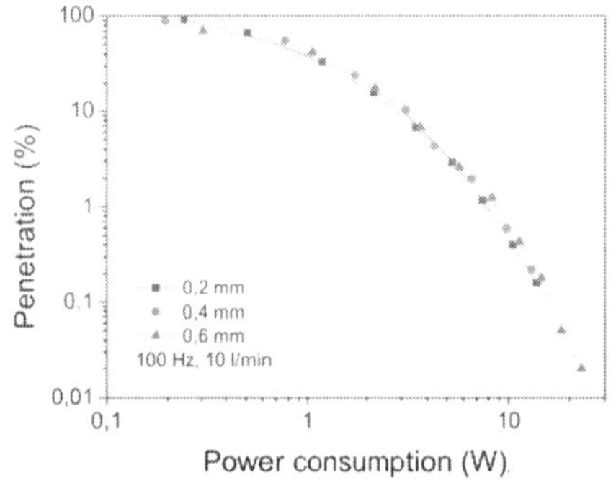

Figure 4. Effect of the wire diameter on the particle penetration of the WST-ESP.

3.2.2. Effect of the ground electrode Width Figure 5.a shows the particle penetration of the WST-ESP as a function of the power consumption for different ground electrode widths at 18 kV. Results show clearly a correlation between the performance of the ESP and the ground electrode width.

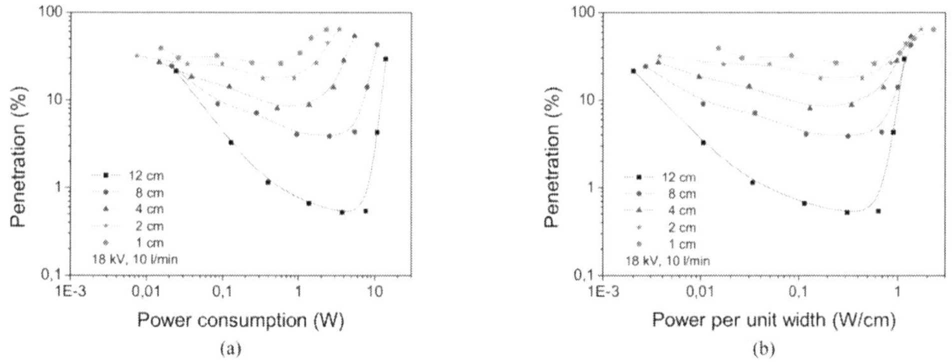

Figure 5. Effect of the ground electrode width on the particle penetration of the WST-ESP.

In figure 5.b the penetration is represented as a function of the power per unit width (Power consumption divided by ground electrode width). The performance of the WST-ESP increases with the ground electrode width. This is due to the increase of the residence time of the particles inside the collection volume.

3.2.3. Discretization of the ground electrode The effect of the ground electrode discretization on the collection efficiency of the WST-ESP is studied in this section. The initial ground electrode is divided into four strips of 20 mm width (Figure 6). The distance between two successive strips is set at 4 cm. Figure 7 shows the effect of the ground electrode discretization on the performance of the WST-ESP. The new ground electrode shape improves the efficiency of the WST-ESP at the same input power. This could be due to the tape edge effect. By increasing the number of edges, more power is injected in the inter-electrode gap and the areas where the electric field strength is intensified (between the wire and the tape edge) are more numerous. In addition, the residence time of particles between the first and last edges increases by discretizing the ground electrode.

3.2.4. Section of the tube Experiments were performed with three glass tubes with different internal sections (S_{T1}=1×1 cm^2, S_{T2}=1.8×1.8 cm^2 and S_{T3}=2.5×2.5 cm^2). In order to have the same air flow velocity inside the three ESPs (about 0.267 m/s), the tests were carried out at different flow rates: 1.6 l/min for the first, 5.1 l/min for the second and 10 l/min for the last one.

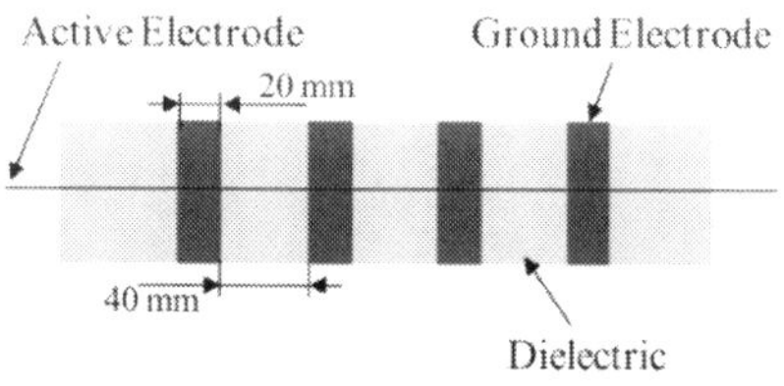

Figure 6. Side view of the WST-ESP with discretized ground electrode.

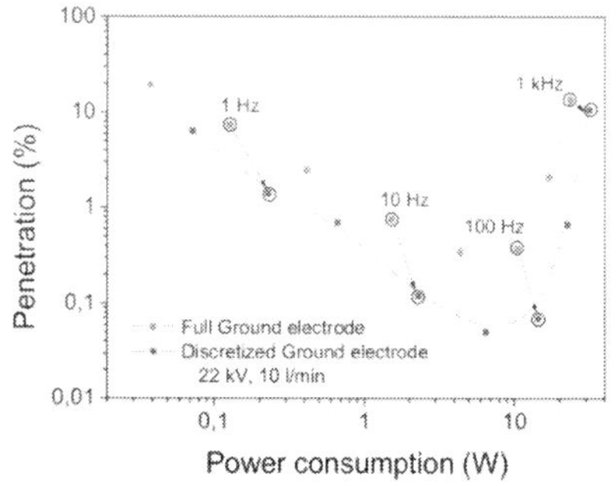

Figure 7. Effect of the ground electrode discretization on WST-ESP performance.

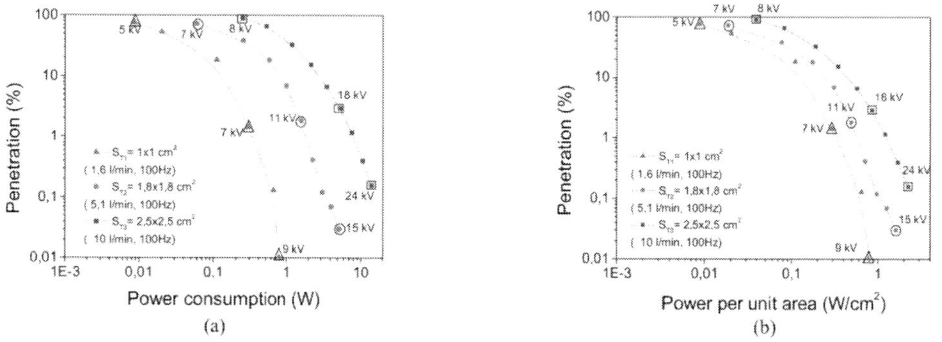

Figure 8. Effect of the section of tube on the performance of the WST-ESP.

The particle penetration varies with the power consumption (figure 8.a). Considering the fact that the volume of gas treated by the three ESPs during the same time interval is different, the particle penetration (figure 8.b) is represented as a function of the power per unit area (Power consumption divided by the section of the tube). The efficiency of the WST-ESP seems to be increased by using smaller sections. This result confirms the interest of clustering several WST-ESPs in parallel.

4. Conclusion

The collection of submicron particles by using a lab-scale barrier discharge ESP was investigated in the case of wire-to-square tube configuration. The parametric study of the WST-ESP shows that using bigger wire diameter does not deteriorate the ESP performance. The penetration decreases with larger or discretized ground electrode. The collection efficiency increases by using smaller sections.

References

[1] M. Jedrusik, J. B. Gajewski, A. J. Swierczok 2001 *J. Electrostat.* **51-52** 245

[2] B. Dramane, N. Zouzou, E. Moreau and G. Touchard 2009 *IEEE Trans. Dielectr. Electr. Insul.* **16** 343.

[3] B. Dramane, N. Zouzou, E. Moreau and G. Touchard 2009 *J. Electrostat.* **67** 117

[4] A. Zukeran, P.C. Looy, A. Chakrabarti, A.A. Berezin, S. Jayaram, J.D. Cross, T. ITo, and J. S. Chang 1999 *IEEE Trans. Ind. Appl.* **35** 1184

Nonwoven filtration mat production by electrospinning method

M. Lackowski, A. Krupa, A. Jaworek

Institute of Fluid Flow Machinery, Polish Academy of Sciences, Fiszera 14, 80-952 Gdansk, Poland

E-mail: mala@imp.gda.pl

Abstract. The filtration of nanoparticles and submicron particles is an important problem in industry and health protection. One of the methods which can be used to solve this problem is to use nonwoven nanofibrous filters. The process of producing filtration mats of different thickness by electrospinning is presented in the paper. The experimental results on filtration properties of nanofibrous filter mat, including the efficiency of removal of cigarette smoke particles from a gas are also presented.

1. Introduction

The filtration of submicron particles, including nanoparticles, is an important problem in industry, environment and health protection. One of the methods which can be used to solve this problem is that which uses the nonwoven nanofibrous filtration mat. The most effective method of the nanofibrous mat production is electrospinning [1, 2]. Mats produced by electrospinning are of uniform density and built from fibers of nearly the same diameter. Most of the papers published in the subject of electrospinning are intended for the production of nanofibers for various nanotechnology applications, with increasing number of applications in biotechnology [3], and recently, explore the problems of fabrication of nanocomposite membranes for gas cleaning applications [4 - 8].

The physical background of electrospinning lies in the utilization of electrical forces for generating shear stress on the surface of a viscous liquid, usually a polymer solution, flowing from a capillary nozzle. Under this stress, the jet becomes thinner, and finally a thin fiber is formed after solvent evaporation. The method has been applied for the production of nanofibrous mats made from polymer material. The goals of this paper are to determine the morphology of the mats produced by the developed technology and study the efficiency of removal of micro and nanoparticles. This type of mat could be used for the production of masks and filters.

In this paper, we demonstrate the electrospinning technology used for the production of nanofibrous filtration mat in the form of non-woven fabric, and the efficiency of removal of cigarette smoke particles from a gas.

2. Experimental set-up

A photograph of the experimental set up is shown in Figure 1. A thin metal nozzle made from a hypodermic needle of outer diameter 0.45 mm was designed for the production of nanofibers. During the process of electrospinning, the polymer jet flowing out from the capillary nozzle is elongated, forming a thin thread, which becomes thinner due to the solvent evaporation. By this method, a nanofibrous, polymer filtration mat can be produced. In the experiments, the electrospun fibers were deposited onto a horizontally rotating grounded plate with a thin metal grid stretched on metal frames (dimensions 15 mm x 15 mm). The distance between the nozzle tip and the table was about 120 mm. The electrospinning of nanofibers was carried out for the poly(vinyl chloride) (PVC) and poly(vinylidene fluoride) (PVDF) polymers. The PVC was dissolved in 1:1 dimethylformamide (DMF) and tetrahydrofuran (THF) mixture by stirring at room temperature to obtain 9% solution. The 15% PVDF solution was obtained by dissolving 0.9 g PVDF in 2.48 g DMAC (N, N Dimethylacetamide) and 2.84 g acetone. These polymers were chosen as the mat material because they are chemically resistant and have good mechanical properties. The spinning was carried out at room temperature and humidity of 45-50%. The flow rate of the polymer solutions was 1 ml/h, and the voltage was: 12 kV for PVC, and 14 kV for PVDF.

The morphology of the nanofibrous filtration mat produced by electrospinning was tested under a scanning electron microscope Zeiss Evo 40. The diameter of the PVC fibers was in the range of 600 to 800 nm, and PVDF 400 to 600 nm.

The fibers were deposited on the grid substrate for times of 8, 16, and 24 min in order to produce mats of various thicknesses. After the electrospinning process was completed, each electrospun mat was dried in a vacuum chamber.

Figure 1. Stand for the production of nonwoven filtration mat.

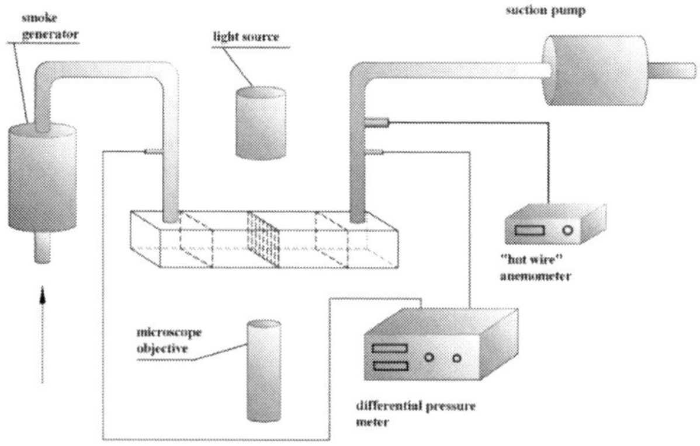

Figure 2. Schematic diagram of experimental stand.

In the paper, the investigations of mechanical properties of nanofabrics were carried out by the optical method. Deformation of nanofibrous mat mounted in a small channel with cross section of 15mm x 15mm, under the air flowing through the channel was observed using microscope NIKON Eclipse TS-100F (Figure 2). The pressure drop on the nanofibrous mat filter of different thicknesses was measured for different air flow rates. The face velocity of the air was measured by hot-wire anemometer TSI 8455.

The efficiency of removal of nanoparticles was measured in a small laboratory channel. The method of measurement of collection efficiency was based on the determination of concentration of particles in the channel before and after the filtration mat by the optical microscope technique. This method was based on light extinction measurement by microscope camera DS1 produced by NIKON. Cigarette smoke was used for testing the removal efficiency of smoke particles. Before and after the experiment, the filtration mat was tested under scanning electron microscope ZEISS EVO 40 in order to estimate qualitatively the effect of particles deposited onto the fibrous filter morphology.

3. Results

The SEM micrographs of PVC polymer mats produced by electrospinning method are shown in Figure 3. Figure 3 (a) presents an image of a clean fibrous filter, and Figure 3 (b) an image of the filter after cigarette smoke treatment. Because the smoke particles are mainly composed of tar matter, the fibers are covered by a thin film of tar which wetted the fiber material. Some thin membranes, which bridge the fibers can also be observed.

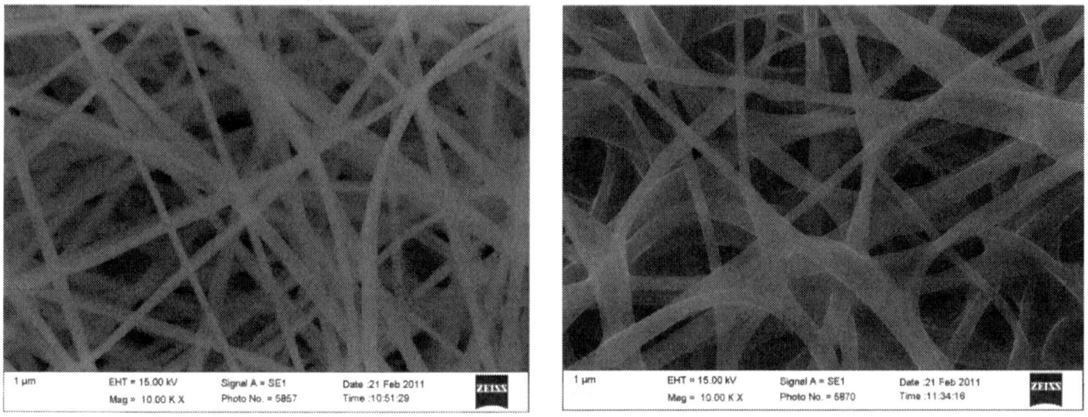

a) b)

Figure 3. PVC nanofibrous polymer mats. (a) Clean filter, (b) Filter after cigarette smoke treatment

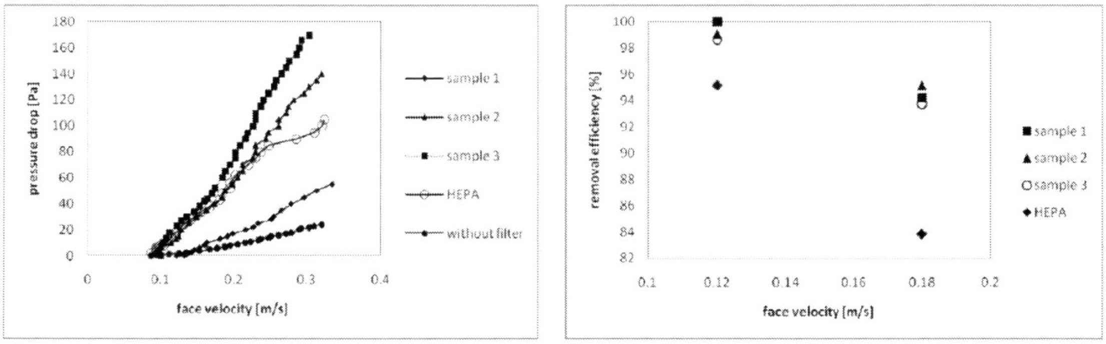

a) b)

Figure 4. Pressure drop (a) and cigarette smoke removal efficiency (b) measured by the optical method. PVC nanofibrous polymer mats. Sample 1: 8 min deposited on the grid substrate, Sample 2: 16 min, and Sample 3: 24 min.

The pressure drop on the filter is a linear function of gas velocity that indicates the viscous flow of air through the filtration mat. The collection efficiency of the nanofibrous mat is higher than for a HEPA filter (Figure 4b) by similar pressure drop that is very important in air conditioning systems. The difference in collection efficiency between nanofibrous filter and HEPA filter increases with increasing face velocity.

4. Summary

The experimental results on filtration properties of nanofibrous filter mat, including the efficiency of removal of cigarette smoke particles from a gas are presented in the paper.

The nanofibrous filtration mat produced by electrospinning method has fibers of diameter in the range from 400 to 800 nm and is of high porosity. Mats produced by electrospinning are of uniform density and built from fibers of nearly the same diameter that allows a uniform pressure drop distribution to be obtained over the entire surface of the filter.

The experimental results presented have shown that a nonwoven nanofibrous filtration mat has a good filtration efficiency for nano and submicron particles, better than HEPA filters. Pressure drop on the nanofibrous filtration mat is similar to HEPA filters

Acknowledgement

This research work was supported by the Polish Ministry of Science and Higher Education within the Project No.4169/T02/2009/37.

References

[1] Doshi J, Reneker D H 1995. J. Electrostatics **35** 151
[2] Ramakrishna S, Fujihara K, Teo W E, Lim T Ch, Ma Z, 2005 World Scientific, Singapore
[3] Teo W E, Ramakrishna S, 2006 Nanotechnology **17,** R89
[4] Ramakrishna S, Fujihara K, Teo W E, Yong T, Ma Z, Ramaseshan R, 2006 Materials Today **9** March, 40
[5] Jaworek A, Krupa A, Lackowski M, Sobczyk A T, Czech T, Ramakrishna S, Sundarrajan S, Pliszka D, 2009 J. Electrostatics **67** 435
[6] Jaworek A, Krupa A, Lackowski M, Sobczyk A T, Czech T, Ramakrishna S, Sundarrajan S, Pliszka D, 2009 Fibers and Textiles in Eastern Europe **17** 77
[7] Sundarrajan S, Pliszka D, Jaworek A, Krupa A, Lackowski M, Ramakrishna S, 2009 Journal of Nanoscience and Nanotechnology **9** 4442
[8] Sundarrajan S, Ramakrishna S, 2007 Journal of Materials Science **42** 8400

Electrostatic application of antimicrobial sprays to sanitize food handling and processing surfaces for enhanced food safety

Shawn M Lyons[1], Mark A Harrison[2] and S Edward Law[3]

[1]MS Student and [2]Professor – Food Science and Technology Department, University of Georgia, Athens, GA, USA 30602-2610

[3]Professor – Biological and Agricultural Engineering Department, Applied Electrostatics Laboratory www.ael.engr.uga.edu , University of Georgia, Athens, GA, USA 30602-4435

E-mail: edlaw@engr.uga.edu

Abstract. Human illnesses and deaths caused by foodborne pathogens (*e.g., Salmonella enterica, Listeria monocytogenes, Escherichia coli* O157:H7, *etc.*) are of increasing concern globally in maintaining safe food supplies. At various stages of the food production, processing and supply chain antimicrobial agents are required to sanitize contact surfaces. Additionally, during outbreaks of contagious pathogenic microorganisms (*e.g.,* H1N1 influenza), public health requires timely decontamination of extensive surfaces within public schools, mass transit systems, *etc.* Prior publications verify effectiveness of air-assisted, induction-charged (AAIC) electrostatic spraying of various chemical and biological agents to protect on-farm production of food crops...typically doubling droplet deposition efficiency with concomitant increases in biological control efficacy. Within a biosafety facility this present work evaluated the AAIC electrostatic-spraying process for application of antimicrobial liquids onto various pathogen-inoculated food processing and handling surfaces as a food safety intervention strategy. Fluoroanalysis of AAIC electrostatic sprays (-7.2 mC/kg charge-to-mass ratio) showed significantly greater ($p<0.05$) mass of tracer active ingredient (A.I.) deposited onto target surfaces at various orientations as compared both to a similar uncharged spray nozzle (0 mC/kg) and to a conventional hydraulic-atomizing nozzle. Per unit mass of A.I. dispensed toward targets, for example, A.I. mass deposited by AAIC electrostatic sprays onto difficult to coat backsides was 6.1-times greater than for similar uncharged sprays and 29.0-times greater than for conventional hydraulic-nozzle sprays. Even at the 56% reduction in peracetic acid sanitizer A.I. dispensed by AAIC electrostatic spray applications, they achieved equal or greater CFU population reductions of *Salmonella* on most target orientations and materials as compared to uncharged sprays and conventional full-rate hydraulic-nozzle sprays.

1. Introduction

Foodborne pathogenic microorganisms in the USA annually cause ~7.6 million acute illnesses, ~5 thousand deaths, and an associated US$152 billion cost in healthcare, workplace losses, *etc.*; nearly 25% of this public-health problem is attributable to unprocessed fresh fruits and vegetables [1]. Decontamination of food-contact surfaces and equipment in food handling and processing facilities presents challenges due in part to lack of effective cleaning and sanitizing options. Inefficient application of antimicrobial chemicals by conventional high-volume sprays often does not evenly cover and protect all surfaces…especially the hard to reach and hidden target areas. Additionally, high liquid volumes are logistically problematic and tend to detrimentally soak cardboard packing cartons carrying the food product. As hypothesized based upon prior work [2], a promising alternative is offered by the air-assisted, induction-charged (AAIC) electrostatic-spraying process in which a reduced liquid-carrier volume of finely atomized, highly charged droplets is aerodynamically propelled (AA..) into the target vicinity where electric forces of attraction efficiently deposit the charged droplets uniformly onto all target surfaces, both directly exposed and obscured. Especially appropriate here is droplet charging by electrostatic induction (..IC) within the nozzle because it accommodates the wide range of target materials and resistances to earth encountered in food handling/processing operations, in contrast to industrial coating by the ionized-field "corona" droplet charging process which demands electric current flow to earth through well grounded targets.

Both increased mass-transfer efficiency and improved biological-control efficacy are reported for AAIC electrostatic spraying of conductive liquids [3, 4]. The highly charged droplets (*e.g.*, 30 μm dia. @ 5-10 mC/kg) are strongly attracted to opposite polarity charge naturally induced into the target and provide uniform coverage of its surface including wrap-around onto backsides [5, 6]. AAIC electrostatic-spraying has been documented successful for applying both biological and chemical pesticides to crops [7, 8]; antifungal preservatives to fresh produce (*viz.*, bananas) [9]; and decontaminant and tanning agents to human skin [10]. The electrostatic deposition benefit for these chemical and biological control agents was increased typically 2-5 fold while using much lower volumes of carrier liquid. Consequently, effective pest and pathogen control can be maintained while dispensing reduced amounts of active ingredient from the nozzle.

The objective of this present food-safety work was to experimentally evaluate the mass-transfer efficiency and antimicrobial efficacy of the AAIC electrostatic-spraying process for applying sanitizer liquids to various surface orientations and compositions commonly encountered in food handling and processing operations…specifically investigating whether the lower volume of spray-carrier liquid (with higher concentration of sanitizer active ingredient) deposited on the target surface would provide sufficient contact time for microbial cell death, as well as determining adequacy of target charge-transfer for various electrically insulating *vs.* conductive food-contact materials.

2. Experimental methods and materials

Within an air-exhausted metal biosafety chamber (1.5 m high x 1 m wide x 1 m deep) an AAIC electrostatic-spraying nozzle [3], commercially provided by Electrostatic Spraying Systems, Inc. www.maxcharge.com and herein designated ESS, was operated with charging ON (-7.2 mC/kg @ 1.2 kV, 207 kPa atomizing air) and charging OFF (0 mC/kg @ 0 kV, 207 kPa) and compared to a conventional hydraulic-atomizing nozzle (TeeJet® #TP40015E @ 295 kPa liquid pressure) to determine mass transfer of active ingredient (A.I.) deposited onto target coupons (23 mm x 90 mm x 1 mm thick). These three spray-nozzle treatments were similarly evaluated for antimicrobial efficacy of sanitizer (peracetic acid $C_2H_4O_3$) they achieved on stainless steel, PVC conveyor belting, and waxed cardboard expressed as reduction in the population of colony forming units (CFU) of *Salmonella enterica* previously inoculated onto these target materials – all assessed to a detection limit of 80 CFU/coupon face using standard microbiological practices. Fluorometric analysis detailed previously [11] was used to quantify mass transfer of an aqueous suspension of fluorescent tracer powder (tracer analogous to mass of A.I.) deposited by each spray-nozzle treatment onto target coupons having surfaces positioned in two orientations with respect to the incoming spray vector: surface perpendicular (frontside, backside) *vs.* parallel. A bactericidal equivalence point was experimentally determined for peracetic acid sanitizer at which this lower than conventionally recommended full-rate mass of A.I. when dispensed from a charged electrostatic nozzle achieved the same population reduction of *Salmonella* as did the full-rate when dispensed from the conventional hydraulic nozzle onto stainless steel frontside surfaces. The equivalence point mass of A.I. was then used to determine population $\log_{10}$ reduction on the three target-material types in the two orientations.

Because the hydraulic nozzle dispensed 6-times more carrier-liquid volume than did the electrostatic nozzle (Hydr.@ 600 ml/min *vs.* ESS @ 100 ml/min), an unbiased comparison having equal mass of A.I. dispensed toward the targets by all nozzle treatments required the A.I. concentration of liquid sprayed from the ESS nozzle to be 6-times that from the hydraulic nozzle. Furthermore, nozzle-to-target spacing (Hydr. @ 42 cm *vs.* ESS @ 70 cm) provided equal width spray swaths (30.5 cm) traversing the target coupons ensuring an identical incoming mass-flux density of A.I. (mg cm^{-2} s^{-1}) for all spray-nozzle treatments as a robotic-arm apparatus [7, 12] accurately replicated a back and forth dual traverse at 76 cm/s (±1%) along the target arc holding four coupons.

3. Results and discussion

3.1. Mass-transfer efficiency

Figure 1 plots the fluorometrically determined treatment mean values for amount of active ingredient tracer deposited (ng/cm^2) by spray applications using the conventional hydraulic-atomizing nozzle compared with the ESS ON (Ch.) and ESS OFF (Unch.) electrostatic nozzle. Statistical analysis (ANOVA) by Tukey's multiple comparison method indicates spray-nozzle treatment mean values within any plotted grouping-of-three differ significantly ($p<0.05$).

As seen where air-assist velocity was strongest onto perpendicular target frontsides, charged spray increased deposition of A.I. to a value 1.2-times greater than uncharged spray and 9.1-times greater than hydraulic-nozzle spray; for backsides, the charged-spray increases were respectively 6.1-times and 29.0-times. For target surface parallel to the spray vector's air-assist velocity, the electrostatic-deposition benefit was 2.9-times *vs.* uncharged and 6.1-times *vs.* hydraulic sprays.

3.2. Antimicrobial efficacy

For the 0.44 dosage equivalency point experimentally determined for peracetic acid sanitizer (Oxonia Active® by EcoLab), Figure 2 plots results of subsequent surface-sanitizing efficacy achieved on the indicated target materials and orientations when dispensing manufacturer's recommended full-rate active ingredient by the hydraulic-nozzle spray *vs.* 0.44-rate by the ESS ON (Ch.) and ESS OFF (Unch.) electrostatic nozzle. On logarithmic scales, the treatment mean values (n = 3 reps., standard error bars) document the population reductions of *Salmonella* expressed as $\log_{10}$ CFU/coupon face.

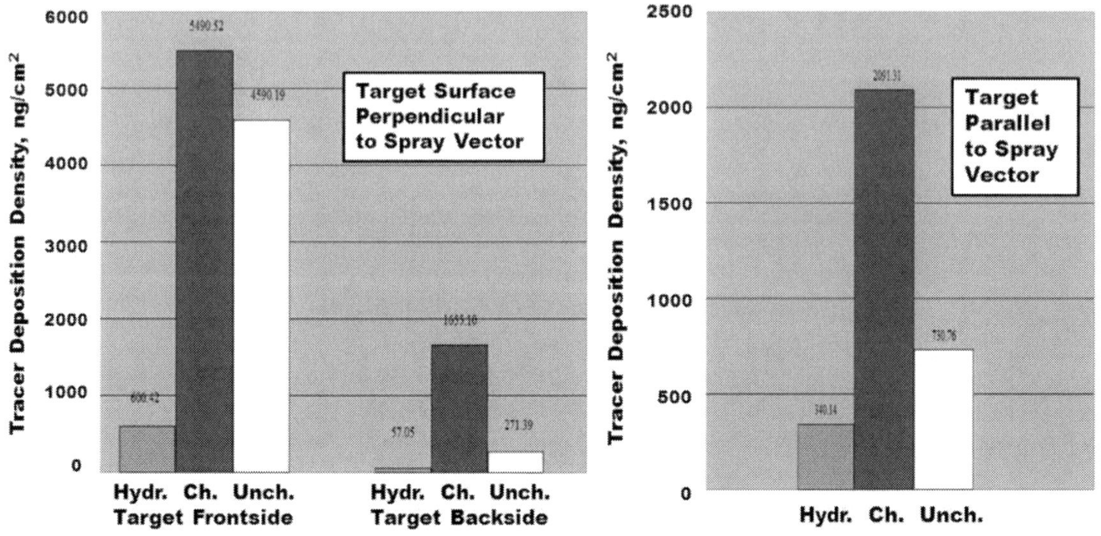

Figure 1. Mean values of areal density of active ingredient (*i.e.*, fluorescent tracer) deposited by the three spray-nozzle treatments onto stainless steel target coupons for the indicated target orientations (n = 3 reps.; all nozzles dispensed equal mass of A.I. toward targets).

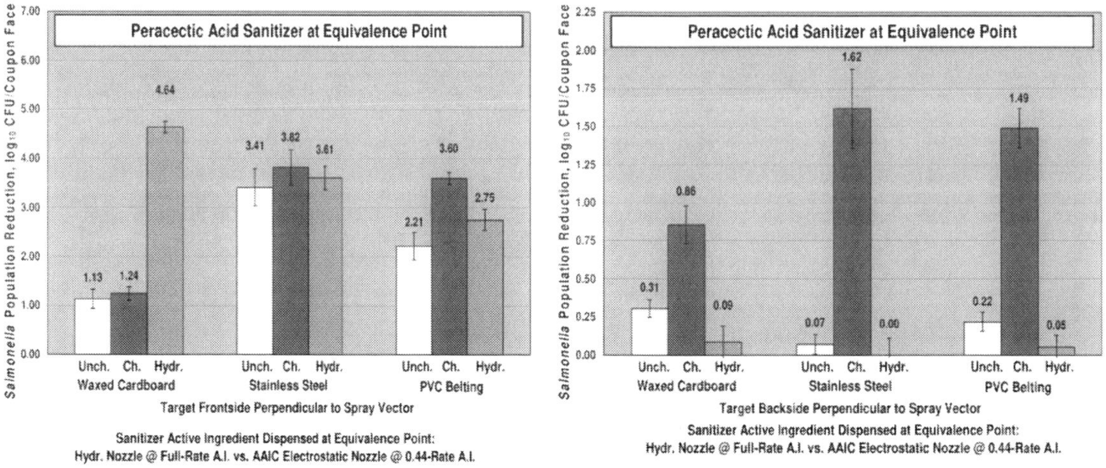

Figure 2. Mean values for $\log_{10}$ reductions of *Salmonella enterica* surface contamination achieved using peracetic acid applied by the three spray-nozzle treatments onto indicated target materials and orientations (n = 3 reps.; A.I. dispensed full-rate Hydr. *vs.* 0.44-rate electrostatic nozzles).

As seen in Figure 2 even at its 0.44-rate of sanitizer A.I. dispensed, the air-assisted charged spray provided equal or greater population reduction of *Salmonella enterica* on all target materials and orientations...except for the vertical frontside waxed cardboard which exhibited hydrophobicity impeding the filming of the reduced-volume spray applications.

3. Conclusion

This study showed significant ($p<0.05$) increases in both spray-deposition efficiency and surface-sanitizing efficacy were provided by reduced carrier volumes of finely atomized, air-assisted, induction-charged (-7.2 mC/kg) antimicrobial spray as compared to similar uncharged spray and to conventional high-volume, hydraulic-atomized spray. Electrostatic deposition benefit up to 29-times A.I. was measured on backside surfaces. For most target materials and orientations tested, 44%-rate A.I. peracetic acid sanitizer dispensed in 1/6th the spray volume achieved equal or greater population reductions of *Salmonella enterica* as did full-rate A.I. applied by conventional hydraulic-nozzle spray.

Acknowledgment

USDA Natl. Inst. of Food and Agrl. food-safety grant #2009-51110-20161 partially funded this work.

References

[1] US Dept. HHS, CDC, Food Safety Office 2008 http://www.cdc.gov/foodsafety
[2] Univ. Ga. Applied Electrostatics Laboratory http://www.ael.engr.uga.edu
[3] Law S E 1978 *Trans. ASAE* **21** 1096-1104
[4] Law S E and Mills H A 1980 *J. Amer. Soc. Hort. Sci.* **105** 774-777
[5] Law S E and Cooper S C 1989 *Trans. ASAE* **32** 1169-1172
[6] Evans M D, Law S E and Cooper S C 1994 *Appl. Engr. in Agrl.* **10** 441-447
[7] Law S E and Scherm H 2005 *J. Electrostatics* **63** 399-408
[8] Giles D K and Blewett T C 1991 *J. Agric. Food Chem.* **39** 1646-1651
[9] Law S E and Cooper S C 2001 *IEEE Trans.* **IA-37** 1597-1602
[10] Law S E, Cooper S C and Harrison M A 2004 *J. Phys.: Conf. Ser.* **178** 331-336
[11] Law S E, Cooper S C and Law W B 1999 *J. Phys.: Conf. Ser.* **163** 243-248
[12] Law S E and Wetzstein H Y 2000 *IEEE Trans.* **IA-36** 998-1009

Dispensing and bio-functionalization of giant unilamellar vesicles on a chip

K V I S Kaler and R Prakash

2500 University Drive NW, Calgary, AB – T2N 1N4, CANADA
kaler@ucalgary.ca

Abstract. Giant unilamellar vesicles (GUVs) [1], typically >10 μm in diameter, can serve as interesting models for mimicking biological membranes and furthermore can be usefully leveraged in drug delivery and bio-sensor applications. In this paper, we have demonstrated a chip based vesicle dispensing scheme that utilizes liquid dielectrophoresis (L-DEP) to dispense, precisely positioned GUVs over a range of diameter 30 – 100 μm, a capability that is not readily achieved by either conventional or microchannel technology. These GUVs can also be individually addressed and tailored for a variety of applications such as bio-sensing and cell-on-a-chip.

1. Introduction

Giant unilamellar vesicles or, GUVs [1] are lipid bilayer vesicles typically with diameter in range of 10-100 μm. There is significant interest in the formation and bio-application of GUVs for numerous reasons. Size of GUVs makes them optimal for mimicking cell behavior and creating artificial cell models [2], often known as cell-on-a-chip. GUVs furthermore may be used to encapsulate intact cells and other bio-molecules in a controlled environment [3] and thus utilized for tailored bio-sensing applications [4].

There are very few standard protocols for producing GUVs and the conventional vesicle formation schemes such as ultrasonication and phase inversion methods are not suitable for forming large mono-dispersed lipid vesicles. The two more popular schemes for forming GUVs are: spontaneous swelling and electroformation [1]. Both of these methods require dry lipid powders on solid substrates with specific solvents in order to produce large vesicles. But, these schemes also have limitations in terms of the population and potential screening of the dispensed GUVs. The evidence clearly demonstrates the need for a more compact and integrated, on-chip GUV dispensing scheme that can precisely dispense arrays of large vesicles and furthermore manipulate the GUVs by individually addressing them in parallel, automated steps. We have previously demonstrated that L-DEP based precision emulsion dispensing scheme can be leveraged to dispense large emulsion droplets (dia. ~ 30-100 μm) [5]. In this work, we utilize a simple integration of L-DEP based dispensing [6, 7] and electrostatic droplet manipulation schemes (or, D-DEP) [8] to dispense and manipulate GUVs, encapsulating cell-sized polystyrene microbeads (dia: 10 μm) with bio-functionalized surface. The GUVs are furthermore utilized to perform a standard DNA-bead binding assay to demonstrate the potential features of our chip based dispensing and bio-functionalization of large phospholipid bilayer vesicles.

2. Materials and Methods

2.1 Materials

Two phospholipid samples, namely POPC (1-palmitol-2-oleoyl-sn-glycero-3-phosphocholine) and NBD-PC (1-acyl-2-{6-[7-nitro-2-1,3-benzoxadiazol-4-yl)amino]hexanol}-sn-glycero-3 phosphocholine), were purchased from Avanti Polar Lipids as 99% pure chloroform solutions. The later (NBD-PC) has a truncated fatty acid chain to incorporate a fluorophore (λex./em.: 460/534 nm; green emission) and was used to facilitate better observation of formed lipid membranes. First a mineral oil dispersion of the two lipid samples was prepared. A combination of 90 μL POPC and 10 μL NBD-PC stocks was spread in a pyrex glass bottom and the chloroform solvent was evaporated under air flow to obtain a thin, dry mixture of the two phospholipids. 10 mL of mineral oil (Crystal Plus 70fg; viscosity ~ 12.7 cSt; STE Oil Company) was then added to the glass bottom and the mixture was ultra-sonicated for approximately 60 min. and then left overnight to prepare uniform, stable lipid dispersion in mineral oil.

A 5 bead/nL sample containing 10 μm diameter polystyrene microbeads and a single-stranded DNA (ssDNA) oligonucleotide with a specific fluorophore (TAMRA™ : λex/em: 546/574 nm; red emission) sample was also prepared in de-ionized water (conc. ~ 2 μM).

2.2 L-DEP based dispensing of GUVs

Device fabrication protocols and dispensing of single emulsion droplets leveraging L-DEP has been reported in [9]. A ~ 1 μL droplet of polystyrene microbead suspension sample (conc. ~ 5 beads/nL) is covered by a comparable sized mineral oil-lipid dispersion sample and placed on one end of the L-DEP electrode scheme (Figure 1(a)). Upon application of an AC voltage (480 Vpp at 100 kHz), an

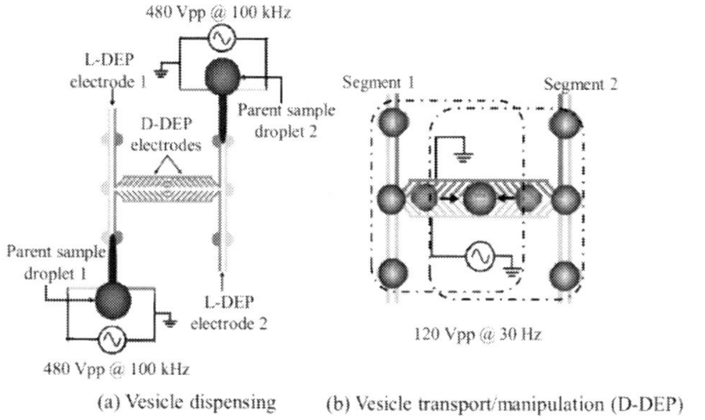

Figure 1: (a) A simple electrode arrangement for L-DEP based vesicle and nanoliter droplet dispensing; (b) Electrostatic manipulation and mixing of droplets and vesicles.

emulsion jet is formed and rapidly propagated to cover the entire L-DEP electrode length [9]. Upon removal of the externally applied voltage, the emulsion jet rapidly disintegrates into precisely positioned single emulsion droplets, containing a lipid monolayer at the aqueous-oil interface (Figure 2(a)). It is important to note that during L-DEP finger actuation, the polystyrene microbeads are positioned at the aqueous-oil boundary and hence coated with functional lipid membrane (Figure 2(b)). This binding of lipid molecules to the bead surface can be enhanced by utilizing specific ligand binding between a biotin molecule (located in lipid bilayer) to avidin (immobilized on the bead surface) [10]. The experimental results of the stable avidin-biotin conjugated DNA-lipid binding will be presented at the conference. At this stage, the L-DEP chip is immersed in lipid-mineral oil dispersion media, partially doped with comparable viscosity silicon oil, to diffuse the outer oil boundary and suspend the uniform monolayer lipid vesicles (vol. ~ 1 nL; dia. ~ 60μm) into an oil bath (Figure 2(c)). The bilayer is assembled by introducing a third aqueous phase (viscosity controlled by increasing glycerol concentration) on top of the oil bath. The sinking interface consists of a second lipid monolayer which results in the bilayer formation over the dispensed daughter droplet, encapsulating functionalized microbeads (Figure 2(d), 2(e)).

2.3 ssDNA binding to the functionalized microbead surface

Large fatty acids in the hydrophobic tail of lipid molecules provide a soluble molecule with suitable binding sites for bio-molecules including DNA, RNA and various proteins [3]. In this work, we have utilized an ssDNA oligonucleotide, containing a fluorophore (TAMRA™) to demonstrate binding between lipid molecules and the ssDNA molecule. An integrated electrode scheme was used (see Figure 1(b)), where electrostatic droplet actuation electrodes (or, D-DEP electrodes) are suitably

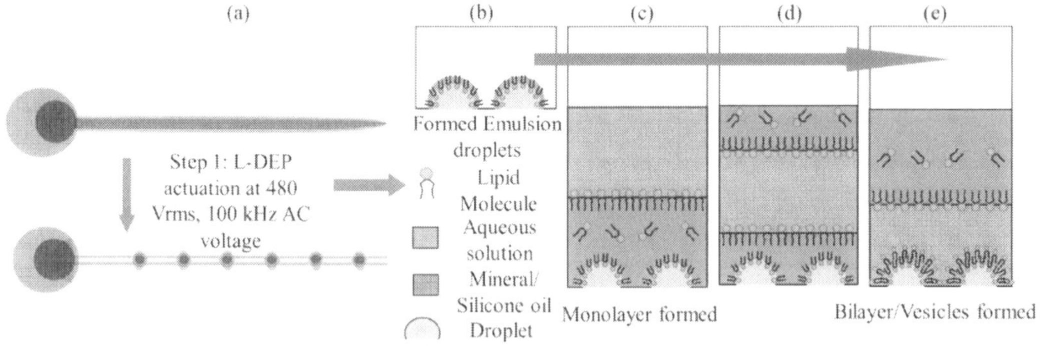

Figure 2: Schematic diagrams showing the experimental steps: (a) L-DEP actuation and formation of single emulsion droplets; (b) Dispensed emulsion droplets with lipid monolayer (Radius.: 40-60 μm); (c) Lipid monolayer droplets formed in mineral/silicone oil bath; (d) high viscosity aqueous glycerol soln. (~ 10 cSt) sinks in the lighter oil bath; (e) Lipid bilayer/vesicles formed.

integrated with the L-DEP electrode scheme, to facilitate transportation of dispensed daughter droplets and vesicles to specific reaction/mixing sites. For this assay, first L-DEP electrode scheme was used to dispense a single array of lipid monolayer vesicles with 5-6 lipid coated beads per droplet. On another L-DEP electrode, arranged in parallel to the first set, a second parent droplet of ssDNA sample (conc. ~ 2 μM) was actuated, under the oil bath. A pair of monolayer lipid vesicle and ssDNA daughter droplet was then transported and mixed using fishbone shaped D-DEP electrode scheme (Figure 1(b)). The mixed vesicle and unmixed droplets are shown in Figure 3(a-f). Since the microbeads have a lipid membrane sitting on their surface, they facilitate binding of ss-DNA molecules. Finally, the GUV is formed from the mixed lipid mono-layered vesicles, as explained earlier in section 2.2.

3. Results and Discussion

The reported experiments demonstrate the versatile applications of GUVs in housing and transporting bi-sensors. In absence of a lipid membrane, DNA or other bio-molecules have a lower affinity to bind to polystyrene bead surface. However, as discussed earlier, the presence of a stabilized lipid molecule at the interface provides sites for interaction between the DNA molecule and the microbead. The results of such a DNA binding assay are shown in Figure 3. Figure 3(a-f) shows unmixed lipid

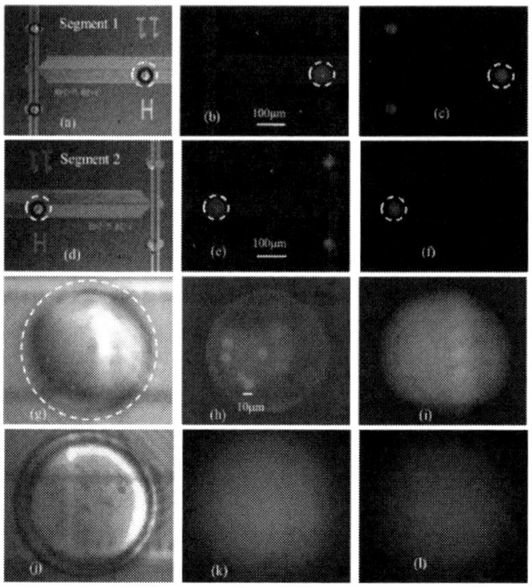

Figure 3: Micrographs showing experimental results. (a, b, c) unmixed ssDNA droplets and mixed monolayer vesicle observed under white, blue (λ_{blue} = 460-480 nm) and green light (λ_{green} = 510-540 nm); (d, e, f) unmixed and mixed lipid monolayer vesicles under white, blue and green light; (g) bilayer vesicle at mixing site encapsulating functionalized microbeads ; (h, i) bilayer vesicle at mixing site with coated bio-functional microbeads showing fluorescence when exposed to blue and green light (resolved microbeads); (j) bilayer vesicle at mixing site encapsulating non-functionalized microbeads (Triton X-100) ; (k, l) bilayer vesicle at mixing site with non-coated microbeads showing only bulk fluorescence when exposed to blue and green light (unresolved microbeads).

vesicles and DNA daughter droplets on L-DEP electrodes, along with the mixed vesicle (centre). Both green and red fluorescence emission was observed from the mixed vesicle and the surface of coated microbeads was resolvable from the bulk emission resulting from the unbound DNA molecules.

Finally, the second lipid layer was self-assembled [5] to form bilayer vesicles (Figure 3(g, i)) and the bio-functional microbeads were encapsulated inside the vesicles. As a control, 0.1 % of Triton X-100 was added to the first parent droplet (containing microbeads) which resulted in degradation of formed lipid membranes and in this case, minimal to no DNA binding was observed. However, since Triton X-100 was only present in the inner sample, the bilayer formation is unaffected and the GUV was again formed for this droplet (Figure 3(j-l)). This effect has been demonstrated in Figure 3(k, l) where due to minimal DNA binding to bead surface, the microbeads cannot be resolved under either blue (for green emission) or green (for red emission) light exposure and entire signal was recorded from the bulk of the GUV.

4. Conclusion

This work demonstrates that on-chip integrated electrode schemes such as L-DEP and D-DEP can be practically leveraged to achieve precision dispensing and subsequent manipulations of functionalized lipid vesicles, which can either themselves serve as bio-sensor or can be leveraged to create artificial test beds for multiple bio-assays in a rapid, automated fashion. The superiority of vesicular bio-sensors in comparison to solid surface based bio-sensors is yet to be fully confirmed.

5. Future Work

We intend to demonstrate the superiority of vesicle based bio-sensors when compared to the conventional bio-sensors [4] and utilize specific lipid and DNA molecules that can provide stable and selective linkage, such as the biotin-avidin linkage resulting in more stable and irreversible binding of DNA or other bio-molecules for preparing tailored bio-sensors. Quantum Dots can replace micro-beads in order to provide enhanced quantum efficiency and more stable photo-response in future bio-sensing applications [11].

6. Acknowledgements

The authors acknowledge the financial support received from National Science and Engineering Research Council of Canada (NSERC) and CMC Microsystems during this work.

References

[1] Pott T. *et. al.*, 2008, "Giant unilamellar vesicle formation under physiologically relevant conditions", *Chemistry and Physics of Lipids*, 154, 115–119.
[2] Loose M. and Schwille P., 2009, "Biomimetic membrane systems to study cellular organization", *Journal of Structural Biology*, 168, 143-151.
[3] Bunge et. al., 2007, "Lipophilic Oligonucleotides Spontaneously Insert into Lipid Membranes, Bind Complementary DNA Strands, and Sequester into Lipid-Disordered Domains", *Langmuir*, 23, 4455-4464.
[4] Scheller F. W., 2001, "Research and Development in Biosensors", F. W. Scheller, *Current Opinion Biotechnology*, 12, 35-40.
[5] Prakash R. and Kaler K. V. I. S., 2010, "Chip based unilamellar vesicle formation and dispensing using dielectrophoresis", *Proc. of MicroTAS, Groningen*, 1, 417-419.
[6] Jones T.B., Gunji M., Washizu M., Feldman M.J., 2001, "Dielectrophoretic liquid actuation and nanoliter droplet formation", *J. Appl. Phys*, 89(3), 1–8.
[7] Chugh D. and Kaler K. V. I. S., 2010, "Integrated liquid and droplet dielectrophoresis for biochemical assays", *Microfluid Nanofluid*, 8, 445-456.
[8] Gunji M., Nakanishi H. and Washizu M., 2004, "Droplet actuation based on single-phase electrostatic excitation", *Proc. of MicroTAS, Malmo*, 1,168–170.
[9] Prakash R. and Kaler K. V. I. S., 2009, "DEP Actuation of Emulsion Jets and Dispensing of Sub-Nanoliter Emulsion Droplets", *Lab Chip*, 9, 2836–2844.
[10] Lasic D. D. and Papahadjopoulos, 1995, "Liposomes Revisited", *Science*, 267, 1275-1276.
[11] Murcia et. al., 2008, "Design of Quantum Dot-Conjugated Lipids for Long-Term, High-Speed Tracking Experiments on Cell Surfaces", *J Am Chem Soc.*, 130(45): 15054–15062.

Method for improving the electrostatics perforation pattern using power controlled discharges

C Garzón[1], E Miranda[1], C Martínez-Cisneros[2], J Alonso[2] and J García-García[1]

[1] Departament d'Enginyeria Electrònica, Universitat Autònoma de Barcelona, 08193 Bellaterra, Barcelona, Spain
[2] Departament de Química Analítica, Universitat Autònoma de Barcelona, 08193 Bellaterra, Barcelona, Spain

E-mail: carolina.garzon@uab.cat

Abstract. The aims of this work are to show the influence of adding a series resistance at the output of a discharge generator circuit and to point out that this component can be used to control the spark energy in electrostatic perforation systems. Analysis of the experimental results reveals that there exists a close connection between the resistor value and the obtained perforation pattern both in hole density and size. The use of a series resistor has a strong influence on the material porosity, which is an important industrial parameter for assessing the pattern perforation quality.

1. Introduction

Electrical Discharge Machining (EDM) is used to drill and mechanize a wide variety of rigid materials that are difficult to cut or shape by more standard tools because of their hardness or complex shape. The EDM operating principle is based on the interaction between solid matter and the electric arc discharge. The severe physical conditions around the spark can be used to mill and drill nearly all kinds of material [1]. Electro-perforation is a particular case of EMD that allows holes to be made in permeable materials such as paper webs or biological membranes so as to obtain a drilling pattern with controlled porosity. Depending on the application considered it is sometimes necessary or convenient to achieve a specific hole size target.

Hole patterns must meet certain specifications to be considered appropriate for industrial applications. One of the most important parameters that needs to be taken into account to evaluate the quality of the perforation pattern is the material porosity. Quite often holes must be imperceptible to the human eye.

In previous work [2], the authors have investigated the relationship between the electrode distance and the perforation density and defined the drag distance as the minimum separation needed to generate a second hole in a specific area.

2. Experimental Results and Analysis.

The experimental setup consists of an array of two programmable step motors Zaber T-LSQ075D [3]. The motors are fixed to an optic table and move the sample in horizontal and vertical sweeps. The discharges take place between two tungsten electrodes (see Figure 1). The motors have 75 mm travel range, 550mm/s maximum speed and an accuracy of ±20μm. The gap in between the electrodes is around 3 mm. The custom-built equipment generates discharge trains up to 100 Hz. Work is in progress to generate higher frequency ranges using IGBTs-based electronics. The tests were made controlling the power discharge of the spark generator circuit. At this stage, the control simply consists iof adding a series resistor in the output circuit but more elaborate electronics is underway.

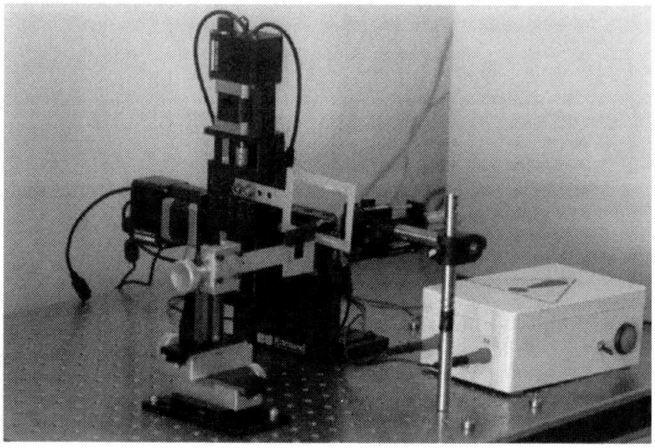

Figure 1. Experimental setup for the electroperforation.

Figures 2 and 3 illustrate the power control effects on the perforation pattern with and without the inclusion of the series resistance in the output circuit, respectively. In order to assess the quality of the perforation process it is convenient to use statistical tools such as the Spatstat package of R language [4].

It is worth noting that the incorporation of a series resistance yields a 60% increase in the number of points per unit area and that the average hole size is reduced by of about 53%. It is clear from these observations that the proposed method can be used to control the perforated area visibility in the industrial process.

Figure 2. Perforation pattern with power discharge control (average hole size 0.22 mm)

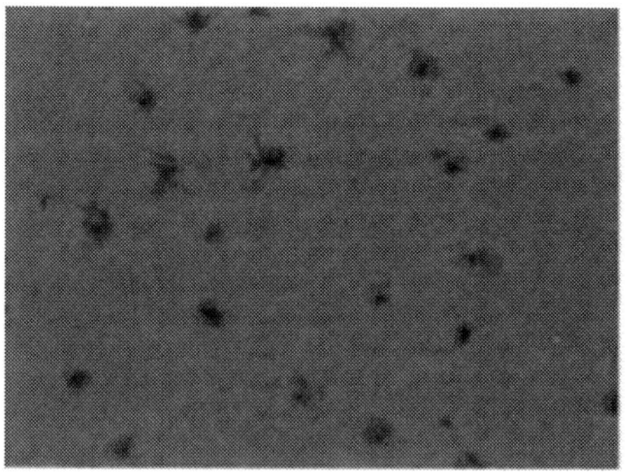

Figure 3. Perforation pattern without power discharge control (average hole size 0.48 mm)

Additionally, the mean drag distance reduces from 0.27mm to 0.21 mm in patterns generated with controlled discharges (Figure 2), i.e. when the resistor (with different values between 200Ω to 1.5kΩ) is introduced in the output circuit. These estimations are obtained using statistical tools such as the cumulative distribution function (CDF) for the nearest neighbour distance [4]. With CDF (Figure 4) it is possible to estimate the most frequent distance between nearest neighbour holes and to describe the

Journal of Physics: Conference Series **301** (2011) 012016 doi:10.1088/1742-6596/301/1/012016

probability distribution for each distance value. Figures 5 and 6 show two histograms for the distance to nearest neighbours in patterns made with and without controlled discharges.

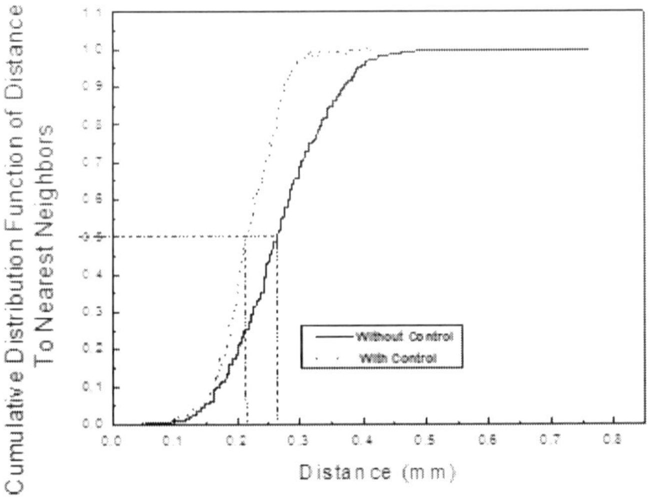

Figure 4. Cumulative distribution function of distance to nearest neighbours.

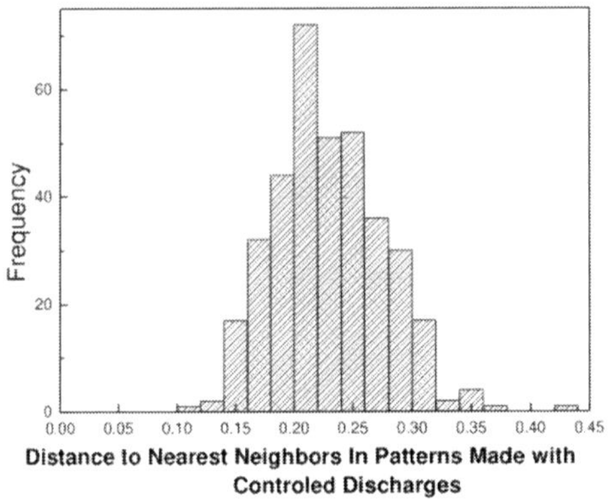

Figure 5. Distance to nearest neighbour in patterns made with controlled discharges (mm)

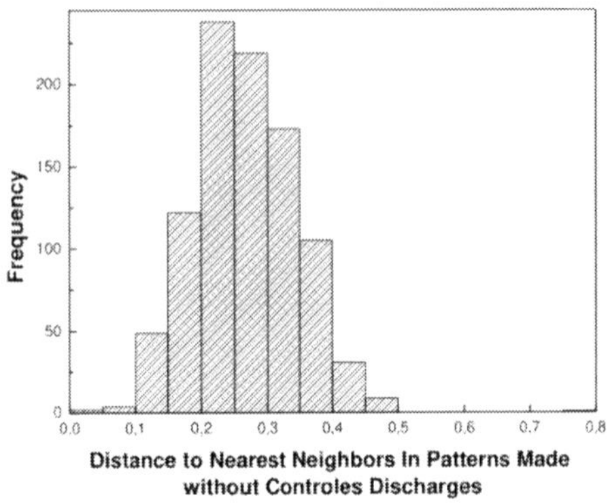

Figure 6. Distance to nearest neighbour in patterns made without controlled discharges (mm)

The porosity is another parameter that is affected by the implementation of the output resistor. Several measurements were made using different resistor values and the results are shown in Figure 7. There is a porosity increase linked to an increased resistance value and its trend line can be described by a 3rd order polynomial expression.

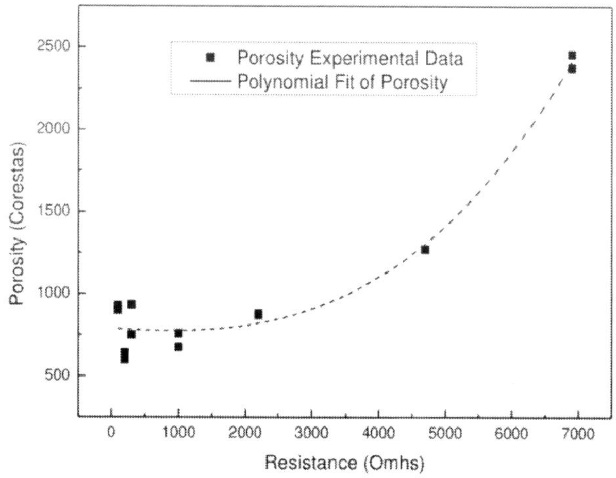

Figure 7. Porosity trend of paper webs perforated with different values of output resistors.

3. Conclusions

In summary, it is shown that it is possible to adjust different parameters of the perforations by using resistors in a typical electro-perforation process.

The porosity increment is the most important parameter for industry and the obtained results show that it is possible to achieve high porosity making holes small enough to reduce the visibility of the perforated areas in an industrial process. This is possible by reducing the drag distance and the consequent increase of holes per unit area due to the power controlled discharges.

References

[1] Ho, K.H.; Newman, S.T. "State of the art electrical discharge machining (EDM)" 2003 Int. J. Mach.Tools Manuf. 41. 1257-1300.

[2] Garcia-Garcia, J; Miranda, E; Martinez-Cisneros, CS, et al. "Porosity enhancement by the utilization of screening patterns in electroperforated paper webs",2010 J. Electrostatics 68. 196-199

[3] http://www.zaber.com/products/product_detail.php?detail=TLSQ075D&tab=Detailed%20Specs

[4] http://www.r-project.org/

Characteristics of sliding discharge in a multi-rod reactor

H Wedaa[1], M Abdel-salam[1], A Ahmed[1] and A Mizuno[2]

[1]Electrical Engineering Department, Assiut University, Assiut, Egypt.
[2]Department of Environmental and Life Sciences, Toyohashi University, Toyohashi, Japan.

E-mail: h_wedaa@yahoo.com

Abstract. This paper is aimed at investigating the characteristics of a sliding discharge (SD) including the onset voltage (V_O), spark voltage (V_S), and current-voltage (I-V) relationship in a multi-rod reactor stressed by sinusoidal AC or pulse voltage. The effects of various parameters (the voltage amplitude, frequency, gas flow rate, and voltage type) on the characteristics of the reactor sliding discharge (V_O, V_S and I-V relationship) have been studied experimentally. It has been found that the DC onset and spark voltages increase with the increase of the gas flow rate, while the effect of the frequency on them is not pronounced. The onset and spark voltages of the stressed reactor for sinusoidal AC voltage are lower than those obtained under a pulse voltage of the same peak value. Subsequently, the sliding current increases with the increase of the sinusoidal AC high voltage, the frequency, and the negative DC voltage, while, it decreases with the increase of the flow rate. It is observed that stressing the reactor with sinusoidal AC voltage gives higher values of sliding current than those obtained using a pulse at the same peak voltage. Stressing the reactor with sinusoidal AC voltage gives higher values of the NO removal efficiency than those obtained using pulse voltage.

1. Introduction

Dielectric barrier discharges (DBD) are characterized by the presence of at least one insulating layer (dielectric barrier) between the electrodes. The dielectric barrier limits the discharge current, preventing the transition to an arc. This stable discharge that can be generated under atmospheric pressure is suitable for ozone generation [1], NOx removal [2, 3] and SO_2 removal [4].

The sliding discharge is a discharge generated on a dielectric surface by a combination of AC or pulse and DC stresses in three-electrode geometry [5, 6]. The resulting DBD with the AC or pulse stress slides along the dielectric surface by the effect of the DC stress applied to a third electrode. The sliding discharge was generated on flat dielectric surfaces [5] or along the inner surface of glass capillaries [6]. The present study is aimed -for the first time- to propose a new design of DBD reactor for NO removal seeking improved efficiency using sliding discharges. The effects of the voltage amplitude, frequency, gas flow rate and the voltage type on the characteristics of a sliding discharge (V_O, V_S and I-V relationship) are examined with experimental confirmation of NO removal efficiency.

2. Experimental setup

The details of the multi-rod DBD reactor used in the present study are illustrated in figure 1. It consists of an internal cylindrical glass tube (inner diameter: 19.5 mm, outer diameter: 22.7 mm, length: 87 mm) and an external cylindrical glass tube (inner diameter: 34.8 mm, outer diameter: 40 mm, length: 270 mm) which was used as the dielectric material. Stainless tubes (diameter: 6 mm, length: 66 mm) were inserted between the two glass tubes as discharge rods. Two grounded stainless meshes were wrapped outside the external glass tube and inside the internal glass tube. To generate the sliding discharge, high sinusoidal AC or pulse voltage with high frequencies and high negative DC voltage are applied to the discharge rods alternately. At a given high sinusoidal AC or pulse voltage, the onset voltage of the sliding discharge is defined as the minimum value of DC voltage where the sliding current starts to flow, while the spark voltage is the corresponding DC voltage leading to spark conditions.

Figure 2 shows a schematic diagram of the experimental setup. A sinusoidal AC high voltage power supply (Function generator (KENWOOD AG-204D) + high voltage amplifier (TREK Model 20/20C)) with variable frequency was used. The net output high voltage applied directly to the DBD reactor was measured using a 1000: 1 high voltage probe (Tektronix P6015A). A 0.1 μF capacitor was connected in series with the plasma reactor to measure the discharge power using Lissajous figures [3]. The waveform of the discharge current was recorded using a 1 kΩ resistor. All waveforms were displayed on a digital oscilloscope (Tektronix TDS 1012B). To generate the sliding discharge, a high voltage DC power supply (Matsusada Precision Inc.) was used and the sliding current was measured using a digital multi-meter (CDM- 12D, Custom Co., Ltd.) with a limiting resistor of 5 MΩ.

A pulse generator (ECG-KOKUSAI model PPS-5000) with variable frequency was used for applying the pulsed voltage and the discharge current was measured using a Pearson current monitor (model 2877), which was located in the return current path to the ground.

The experimental measurements were carried out at room temperature under normal pressures with a constant relative humidity (8%). Simulated gas (dry air + NO, N_2 base) was fed through the plasma reactor with the NO initial concentration of 200 ppm. FTIR (Fourier Transform Infra-Red) spectroscope gas analyzer (model SESAM 3-N) was inserted to measure the concentrations of NO.

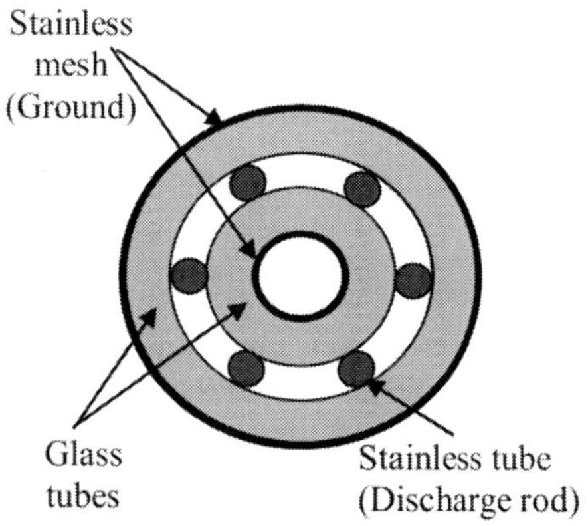

Figure 1. The cross-section of the suggested DBD reactor.

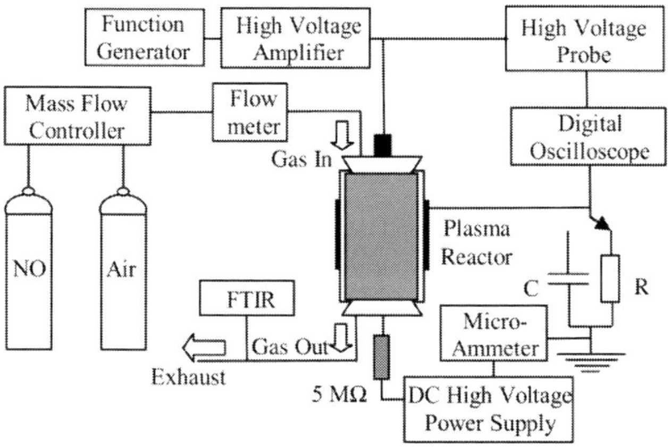

Figure 2. A schematic diagram of the experimental setup.

3. Results and discussion

3.1. Sliding discharge characteristics

3.1.1 Effect of AC high voltage amplitude. Figure 3 shows the relation between the sliding current and the negative DC voltage for various sinusoidal AC peak voltages (5.5, 6, 6.5 kV) at a frequency of 3 kHz and flow rate of 5 L/min. It is clear that the sliding current increases with increasing DC voltage for all cases and the sliding current increases with the AC high voltage.

3.1.2 Effect of the frequency. It is observed that the onset and spark voltages decrease with increasing sinusoidal AC peak voltage irrespective of the frequency, while the effect of the frequency on Vo and

Vs is not clear. Figure 4 shows the sliding current as a function of the negative DC voltage for various frequencies (2, 3, 4 kHz) at a sinusoidal AC peak voltage of 6 kV and flow rate of 5 L/min. It is clear that the sliding current increases with increasing the frequency.

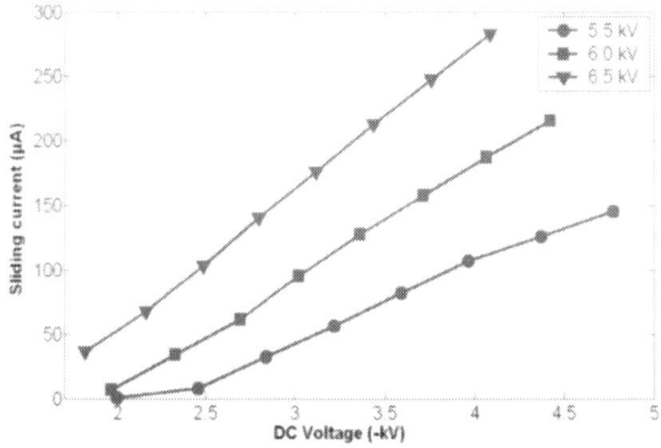

Figure 3. Sliding current as a function of the negative DC voltage for various sinusoidal AC peak voltages at 3 kHz and 5 L/min.

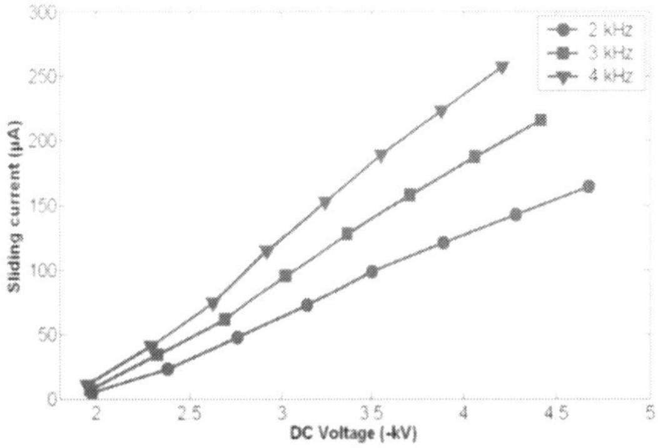

Figure 4. Sliding current as a function of the negative DC voltage for various frequencies at 6 kV and 5 L/min.

3.1.3 Effect of the gas flow rate. It is observed that the onset and spark voltages increase with increasing flow rate due to the retarding effect of the gas flow on the discharge ion motion. Subsequently, the sliding current decreases with increasing gas flow rate as shown in figure 5, which shows the relation between the sliding current and the negative DC voltage for various gas flow rates (2, 5, 10 L/min) for a sinusoidal AC peak voltage of 6 kV and frequency of 4 kHz.

3.1.4 Effect of the voltage type. Figure 6 shows the relationship between negative DC voltage and peak voltage at onset and spark conditions for various voltage types (sinusoidal AC and oscillatory pulse) at frequency of 4 kHz and flow rate of 5 L/min. The first oscillation of the applied pulse has a rise time of about 235 ns and tail time (time up to 50% of the peak value) of 1550 ns. The duty cycle being defined as the ratio between on-time of positive-going of the first oscillation and the pulse period (1/4000= 0.25 ms) is 0.014. It is clear that the onset and spark voltages decrease with the peak voltage irrespective of the voltage type. The AC voltage gives lower onset and spark voltages when compared with the pulse voltage at the same voltage due to the shorter stress time with the application of pulse voltage compared with AC voltage.

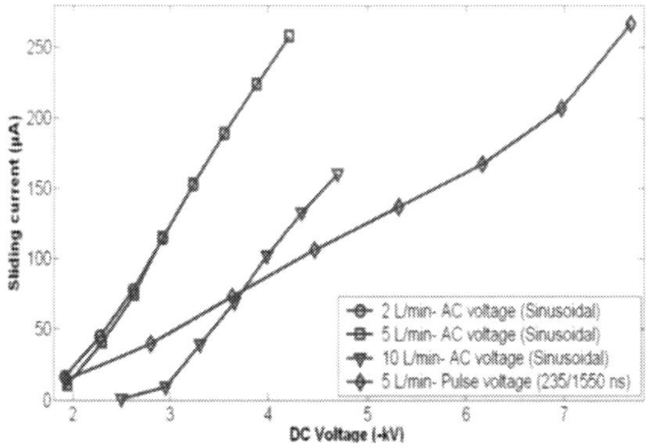

Figure 5. Sliding current as a function of the negative DC voltage for various flow rates and voltage types at 6 kV, and 4 kHz.

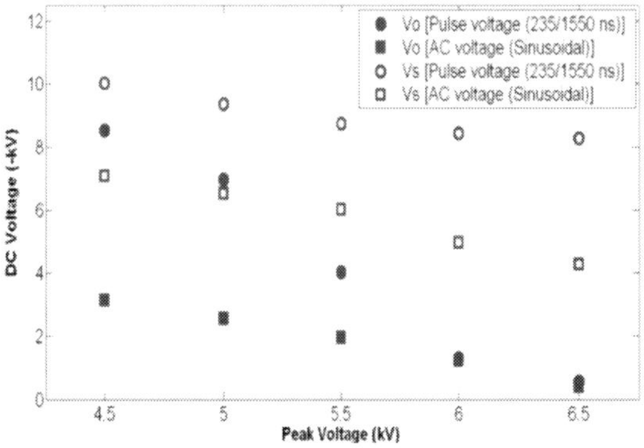

Figure 6. Relationship between DC voltage and peak voltage for various voltage types at 4 kHz, and 5 L/min.

Subsequently, stressing the reactor with AC voltage gives higher values of sliding current when compared with the pulse voltage at the same peak voltage as shown in figure 5, which shows the relation between the sliding current and the negative DC voltage for various voltage types at peak voltage of 6 kV, frequency of 4 kHz and flow rate of 5 L/min.

3.2. Removal of Nitric Oxide (NO)

Figure 7 shows the temporal variation of the NO concentration in ppm at sinusoidal AC peak voltage of 6.5 kV, frequency of 3 kHz, flow rate of 5 L/min and DC voltage of -2.275 kV. It is clear that using SD with DBD reduces the concentration more than by using DBD only due to the increased current.

Figure 8 shows the NO removal efficiency (=((NO_i-NO_f)/NO_i)×100, where NO_i and NO_f are the initial and final concentrations of NO in ppm, respectively) as a function of the energy density (=(P_d×60)/Q, where E_d, P_d, and Q are the energy density (J/L), discharge power (W) and gas flow rate (L/min), respectively) for various voltage types (AC and pulse) at a peak voltage of 6.5 kV, frequency of 3 kHz, flow rate of 5 L/min and variable negative DC values. The NO removal efficiency increases with the increase of energy density irrespective of the voltage type. However, stressing the reactor with sinusoidal AC voltage gives higher values of removal efficiency as compared with the pulse voltage. The removal efficiency expressed as a percentage of the energy density is high at first but decreases with increasing energy density irrespective of the voltage type as shown in figure 8.

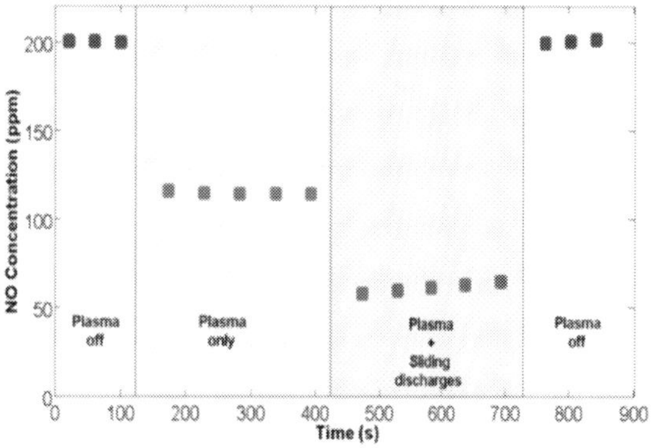

Figure 7. Temporal variation of NO concentration at 6.5 kV, 3 kHz, 5 L/min, and DC voltage of -2.275 kV.

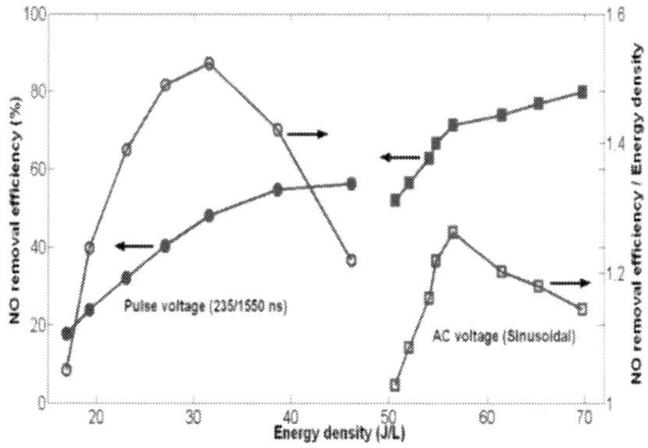

Figure 8. NO removal efficiency as a function of the energy density for various voltage types at 6.5 kV, 3 kHz, 5 L/min, and variable DC values.

4. Conclusion

The generation of sliding discharges inside the multi-rod DBD reactor by a combination of AC or pulse and DC stresses was investigated. The sinusoidal AC high voltage and corresponding DC high voltage at sliding-discharge onset and spark condition in the DBD reactor were measured. From the electrical measurements, it has been shown that the sliding discharge generation depends on the voltage amplitude, frequency, gas flow rate and the voltage type (sinusoidal AC or pulse). Preliminary results showed the effectiveness of the sliding discharge generated in the DBD reactor for removing NO but further investigation is necessary to develop more effective NO and NO_x removal reactors.

References

[1] Fang Z, Qiu Y, Sun Y, Wang H and Edmund K 2008 *Journal of Electrostatics* **66** 421
[2] Abdel-Salam M, Hashem A, Yehia A, Mizuno A, Turky A and Gabr A 2003 *Journal of Physics D: Applied Physics* **36** 252
[3] Toda K, Takaki K, Kato S and Fujiwara T 2001 *Journal of Physics D: Applied Physics* **34** 2032
[4] Ma H, Chen P, Zhang M, Lin X and Ruan R 2002 *Plasma Chem. and Plasma Process.* **22** 239
[5] Louste C, Artana G, Moreau E and Touchard G 2005 *Journal of Electrostatics* **63** 615
[6] Hensel K, Sato S, and Mizuno A 2008 *IEEE Transactions on Plasma Science* **36** 1282

Comparison of airflow patterns produced by DBD actuators with smooth or saw-like discharge electrode

Artur Berendt[1], **Janusz Podliński**[1], **Jerzy Mizeraczyk**[1, 2]

[1] Centre for Plasma and Laser Engineering, The Szewalski Institute of Fluid Flow Machinery, Polish Academy of Sciences, Fiszera 14, 80-952 Gdańsk, Poland
[2] Department of Marine Electronics, Gdynia Maritime University, Morska 81-87, 81-225 Gdynia, Poland

E-mail: aberendt@imp.gda.pl

Abstract. In this paper we compare the performance of DBD actuators when either a smooth or saw-like electrode is employed. Two electrode arrangements of DBD actuators are investigated. The first is the classic DBD actuator and the second is a DBD actuator with floating electrode. The usefulness of the saw-like electrode in these two types of DBD actuators is studied.

1. Introduction

The plasma generated by the surface dielectric barrier discharge (DBD) is capable of inducing electrohydrodynamic (EHD) flow for modifying airflow around aerodynamic elements [1-3]. Devices using the DBD for flow control are called DBD actuators. DBD actuators can improve the properties of aerodynamic elements modifying airflow around them and are useful to control flow separation and laminar-turbulent flow transition, to reduce drag and improve the lift of airfoils.

In the past decade many DBD actuator configurations were investigated [3]. The results obtained showed that DBD actuator properties, such as size of the plasma region and generated airflow velocity depend on electrode configuration, dielectric material properties and applied voltage. Usually, a smooth plane electrode was used as the HV electrode. However, other shapes of electrodes were also used to improve airflow generated by DBD actuators e.g. a wire electrode.

In this paper we compare the performance of DBD actuators when either a smooth or saw-like electrode is employed. Two electrode arrangements of DBD actuators are investigated. The first is the classic DBD actuator and the second is a DBD actuator with floating electrode. The usefulness of the saw-like electrode in these two types of DBD actuators is studied.

Also the influence of the floating electrode position on the airflow produced and the discharge current was studied for the DBD actuator with the saw-like floating electrode.

2. Experimental set-up

2.1. Classic DBD actuators

The classic DBD actuator consisted of two electrodes mounted on opposite sides of a 2 mm-thick glass plate. The HV electrode (on the upper side of the actuator) was exposed to ambient air when the grounded electrode was insulated with a polyimide tape to eliminate discharges on the bottom side of the actuator. Two types of classic DBD actuators were studied. The first, without any displacement between the HV electrode and grounded electrode (25 mm wide) (Fig. 1a) and the second in which the displacement between the HV electrode and the grounded electrode (10 mm wide) was 20 mm

Published under licence by IOP Publishing Ltd

(Fig. 1b). Both types of classic DBD actuators were made in two variants, with either a smooth HV electrode or a saw-like HV electrode. Both types of HV electrodes, made of a 50 µm-thick copper tape, were 6 mm wide (Fig. 2).

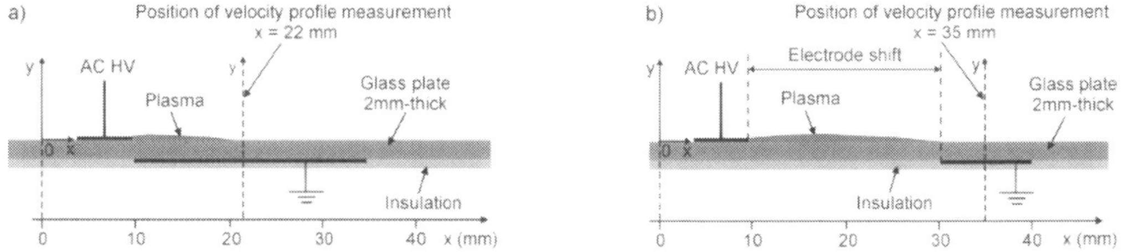

Figure 1. Side view of the classic DBD actuator with an electrode displacement of 20 mm (a) and the classic DBD actuator without electrode displacement (b)

Figure 2. Photo of a smooth (a) and a saw-like (b) electrode used in the classic DBD actuators

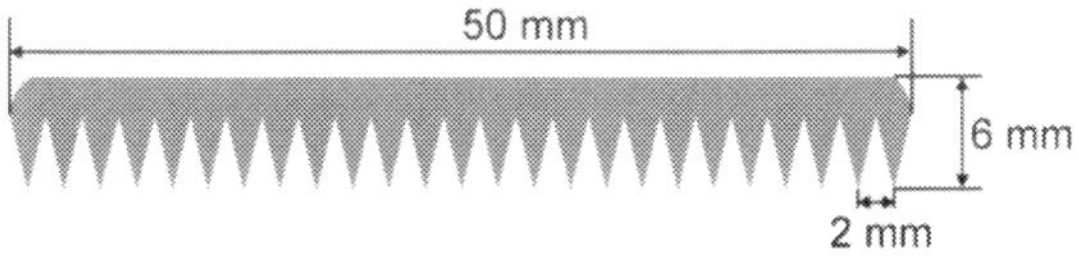

Figure 3. Saw-like floating electrode used for the DBD actuator

2.2. DBD actuators with floating electrode

The first DBD actuator with floating electrode consisted of a HV electrode and a grounded electrode, both fixed on the one side of the 2 mm-thick glass plate and both insulated, and a third electrode at floating potential placed on the opposite side of the glass plate (Fig. 4). The floating electrode was exposed to ambient air. Either the smooth or saw-like electrode was used as the floating electrode. The floating electrode and the HV electrode were 6 mm wide while the grounded electrode was 10 mm wide. The displacement between the HV electrode and the floating electrode was 1 mm, and between the floating electrode and the grounded electrode was 15 mm.

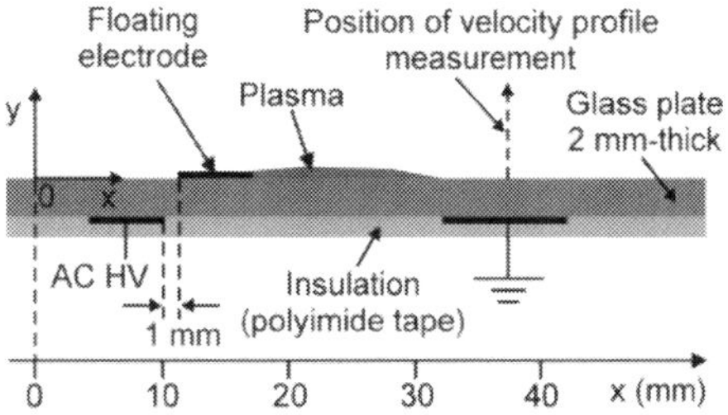

Figure 4. Side view of the DBD actuator with floating electrode

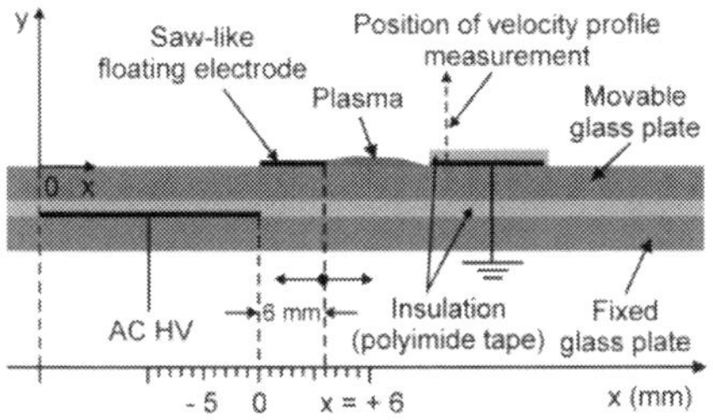

Figure 5. Side view of the DBD actuator used for optimisation of the saw-like floating electrode position

To optimise the floating electrode position, a second DBD actuator with saw-like floating electrode was used (Fig. 5) (floating electrode in x = + 6 position). It consisted of two 2 mm-thick glass plates and three 50 µm-thick copper electrodes. The HV electrode, the grounded electrode and the floating electrode were 20 mm, 10 mm and 6 mm wide, respectively. The length of HV and grounded smooth electrodes was 100 mm. The saw-like floating electrode was 50 mm long (Fig. 3). The first glass plate with the HV electrode was fixed. The second glass plate with the saw-like floating electrode and the grounded electrode, placed on the first one, could be shifted along the x direction. The HV and grounded electrodes were insulated with polyimide tape. The saw-like floating electrode was exposed to ambient air. The distance between the grounded electrode and the saw-like floating electrode was 10 mm, whereas the displacement between the HV electrode and the saw-like floating electrode could be varied.

2.3. *Experimental apparatus*

The experimental set-up consisted of an AC power supply, an oscilloscope, a discharge current probe, a flow channel in which the DBD actuators were placed, and a particle image velocimetry (PIV) [3] equipment for two-dimensional (2D) measurements of airflow velocity fields generated above the DBD actuator.

A sinusoidal high voltage (frequency 1.5 kHz) applied to the DBD actuators was generated by the power amplifier (TREK, model 40/15, U_{pp} – up to 80 kV) which amplified the signal from the function generator METEX MS9150. Discharge current was measured by a Pearson current probe (Rogowski coil). The voltage signal was delivered to the oscilloscope from a high-voltage probe in which the HV power amplifier was fitted.

The experiments were carried out in ambient air at atmospheric pressure in an open-ended flow channel. During the experiment there was no externally forced airflow in the flow channel.

3. Results

3.1. *Classic DBD actuators*

In Fig. 6 an example of the time-averaged flow velocity vector field of the airflow generated by the classic DBD actuator without electrode displacement (saw-like HV electrode).

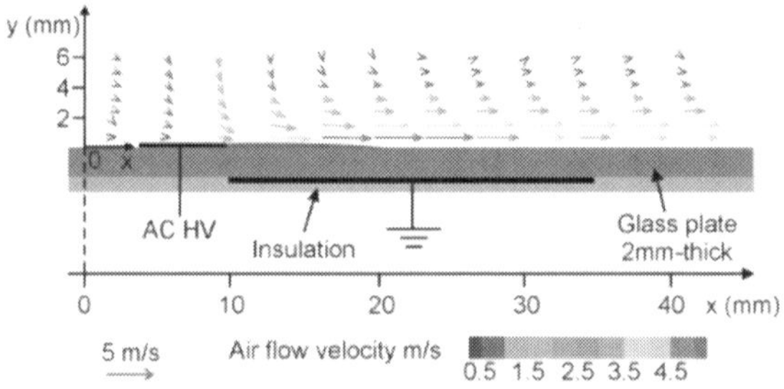

Figure 6. The time-averaged flow velocity vector field of the airflow generated by the DBD actuator (saw–like HV electrode) without electrode displacement. The applied sine-wave voltage was 32 kV_{pp}, the frequency - 1.5 kHz.

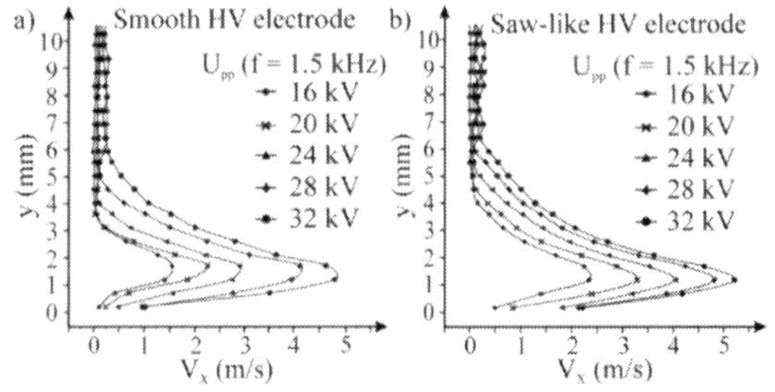

Figure 7. Flow velocity profiles measured for the classic DBD actuators without electrode displacement for smooth (a) or saw-like (b) HV electrode

Fig. 7 shows flow velocity profiles obtained for the classic DBD actuators without electrode displacement (saw-like HV electrode) at different applied voltages. The profiles were determined 12 mm downstream of the active edge of the HV electrode. The profiles clearly show that at the same applied voltage the velocities of the airflow generated by the DBD actuator with the saw-like HV electrode are higher than those for the DBD actuator with the smooth HV electrode. The most remarkable difference in airflow velocities occurs at the lower applied voltages. Similarly, the airflow velocity generated by the classic DBD actuator with 20 mm electrode displacement is higher when the saw-like HV electrode is used.

3.2. *DBD with floating electrode*

The time-averaged flow velocity fields of the airflow generated by the DBD actuator with smooth or saw-like floating electrode are presented in Figs. 8a and 8b, respectively. The vertical airflow velocity profiles measured for DBD actuators with the floating electrode (smooth or saw-like) were determined 20 mm downstream the active edge of the floating electrode (Fig. 9). The results show that velocities of the airflow generated by the DBD with saw-like floating electrode are higher than for the DBD with smooth floating electrode.

To find the optimal position of the floating electrode in the DBD actuator, measurements of the amplitude of DBD current pulses and the flow profiles were carried out for different positions of the floating electrode in respect to the HV electrode. The results obtained show that the maximum amplitude of the discharge current pulses and maximum airflow velocity were observed when the floating electrode was in x = + 1 mm position (where the floating electrode saw teeth were 1 mm downstream of the HV electrode edge). Thus, we concluded that position x = + 1 mm is the optimal position of the floating electrode in the investigated DBD actuator.

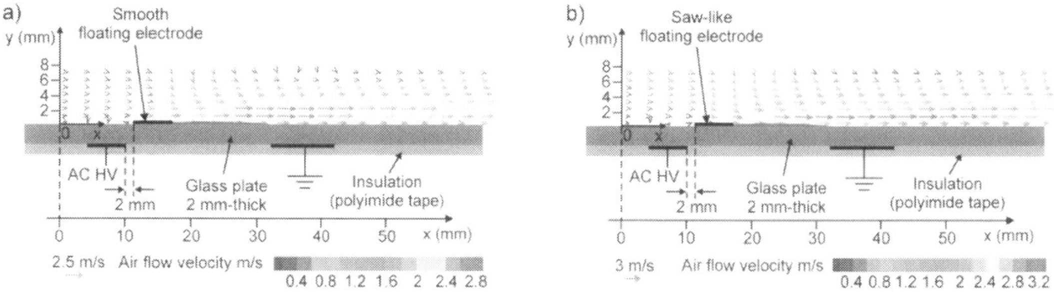

Figure 8. Time-averaged flow velocity vector fields of an airflow induced by the DBD actuator with smooth (a) and saw-like (b) floating electrodes. The applied sine-wave voltage was 60 kV_{pp},the frequency - 1.5 kHz.

Figure 9. Flow velocity profiles measured for the DBD actuator with smooth (a) and saw-like (b) floating electrodes

4. Summary and conclusions

Using the 2D PIV method the influence of the HV electrode shape on the airflow generated by the classic DBD actuator and the DBD actuator with floating electrode was investigated.

The comparison of obtained velocity maps clearly showed that in all cases velocities of the airflow generated by the DBD actuators were higher when the saw-like electrode was used. The most significant differences in airflow velocities were observed when relatively low voltages were applied. Also, when the saw-like electrode is used the the DBD starts at lower voltage.

In order to determine the optimal position of the saw-like floating electrode in the DBD actuator measurements of the DBD current pulses and of the generated airflow were made. The results obtained showed that the optimum position of the saw-like floating electrode was x = + 1 mm i.e. when nearly the whole saw-like floating electrode was above the HV electrode and only its saw teeth were 1 mm downstream of the HV electrode edge. The floating electrode can by useful for forming long surface DBD for actuators [4, 5].

Acknowledgement

The research presented in this paper received funding from the European Community, Seventh Framework Programme FP7/2007-2013 under grant agreement no.: 234201 (PLASMAERO – Useful PLASMas for AEROdynamic control www.plasmaero.eu) and by the R. Szewalski Institute of Fluid Flow Machinery, Polish Academy of Sciences, Gdansk.

References

[1] Roth J R, Sherman D M and Wilkinson S P 1998 *Proc. 36th AIAA Aerospace Sciences Meeting and Exhibit* (Reno, USA) 98-0328
[2] Touchard G 2008 *I.J. PEST* 2, 1
[3] Moreau E 2007 *J. Phys. D: Appl. Phys.* 40 pp 605-636
[4] Berendt A, Podlinski J, Mizeraczyk J 2011 *EPJ AP* (Preprint DOI: 10.1051/epjap/2011100441)
[5] Podlinski J, Berendt A, Mizeraczyk J 2010 *Proc. 12th Int. Symp. on High Pressure, Low Temperature Plasma Chemistry* (Trencianske Teplice, Slovakia) p 74

Journal of Physics: Conference Series 301 (2011) 012019 doi:10.1088/1742-6596/301/1/012019

Non-thermal plasma reactor with back corona discharge electrode

T Czapka[1], **R Kacprzyk**[1]

[1] Institute of electrical Engineering Fundamentals, Wroclaw University of Technology, Wyb. Wyspianskiego 27, 50370 Wroclaw, Poland

E-mail: tomasz.czapka@pwr.wroc.pl

Abstract. The new design of plasma reactor with back discharge electrode is presented and characterized. The laboratory scale plasma reactor was constructed in a plane parallel geometry with a gas permeable low-field electrode system. The low-field electrode was covered with a dielectric layer enabling the appearance of back corona discharges. The total volume of the reactor is equal to 2 dm^3.The discharge properties of the reactor operating at dc voltage in air under normal conditions are given. The results of optimizing the electrical properties are also presented. The influence of back discharges on the discharge current is discussed. The maximum discharge current density obtained during the experiment was equal to about 25 $\mu A/cm^2$.

1. Introduction

Nowadays plasma technology is widely and successfully used in many technological applications such as plasma surface treatment, environmental plasma chemical technologies, reactive magnetron sputtering and more [1]. In general, a plasma is an ionized gas (excited electrically, thermally or from electromagnetic irradiation) containing electrons, ions, radicals and atoms. It is possible to distinguish two types of plasma: thermal and non-thermal ones with completely different features. The energetic level of the particular particles is crucial in this division. Mainly electrical discharges in gases (e.g. corona discharge, dielectric barrier discharge, high frequency discharge) and electron beam irradiation are used for generating a low temperature plasma [2].The most common and still developed non-thermal plasma (NTP) applications are gaseous pollutant removal, water purification, film deposition and surface modification. Decomposition of volatile organic compounds (VOCs) is realized using plasma reactors. So far there were proposed many types of plasma reactors with different discharges generating the plasma (e.g. with corona discharges, dielectric barrier). There is also possibility of increasing the discharge current and, finally, charged particles in the discharge region by utilising the back corona discharge phenomenon.

2. Back corona discharge phenomenon

The phenomenon of back corona discharge was observed for the first time in electrostatic precipitators [3] as a damaging one, causing serious decrease of their collecting efficiency. The electric charge during the corona discharge accumulated on the dielectric layer covering the counter electrode. As a result the electric field across the dielectric layer was intensified. When the accumulated charge exceeded a critical value, breakdown through the dielectric layer appeared from the counter electrode as a back discharge. Charge of opposite polarity was then injected into the inter-electrode region. The phenomenon occurred for layers exhibiting volume resistivity ρ_V higher than $5 \cdot 10^8$ Ωm [4].

Much research was undertaken to describe and understand the back corona discharge phenomenon. Initial investigations described ways of preventing the negative effect in the electrostatic precipitators. Subsequent scientific attempts were focused on applications using this phenomenon such as decomposition of hydrocarbon gaseous contaminants and ozone production [5]. In this paper a new design of plasma reactor utilising the back corona discharge phenomenon is presented. This reactor is intended to be used for VOCs decomposition from air streams.

3. Reactor and measurement system

The cross-section of the laboratory type reactor is shown in figure 1. The reactor was constructed in a plane parallel geometry. The reactor housing consisted of transparent dielectric side walls (1) tightened by clamping screws (2).

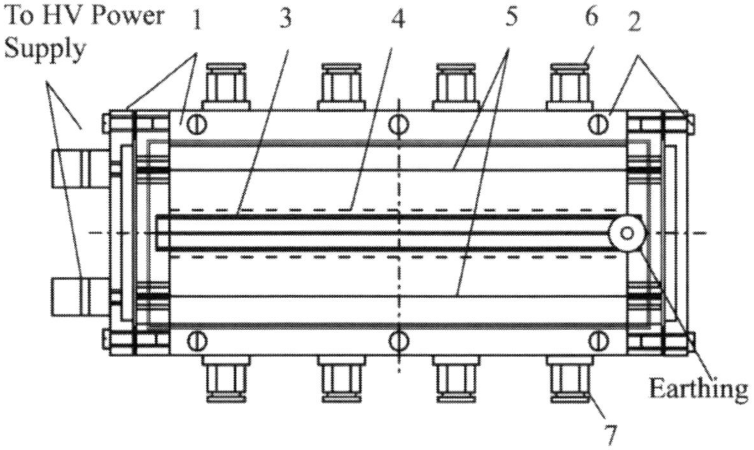

Figure 1. Scheme of the plasma reactor with gas permeable low field electrode and plane parallel geometry.

The earthed low-field electrodes (3) were made of perforated metal sheet (perforated stainless steel with pores diameter $\varnothing = 0.5$ mm, distance between particular pores equal to 1.09 mm) covered with dielectric layer (4). Perforation of the electrodes should enable the processed gas to flow in parallel to the vector of the electrical field in the discharge region. The reactor was also equipped with inlet (6) and outlet (7) pneumatic connectors.

A ceramic glaze was used as the dielectric layer. Permittivity, volume resistivity and thickness of the applied layer were ε=6.8, ρ_v=$7 \cdot 10^{12}$ Ωm and d=30 μm respectivelyMetal strings (Cu-Ni alloy wires $\varnothing = 0.3$ mm) mounted in parallel, operated as the corona electrode (5). The total volume of the reactor was equal to 2 dm^3.

The reactor discharge system was supplied from two stabilized dc high voltage power supplies – Glassman PS/EW40P150YQ2 and PS/EW40N150YQ1 types, regulated in the range of 0 to ± 20 kV. The discharge current was measured by an analogue dc current meter WUT PA 100.

4. Results and discussion

All of the measurements were carried out with the reactor filled with the air under atmospheric pressure and at room temperature. The current-voltage (*I-V*) characteristics for dc corona discharge (low field electrode without dielectric layer) and back corona discharges (low field electrode covered by dielectric layer) are presented in figure 2.

Discharge currents obtained for back corona discharges were much higher (even up to 10 times) in comparison to the case with corona discharges only. A similar effect of the back corona was observed for opposite polarity of applied voltage. The influence of polarity of applied voltage on the back discharge current was also shown in figure 2. The results were obtained for a fixed distance between wires and low field electrode *a*, distance between adjacent strings *b* and number of strings *n*.

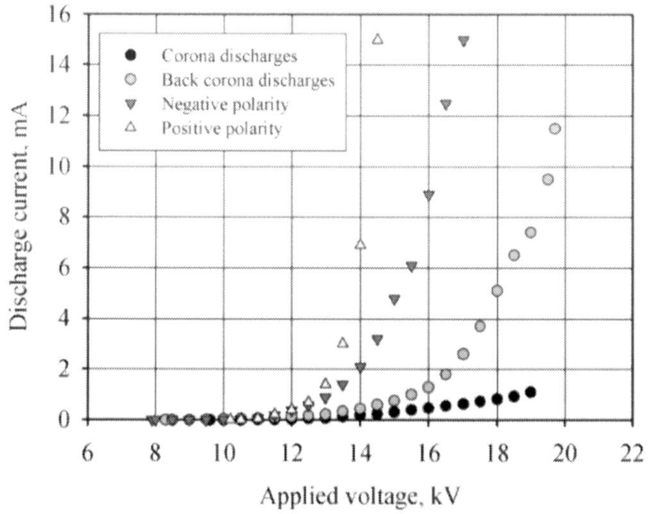

Figure 2. Dependence of discharge current on the type of discharges in the reactor (corona discharges or back discharges for negative polarity, a=25 mm, b= 25mm, n=5) and polarity of applied voltage to corona electrodes (a=20 mm, b=40 mm, n=3).

It was noticed, that the onset (inception) voltage was lower for negative polarity of applied voltage to the strings. On the other hand, the critical value of discharge current was obtained more rapidly for positive voltage (*I-V* characteristics were shifted in relation to each other). The maximum discharge current densities for both polarities were equal to about 26 μA/cm^2. The discharge was very unstable and following increase of voltage led to streamer discharges.

The results of the optimization of the electrical properties are shown in figure 3. During this experiment parameters *a*, *b*, *n* describing the configuration of the electrode system were changed to obtain optimal discharge properties.

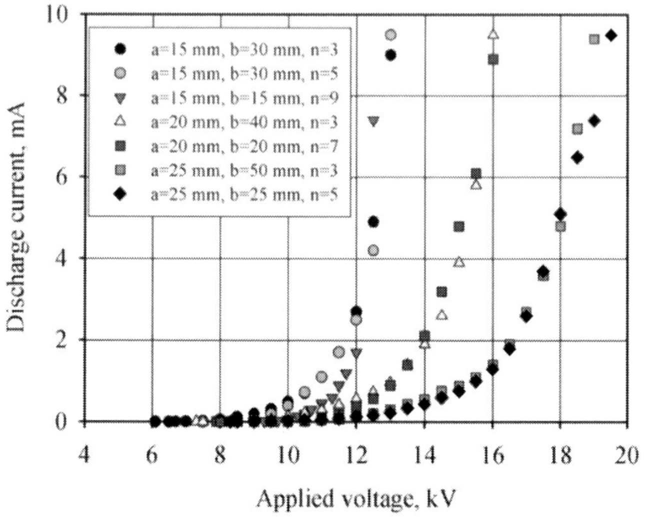

Figure 3. Dependence of discharge current on configuration of corona strings in the plasma reactor. Results obtained for negative polarity of applied voltage to corona strings.

It was found that the distance between the corona electrodes and the low field electrode *a* strongly influenced the discharge current. There was a visible shift between the obtained *I-V* characteristics.

The same value of discharge current was obtained (e.g. 10 mA) for a lower value of applied voltage when the distance *a* decreased. It was also noticed that electrical properties were not dependent on the number of corona strings *n*. The number of used corona strings may be important for expanding the extent of the back discharge occurring on the low field electrode. The distribution of back corona (on the low field electrode) and corona discharges (from strings) in the reactor is shown in figure 4.

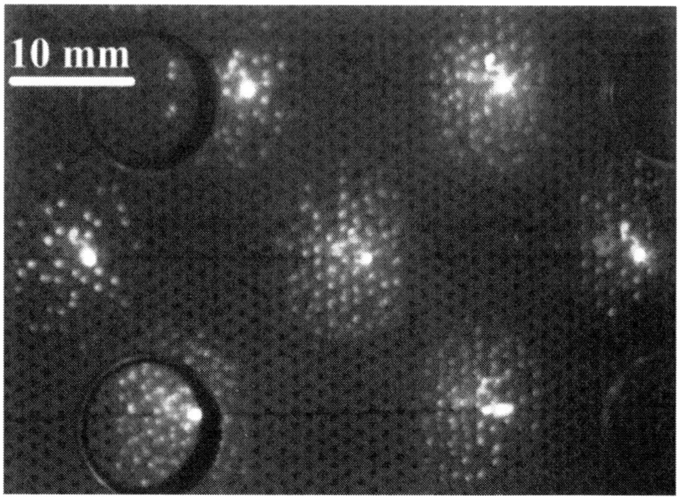

Figure. 4. Corona and back corona discharges observed in the plasma reactor in air in normal conditions for a=15 mm, b=15 mm and n=9. Average current density ca 15 μA/cm^2. The bright points on the corona electrodes were local discharges for negative polarization voltage.

5. Conclusions

Back corona development on the perforated dielectric layer was experimentally confirmed for both polarities of the voltage applied to the corona electrode. The back discharge phenomenon led to reach discharge current density even 10-times higher than the current density measured in the plasma reactor with corona discharges only – figure 4.

The critical value of discharge current density leading to streamer discharge was equal to 26 μA/cm^2. Applying positive voltage to the corona electrode led to a more rapid increase of discharge current and lower onset voltage (see figure 2). This is consistent with theory of electrical breakdown in gases. The maximum density power dissipated in the reactor was about 280 W/dm^3.

The configuration of the electrode system also influenced the discharge properties of the reactor – figure 3.

Acknowledgments

This work was realized under grant No N N510 108138 funded from Ministry of Science and Higher Education.

References

[1] Yamamoto T, Yoshizaki A, Kuroki T and Okubo M 2004 *IEEE Trans. Ind. Appl.* **40** 1220–25
[2] Penetrante B M, Hsiao M C, Merritt B T, Vogtlin G E and Wallman P H 1995 *IEEE Trans. Plasma Sci.* **23** 679–87
[3] White H J 1963 *Industrial Electrostatic Precipitation* (Addison-Wesley Publ. Co., Pergamon Press, Oxford)
[4] Mizuno A 2000 *IEEE Trans. Dielectr. Electr. Insul.* **7** 615–27
[5] Jaworek A, Krupa A and Czech T 1996 *J. Phys. D-Appl. Phys.* **29** 2439–46

Journal of Physics: Conference Series **301** (2011) 012020 doi:10.1088/1742-6596/301/1/012020

Online-characterization of dielectric barrier discharge plasma actuators for optimized efficiency of aerodynamical flow control applications

J Kriegseis[1], D Schröter[1], S Grundmann[2] and C Tropea[1,2]
[1] Institute of Fluid Mechanics and Aerodynamics,
[2] Center of Smart Interfaces,
Technische Universität Darmstadt, D-64287 Darmstadt, Germany
E-mail: kriegseis@sla.tu-darmstadt.de, grundmann@csi.tu-darmstadt.de

Abstract. The impact of fluctuating and transient kinematic and thermodynamic airflow conditions on the performance of dielectric barrier discharge (DBD) plasma actuators is demonstrated. A novel online-characterization and control approach is introduced, revealing the possibility of compensating for impaired discharge performance due to changing airflow scenarios during actuator operation. The goal of controlling the plasma actuator performance online and in situ is achieved and successfully demonstrated.

1. Introduction

Dielectric barrier discharge (DBD) plasma actuators have proven to be an attractive and promising control device for various aerodynamic flow-control applications, as summarized comprehensively by Moreau [1]. In the large number of recent publications on this topic it is commonly assumed that the discharge effects the flow, but not vice versa. The reports of flow-control experiments deal with free stream velocities ranging between several meters per second and supersonic conditions. In contradiction to that, mainly quiescent air experiments have been conducted to characterize actuator performance, for instance the impact of ambient conditions (e.g. temperature, pressure, humidity) or actuator-specific parameters (e.g. geometry, dielectric material) on the wall jet induced by plasma actuators.

However, very few studies quantitatively report on the electrical performance of such actuators, i.e. the discharge intensity, the power consumption and the corresponding electrical efficiency. This is a clear deficiency, since electrical performance should be the basis on which to evaluate the overall control authority of plasma actuators in flow control applications. Kriegseis *et al.* [2] suggested a new diagnostic approach for evaluating the performance of plasma actuators quantitatively, which is based on voltage-charge cyclograms. These Lissajous figures are a powerful and well-established means to quantitatively analyze dielectric barrier discharges [3,4], in the present case yielding the electrical power consumption P_A of plasma actuators.

Based on the previously reported dependency of discharge-light emission on the consumed power by Enloe *et al.* [5], Kriegseis *et al.* [6] demonstrated correlations of power consumption P_A, effective discharge capacitance C_{eff}, (chord wise) plasma length Δx and resulting thrust production F in quiescent air. First (qualitative) reports by Pavon *et al.* [7] indicated the adverse effect of high airflow speed on the discharge intensity in terms of a reduced overall light emission. Recently, Barckmann *et al.* [8] demonstrated and quantified that power consumption and effective discharge capacitance of plasma actuators are considerably affected by changing airflow velocities.

Continuing in the direction of recent investigations [2, 6–8] the present manuscript addresses the above mentioned discrepancies with a new online diagnosis tool for DBDs. Such a monitoring system is essential for optimal flow-control using DBD plasma actuators, since it provides the required information to characterize and quantify the impact of fluctuating or transient airflow conditions on the plasma-actuator performance during operation. A quantitative measure of the performance is fundamentally important for any optimization procedure, since any plasma actuator flow-control system, which has previously been impedance matched for a particular reference Mach number and thermodynamic state, can be considerably de-tuned by simply varying the free stream velocity and/or the altitude (temperature, pressure or density). Therefore, a detailed online knowledge of the actuator's performance is essential to permanently maintain the optimum electrical efficiency and flow-control effectiveness during operation at changing airflow conditions.

2. Experimental Procedure

The experimental setup comprises two measurement systems, as sketched in Figure 1. A plasma actuator as used in [2] of $L = 0.11$m length is mounted at the wall of the wind-tunnel's test section. At the opposite window a CMOS camera (Phantom V12.1, 512×512 pixels, 24fps; Nikon 105 mm, AF Micro NIKKOR f/2.8D) is used to record the spatio-temporal light emission of the discharge during the power-consumption analysis (cp. [6]). The electrical control circuit is built up using a digital oscilloscope (Picotech PicoScope4424, 4CHs, 2500p/Ch, 10MS/s) to record the operating voltage V (Testec HVP-15HF, 1000:1) and the voltage V_p across the charge-probe capacitor $C_p = 22$ nF (LeCroy PP006A, 10:1). The operating voltage V is generated by a high voltage generator (GBS Elektronik, Minipuls2), which is driven by a notebook-controlled laboratory power supply (Voltcraft VSP 2410) and a function generator (GW Instek, SFG-2004, fixed frequency: $f = 12.0$ kHz).

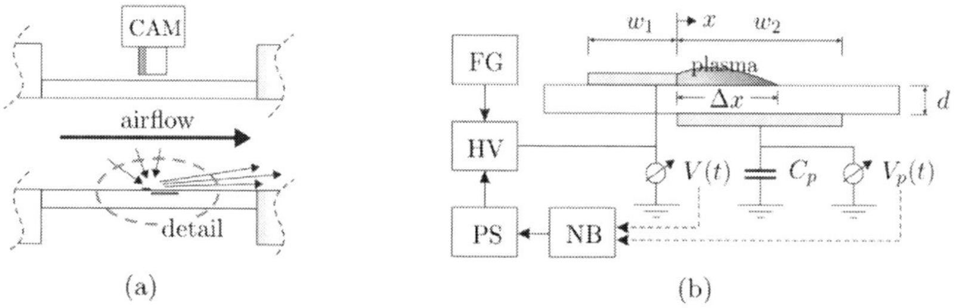

Figure 1. Sketch of experimental setup. (a) Wind tunnel test section and overhead camera (CAM); (b) detailed view of electrical plasma-actuator setup comprising function generator (FG), power supply (PS), high voltage (HV) transformer, notebook (NB) and plasma actuator.

The experiments are conducted in a blow-down type wind tunnel in order to obtain a transient airflow during experimentation, as shown in Figure 2. The static pressure p ranges between $p_{\min} = 0.89$ bar and $p_{\max} = 1.45$ bar during the blow-down. Benard *et al.* [9] and Versailles *et al.* [10] previously reported a favorable and adverse impact on plasma actuator performance at quiescent air conditions under reduced and elevated pressure levels, respectively.

The online-characterization of the voltage-charge cyclograms, i.e. for monitoring or controlling purposes, is based on the diagnostic approach as introduced by Kriegseis *et al.* [2].

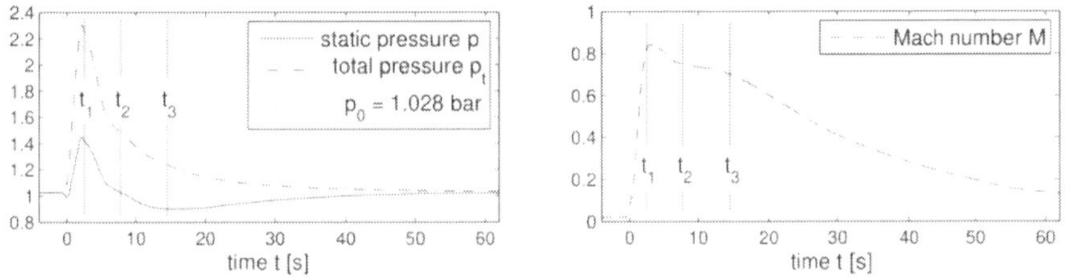

(a) pressure p, p_t [bar] (b) Mach number M

Figure 2. Transient flow conditions. (a) Static and total pressure p, p_t; (b) Mach number M; characteristic times are labeled as t_1, t_2, t_3 (cp. Table 1).

For every time step t^i the algorithm calculates the power consumption

$$P_A^i = f\, E^i = f \oint_{t^{i-1}}^{t^i} Q\, dV \quad \text{with} \quad Q = C_p V_p. \tag{1}$$

When using the diagnostic tool in closed-loop control mode, the algorithm furthermore compares P_A^i with a pre-set power level P_A^* and calculates the control signal $u^{i+1} = u^{i+1}(u^i, P_A^i, P_A^*)$ for the next time step by means of a PD control algorithm, which is then sent to the power supply (see Figure 2(b)). The light-emission data is used to validate the online-diagnostic tool by means of the temporal plasma length evolution Δx (cp. [6]).

3. Results

The pre-set initial conditions of f = 12kHz, V = 10kV and p_0 = 1.028 bar result in an initial power consumption $P_A = P_A^* = 7.2$W for $t < 0$. Three characteristic times are highlighted for $p(t_1) = p_{\max}$, $p(t_2) = p_0$, $p(t_3) = p_{\min}$ with the purpose of distinguishing pressure and air-speed effects on the discharge performance (see Figures 2 and 3).

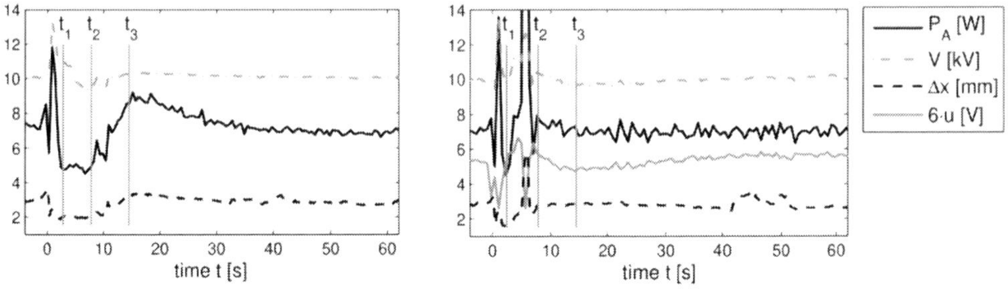

(a) (b)

Figure 3. Results of consumed power P_A, operating voltage V, plasma length Δx and remote signal u for (a) monitoring and (b) controlling modes of operation (cp. Table 1).

The results the monitoring experiment (Figure 3(a)) clearly reveal the impact of the transient flow conditions on the resulting actuator power P_A. Immediately after the wind tunnel valve is opened at $t = 0$ a power peak occurs due to an initial expansion wave passing the test section, which is followed by a significant performance drop ($P_A = 4.8$W) once the blow-down scenario is fully developed at t_1 under adverse pressure conditions at maximum airflow speed (see Table 1). With decreasing Mach number and pressure at t_2 a constantly reduced performance

(P_A = 4.9W) is observed, solely due to the impact of high speed airflow (M = 0.75) at ambient pressure conditions p_0, which agrees with the reports of Barckmann *et al.* [8]. The influence of the minimum pressure at t_3 exceeds the adverse airflow impact (M = 0.69), which results in an increased performance P_A = 8.7W as compared to the initial value P_A^* = 7.2W. Thereafter, all quantities asymptotically return to their initial values again. The plasma length Δx of the simultaneously recorded light-emissions further confirms the correctness of the online characterization of the power consumption.

Table 1. Measured data at characteristic times: ^{m} monitoring, ^{c} controlling (cp. Figure 3).

	p [bar]	M	P_A [W]	V [kV]	Δx [mm]	u [V]
t < 0	1.03	0	$7.2^{m,c}$	$10.0^{m,c}$	$2.9^{m,c}$	0.89^c
t_1 = 2.4[s]	1.45	0.84	4.8^m 5.7^c	10.8^m 10.1^c	1.9^m 1.6^c	0.97^c
t_2 = 7.8[s]	1.03	0.75	4.9^m 7.5^c	9.4^m 10.2^c	2.0^m 2.8^c	0.95^c
t_3 =14.5[s]	0.89	0.69	8.7^m 7.2^c	10.3^m 9.8^c	3.3^m 2.8^c	0.78^c

For identical initial and airflow conditions the results of the closed-loop control experiment are shown in Figure 3(b). The control algorithm fails for the very strong initial power oscillation of the passing expansion wave. Thereafter the performance drop is identified at t_1 and the algorithm counter-acts this drop, as shown be the slope of the control signal u. At t = 5.5s the control algorithm collapses for a single time step, which causes a power overshoot. Apart from this peak the algorithm successfully conducts a closed-loop control of the the above-discussed power variations, which again is confirmed by the results of the plasma-length Δx. Qualitatively, the slope of the remote signal u of the controlled case directly mirrors the slope of the power P_A of the uncontrolled case, which underlines the successful counter-action of the controller even in this simple proof of concept approach.

4. Conlusions

In accordance with literature, the impact of changing kinematic and thermodynamic airflow conditions on the performance of dielectric barrier discharge (DBD) plasma actuators is demonstrated. The necessity of counter-acting these performance fluctuations is met by a novel online-characterization and in-situ-control approach. Based on the measured real-time performance data, the possibility of achieving a constant plasma-actuator performance during operation under fluctuating and transient flow conditions is demonstrated in a simple proof of concept approach. This is an important insight, since beyond the common purpose of favorably manipulating the airflow, any advanced DBD-based flow control system will necessarily require an appropriate closed-loop control of the discharge device.

References

[1] Moreau E 2007 *J. Phys. D: Appl. Phys.* **40** 605–636
[2] Kriegseis J, Möller B, Grundmann S and Tropea C 2010 Capacitance and power consumption quantification of dbd plasma actuators *Journal of Electrostatics (accepted 15.04.2011)*
[3] Kogelschatz U 2003 *Plasma Chemistry and Plasma Processing* **23** 1–46
[4] Wagner H, Brandenburg R, Kozlov K, Sonnenfeld A, Michel P and Behnke J 2003 *Vacuum* **71** 417 – 436
[5] Enloe C L, McLaughlin T E, VanDyken R D, Kachner K D, Jumper E J, Corke T C, Post M and Haddad O 2004 *AIAA Journal* **42** 595–604
[6] Kriegseis J, Möller B, Grundmann S and Tropea C 2011 *AIAA-paper 2011-155*
[7] Pavon S, Dorier J L, Hollenstein C, Ott P and Leyland P 2007 *J. Phys. D: Appl. Phys.* **40** 1733–1741
[8] Barckmann K, Kriegseis J, Grundmann S and Tropea C 2010 *AIAA-paper 2010-4258*
[9] Benard N, Balcon N and Moreau E 2008 *J. Phys. D: Appl. Phys.* **41** 042002 (5pp)
[10] Versailles P, Gingras-Gosselin V and Vo H 2010 *AIAA Journal* **48** 859–863

Effects of a transmission line on the properties of a surface discharge based on a segmented ground electrode design

J Jolibois, K Takashima and A Mizuno

Department of Environmental and Life Sciences, Toyohashi University of Technology Tempaku-cho, Toyohashi, Aichi, Japan

E-mail: mizuno@ens.tut.ac.jp

Abstract. The paper deals with a new way of modifying the properties of a surface discharge. This method consists of influencing the surface potential distribution on the dielectric layer. For this, we used two configurations of plasma device. The first one is a standard surface discharge. While the second one is based on a surface discharge with a segmented ground electrode to which we added a transmission line. For these experiments, the surface discharge is supplied by an AC high voltage of a few kV with frequency ranging between 0.5 to 4.5 kHz, and operates in DBD mode. The velocity of the produced electric wind has been recorded using a pressure probe in a quiescent environment. From these measurements, the maximum velocity has been extracted and used to compare the performance of the different configurations. In addition, the voltage waveform and current-discharge, from which the power can be calculated, have also been recorded. The results highlight the capability to modify the topology of the electric wind produced by the new configuration of plasma device. The highest induced velocity achieved is 15% greater than the standard case with the same power consumed.

1. Introduction

The purpose of this study focuses on optimizing the electromechanical properties and/or increasing the size of the plasma sheet of a surface discharge. Indeed, the performance of these systems, as devices enabling active control of an aerodynamic airflow or reduction of air pollution, are related to the electric wind velocity induced by the discharge and / or by the extension of the plasma above the insulating barrier. Conventionally, the performance improvement of a surface discharge is obtained by modifying two parameters: 1) the shape of the device and 2) the input electric parameters [1-5]. In summary, these enhancements increase the injected power into the plasma.

In the present work, a new way of modifying the properties of surface discharge is proposed. This method consists of influencing the surface potential distribution on the dielectric layer. To achieve this, a surface discharge based on a design of segmented ground electrode is achieved. Then, we have connected a transmission line (i.e. an LC network) to the segmented ground electrode.

From the electrical power measurements and those of induced flow, a comparison of electromechanical behavior has been performed between the new setup and the standard discharge.

2. Experimental part

Figure 1 shows a schematic illustration of the setup of both discharges. The standard surface discharge (named SD # 1) that forms the baseline case for this study, consists of two electrodes asymmetrically flush mounted on each side of a dielectric plate, plus a counter-electrode placed on the top side of the insulating wall. This counter-electrode is separated relatively to the right edge of the electrode #1 by an air gap of 40 mm, named the SD gap. Each electrode is made of aluminium strip. The upper electrodes are 100 mm-long (in spanwise) for 20 mm in width. The lower electrode (electrode #2) is 25 mm wide for 100 mm in spanwise. The dielectric used is an alumina plate 150 × 150 mm^2 and 1.5 mm-thick.

The new design of surface discharge (named SD # 2) is similar to the baseline case, except the bottom electrode is substitued by a segmented electrode (electrode #2') to which we added a tansmission line. This segmented electrode is composed of three 5 mm wide aluminium strips. Each strip is separated by an air gap of 5 mm. This keeps the total area of the combined width of strips and air gaps equal to electrode #2. The transmission line consists of an *LC* network composed of two coils and three capacitors. Depending on the inductance and capacitance values, several SD # 2 configurations have been investigated.

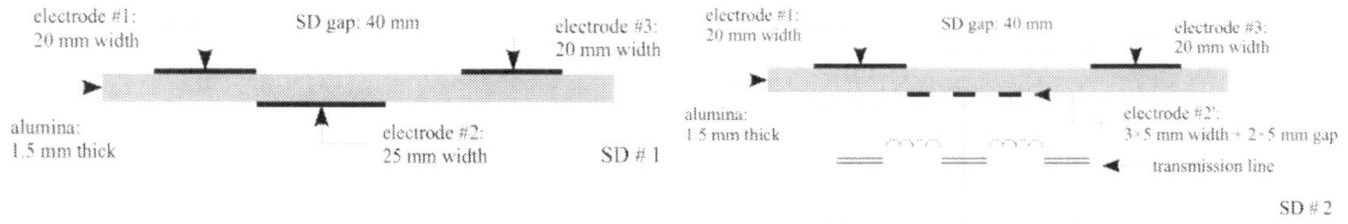

Figure 1. Scheme of two plasma devices used: standard discharge (left) and new configuration (right).

Surface discharges are powered by an AC high-voltage connected to the electrode #1 as shown in Figure 2. The AC high voltage is obtained with the help of a transformer supplied by a power amplifier (NF Corporation, model 4510). The transformer can supply a maximum peak voltage of 20 kV at driving frequencies up to 5 kHz. The two other electrodes are grounded. In this case, the surface discharges operate in DBD mode [4, 5].

The applied voltage is measured by using an HV probe (Tektronix, model P6015A). The current, *i*, is deduced from the voltage across a non-inductive resistor (R = 100 Ω) connected in series between the electrode #2 (or the whole electrode #2' and transmission line) and the earth. Each electrical waveform was recorded using a fast digital oscilloscope (Tektronix, model DPO 2024). From both voltage and currents curves, the total electrical power consumption can be calculated as follows:

$$P=\frac{1}{nT}\int_{t=0}^{nT} v(t)i(t)dt \qquad (1)$$

where $v(t)$ and $i(t)$ are the measured voltage and current versus time, respectively, T is the waveform period, and n is the number of periods. In practice, the time-averaged value computed with only one cycle would not allow a convergent value to be reached because it should vary from one cycle to another. Therefore approximately 10 cycles have been recorded.

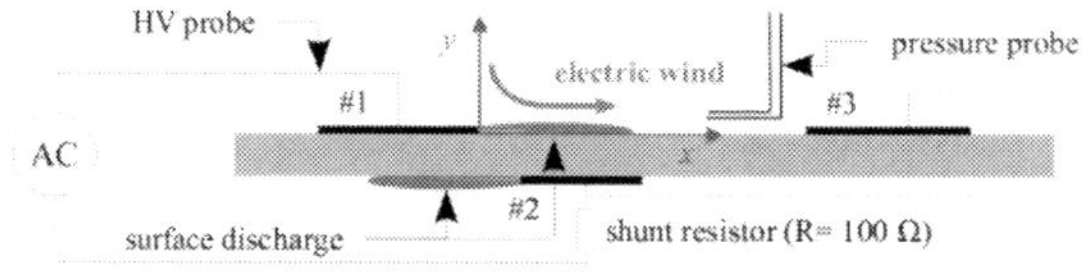

Measurements of the induced velocity above the plasma devices are realized with a total pressure probe made in glass connected to a low differential pressure manometer (Manostar, model W081). The probe is located at x = 10 mm (horizontal axis) and y = 0.2 mm (vertical axis) above the dielectric. The origin of the coordinates corresponds to the right edge of electrode #1 as shown in Figure 2. This setup allows the time-averaged velocity component U to be determined by using the following expression:

$$U=\sqrt{\frac{2(\Delta p)}{\rho}} \qquad (2)$$

where Δp is the pressure difference between the measured pressure p and the atmospheric pressure p_{atm}, and ρ is the air density.

3. Results

3.1. Capacitor value of the *LC* network

In this section, the influence of capacitance variation of the transmission line on the electromechanical behavior of the surface discharge has been analyzed. In this experiment, the coil value of the *LC* network is fixed with $L = 1$ mH. Only the capacitance varies. The capacitor C ranges from 0.1 to 100 μF. A sinusoidal waveform is applied, with the peak voltage fixed at 9 kV.

Figure 3 displays the power consumed as a function of frequency between the SD # 1 and the SD # 2 with several configurations of *LC* network. Compared to the baseline case, it appears that the SD #2 device, without and with the transmission line, has no significant effect on electrical power consumption.

Evolution of the induced electric wind versus power consumed is presented in Figure 4. Several significants results are highlighted:

1. Without the *LC* network, it appears that the plasma device with the segmented ground electrode generates higher electric wind compared to the baseline case. The velocity gain reaches about 10% at a given power consumed.

As shown by other authors [6] for another range of voltage-frequency and a different electrode arrangement, this result confirms that the use of a segmented ground electrode increases the induced flow by the surface discharge.

2. Depending on capacitance value, the electric wind velocity is either decreased or increased. For the capacitors of 10 and 100 μF, we can see that the measured velocity is constantly reduced. This reduction is of the order of 12% on average with $C = 10$ μF. In the case where $C < 10$ μF, the flow produced by SD # 2 is initially reduced. But from a power threshold, a velocity gain is achieved. The highest induced velocity reached 15% greater than the standard case with the same power consumed.

It seems that the electric wind is higher with a low capacitor value, here $C = 0.1$ μF.

These results are interesting because they highlight the capability of modifying the electric wind topology by influencing the surface potential distribution on the dielectric barrier.

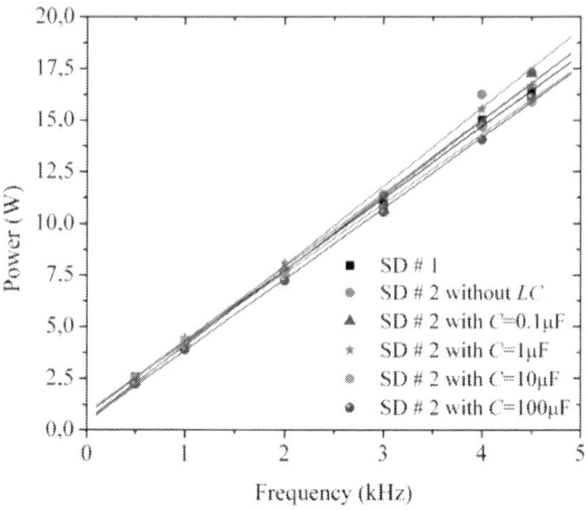

Figure 3. Evolution of the electrical power versus frequency for every configuration.

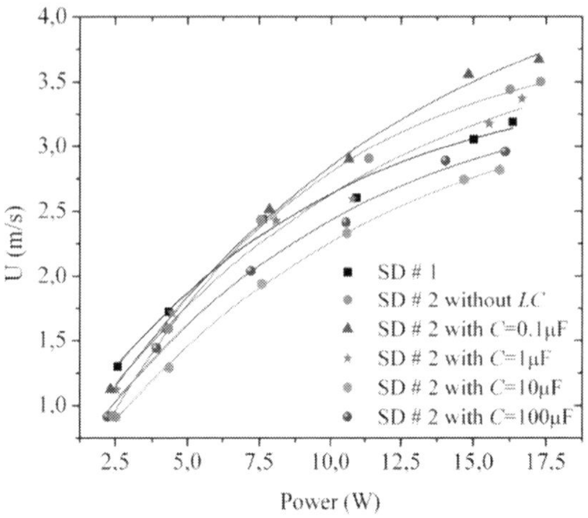

Figure 4. Electric wind velocity versus power consumed for every configuration.

3.2. Inductance variation of the *LC* network

In a second set of experiments, the effect of inductance variation of the transmission line on the properties of the plasma device is investigated. To achieve this, the capacitor value of the *LC* network is fixed with $C = 0.1$ μF, deduced from previous results (see Figure 4). Here, three values of coil are tested, i.e. 1, 10 and 100 mH. The waveform applied is sinusoidal, with the peak voltage fixed at 9 kV.

Electrical power consumption versus frequency between the SD # 1 and the SD # 2 with several configurations of *LC* network is shown in Figure 5. As in the previous section, one can remark that the plasma device with a segmented ground electrode, in the presence or absence of the transmission line, has no effect on the power consumed for the range of voltage-frequency used in this study.

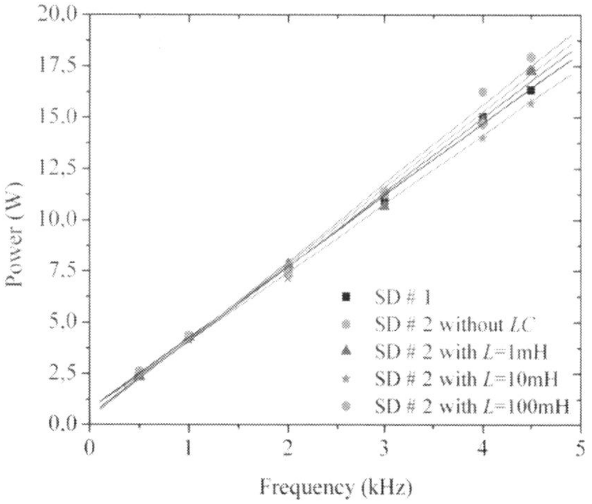

Figure 5. Evolution of the electrical power versus frequency for every configuration.

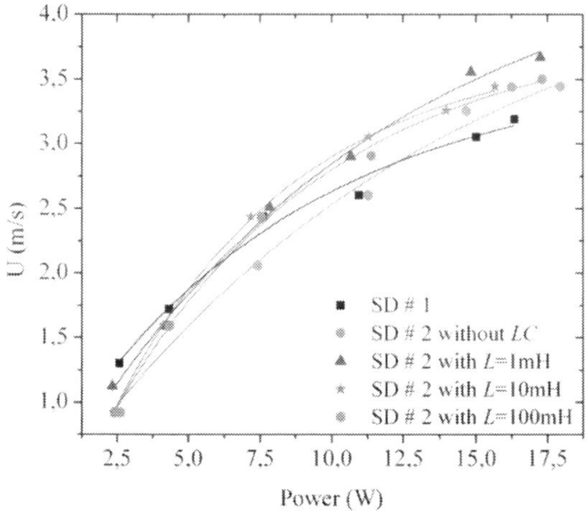

Figure 6. Electric wind velocity versus power consumed for every configuration.

Figure 6 shows the electric wind velocity versus power consumed for every configuration tested. Whatever configuration of the SD # 2 used, the induced flow is constantly reduced up to a power threshold. Beyond this threshold, the velocity is higher compared to the electric wind generated by the baseline case. Moreover, it appears that the coil effect on the electromechanical behavior is better with a low inductance value, i.e. with $L = 1$ mH.

4. Conclusion

In this paper, the results show the capability of modifying the properties of a surface discharge by using a segmented ground electrode to which we connected a transmission line. Firstly, it appears that this electrode arrangement, without or with an *LC* network, has no effect on the power consumed by the discharge. Secondly, the SD # 2 plasma device (in presence or absence of *LC* network) produces an effect upon the induced electric wind velocity. Depending on the values of inductance and capacitance, the electric wind velocity can be reduced or accelerated. This can be explained by the fact that the transmission line modifies the distribution of surface potential by varying the propagation and the shape of the sinusoidal waveform, resulting a variation of the propagation velocity of streamers above the dielectric barrier. From these results, it seems that low values of capacitance ($C = 0.1$ µF) and inductance ($L = 1$ mH) allow the most significant velocity gains to be achieved from a threshold power consumed. It means that the propagation velocity of the sinusoidal waveform into the *LC* network is proportional to the velocity of streamers (about 10^6 m/s).

This study should be completed by future investigations to better point out the physical mechanisms that occur at the surface discharge. First, by highlighting the spatial and time variation of potential at the dielectric surface by adapting the method described in [7] to our setup. And finally, by performing electric wind velocity mapping according to the coordinates x and y.

References

[1] E. Moreau, 2007, *J. Phys. D: Appl. Phys.*, **40**, No 3, pp. 605.
[2] M. Forte, J. Jolibois, J. Pons, E. Moreau, G. Touchard and M. Cazalens, 2007, *Exp. Fluids*, **43**, No 6, pp. 917.
[3] J. Jolibois and E. Moreau, 2009, *IEEE Trans Dielectr. Electr. Insul.*, **16**, Issue 3, pp. 758.
[4] K. Takashima, N. Zouzou, E. Moreau, A. Mizuno and G. Touchard, 2007, *IJPEST*, **1**, No 1, pp. 14.
[5] E. Moreau, R. Sosa and G. Artana, 2008, *J. Phys. D: Appl. Phys.*, **41**, No 11, pp 5204.
[6] C. Hale, R. Erfani and K. Kontis, 2010, *AIAA Paper* 2010-5206.
[7] C.L. Enloe, T.E. McLaughlin, J.W. Gregory, R.A. Medina and W.S. Miller, 2008, *AIAA Paper* 2008-1103

Electric field and charge distribution in back discharge in point-plane geometry

A. Krupa

Institute of Fluid Flow Machinery, Polish Academy of Sciences, Fiszera 14, 80-952 Gdansk, Poland

E-mail: krupa@imp.gda.pl

Abstract. The paper presents experimental and theoretical investigations of electric field and charge distribution on the fly-ash layer covering the plate electrode in the back discharge in point-plane geometry. Two kinds of dielectric layer were used in the experiments: highly resistive fly-ash from an electrostatic precipitator and a dielectric plate (PET) with a small pinhole.

1. Introduction

Back discharge is a type of electric discharge that occurs in electrostatic precipitators, used for removal of particulate matter from coal-fired power plants, when low sulphur coal is burned. Ions generated in the vicinity of the discharge electrode accumulates on the surface of the dielectric layer (fly-ash deposit) and causes an increase in the magnitude of electric field within this layer. This charge causes also a decrease in the magnitude of electric field in the interelectrode space. An increase of electric field within the layer, leads to the breakdown of this layer. Initially the charge accumulated on the layer surface flows through small channels to the plate electrode, and faint, glowing spots can be visible on the layer surface. An increase of the voltage and discharge current leads to an increase in the current density in these channels that causes heating of the material of the layer. When sufficiently high current flows through one such channel, it becomes conducting due to thermal ionisation of the layer material. These ionization processes on the layer surface lead to gaseous discharges such as glow or breakdown streamers to occur and finally the arc discharge can be incepted at lower voltages than without the dielectric layer. Effects of gas composition on characteristics of the back discharge were investigated by Masuda et al. [1, 2]. The current-voltage characteristics of back-discharge were presented in [3] and spectroscopic studies of the discharge in [4]. A back-corona discharge was also tested as a plasma source for decomposition of hydrocarbons [5]. The investigations of the motion of dust particle emitted from the collection electrode were studied in the paper [6]. The paper [7] presents investigations of back discharge occurring in air and flue gases produced by the process of burning of LPG or charcoal. The aim of that work was to determine an effect of back discharge on morphology of fly ash layer and exhaust gas composition. In the present paper, back discharge generated in needle-to-plate electrode configuration, with the plate electrode covered with a dielectric layer, is investigated. The aim of this work was to determine the surface charge density before back corona occurred, and the rate of charge decay on the layer. The distribution of surface charge density deposited on the fly-ash layer was also measured in the vicinity of the crater after a breakdown streamer occurred, and after switching off the high voltage.

2. Experimental set-up

Schematic of the experimental setup is shown in Figure 1. The back discharge is generated in a needle-to-plate electrode configuration with the plate electrode covered with a dielectric layer. The distance between needle tip and plate was fixed at 35 mm. Fly-ash of high resistivity taken from power plant Tarong Energy Australia was used in the experiments. Additionally, a dielectric plate (PET) with a small pinhole in the middle was used in the experiments for reference. The dielectric plate allowed the generation of highly localized and repeatable back discharges. The discharge was generated in air at a fixed temperature in the range of 20 to 120°C. The structure of the fly-ash layer deposited on the collection electrode depends on the discharge current density and the radial position from the needle electrode axis. The surface charge density and the rate of charge decay on the fly ash layer was measured before back corona occurred. The distribution of surface charge density deposited on the fly-ash layer was also measured in the vicinity of the crater after a breakdown streamer, just after the high voltage was switched off. In the experiments a precision electrostatic voltmeter, Trek Model 341B, employing the field-nulling technique for noncontact measurement of voltage was used. The electrostatic voltmeter was equipped with high Temperature Probe (up to 100 °C) Model 3453ST (side-viewing). This permits stationary measurements of charge decay at a fixed point, and measurement of surface charge distribution along the radial coordinate of the dielectric sample via moving the side-viewing probe above the surface of the dielectric layer.

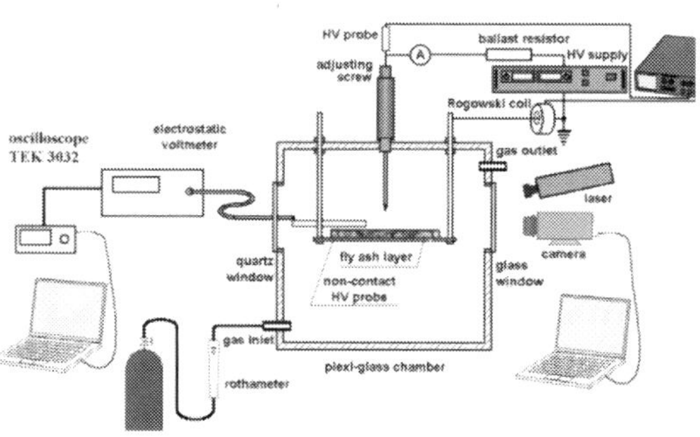

Figure 1. Schematic of experimental setup for back discharge investigation in needle-to-plane electrode configuration

Figure 2. SEM image of highly resistant fly ash from Tarong Energy Australia

3. Results

The surface charge density distributions on the PET plate are presented in Figures 3 and 4. The effect of the diameter of the circular opening in the plate is shown on Figure 3. The magnitude of surface charge density on the PET layer outside the opening is nearly the same, but in the vicinity of the opening the surface charge density decreases. The distribution of charge accumulated on the PET plate without opening and with an opening of diameter 1 mm for the charging voltages 17 kV and 19 kV is presented in Figure 4. Figure 5 shows the radial distribution of surface charge density after the corona discharge. In a zone of about 10 mm from axis of the needle, the surface charge density is nearly constant. Out of this zone, the level of charge density decreases with the distance. This phenomenon is probably caused by the combined effect of charge distribution on the layer following the Warburg law, and differences in density of the fly ash layer. Just below the needle, the layer is highly packed due to ion bombardment. The average dielectric constant is larger and resistivity of the packed layer is lower (because of smaller air voids). As an effect, the deposited charge is higher due to Warburg law in spite of higher conductance in this zone. In the less-densely packed layer the surface charge density is lower due to lower ion current density flowing to the layer. In Figure 5b, the distribution of surface charge density on the fly-ash layer with a crater remaining after the discharge is presented.

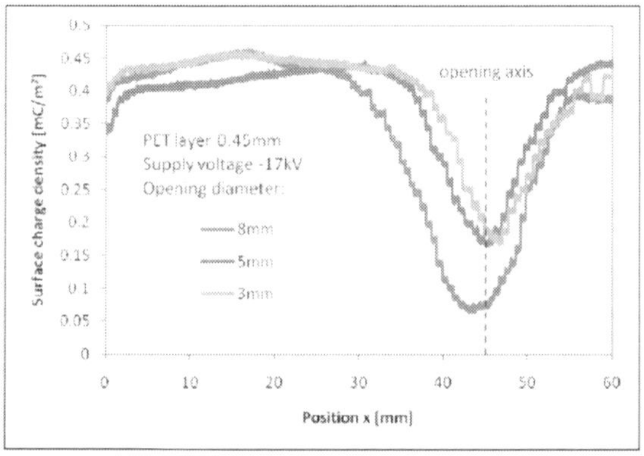

Figure 3 Surface charge density distribution on the PET plate for various opening diameters.

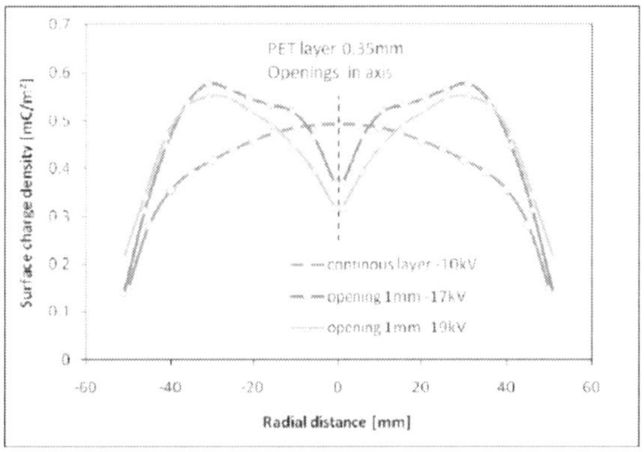

Figure 4. Surface charge density distribution on the PET plate with opening of diameter 1mm for various charging voltages.

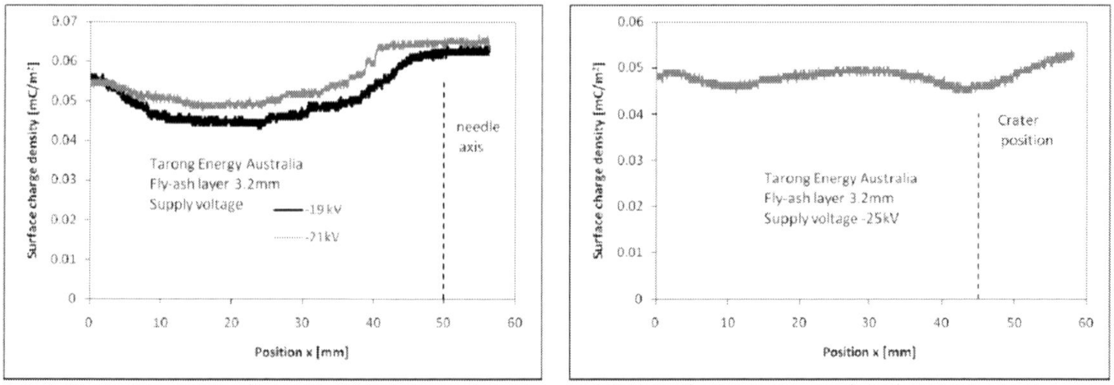

a)continous layer b)back discharge crater in the position 45 mm

Figure 5. Surface charge density distribution on the highly resistive fly-ash layer. The discharge needle axis in the position 50 mm

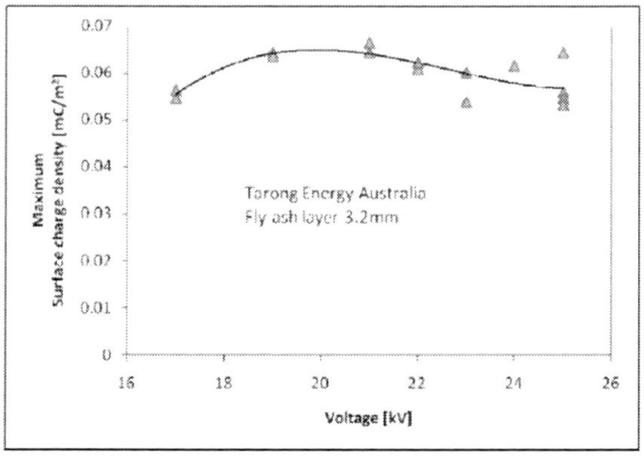

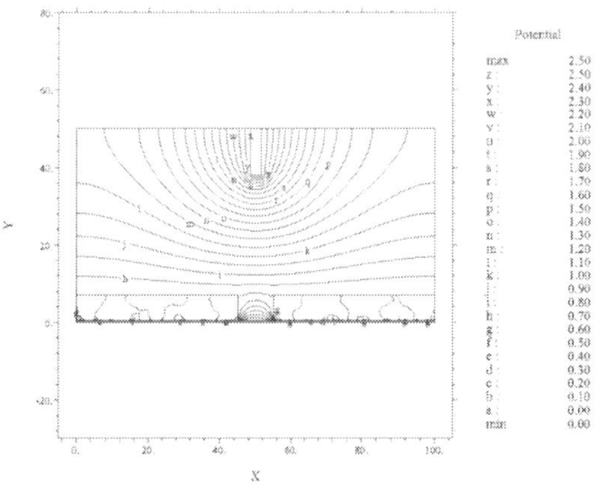

Figure 6 Maximum surface charge density (near the needle axis) on the highly resistive fly-ash layer vs. discharge voltage

Figure 7 Electric field distribution in the needle to plate geometry with back discharge crater in dielectric layer. Discharge voltage 25 kV.

The maximum surface charge density on the fly ash layer (near the axis) vs. charging voltage is shown in Figure 6. The charge density increases with the voltage up to a certain maximum, and with higher charging voltage, it starts to decrease. This effect can be explained by the results of measurements presented in Figure 8, which show the time resolved decay of charge on the layer. For lower voltages (Figure 8a), the charge decreases slowly and smoothly due to conduction current through the layer and/or the surface current. Only at the beginning of the charge decay, a small jump of charge can be noticed. In the case of higher voltages (Figure 8b), the charge decay is stepwise. The few jumps in charge density are the result of electric breakdown through the layer leading to local charge neutralization in the breakdown points. This stepwise decay disappears when the surface charge density decreases below 0.05 mC/m^2. From the value of charge density the electric field through the layer can be determined. The estimated electric field magnitude is of the order of 2 MV/m and is close to the breakdown field of air (3 MV/m). The breakdown results in a shorter average time constant for charge decay at higher charging voltages. Numerically obtained electric field distribution in the needle to plate geometry with a back discharge crater in the dielectric layer in the axis is presented in Figure 7. It can be seen from this simulation that the magnitude of the electric field within the crater increases similarly to the needle tip and can lead to ionization processes.

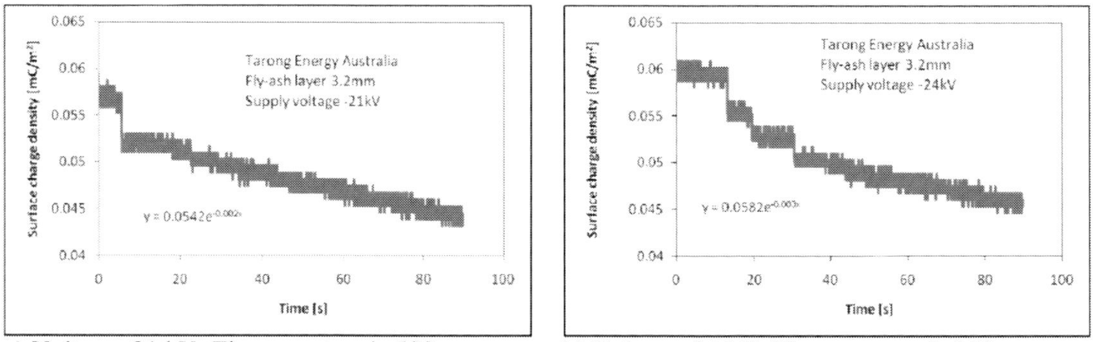

a) Voltage -21 kV. Time constant is 500 s. b) Voltage -24 kV. Time constant is 330 s.

Figure 8 Surface charge decay on the highly resistive fly-ash layer in the position 5 mm off the axis.

4. Summary

The paper presents experimental and theoretical investigations of electric field and charge distributions on the fly-ash layer covering the plate electrode in a back discharge in point-plane geometry. The experiments were carried out for highly resistive fly-ash from Australia and dielectric plate (PET) with a small pinhole, in the middle. The presented results of charge distribution measurement provide information on the mechanisms of back discharge onset and the effect of current density, resistivity of fly-ash on the discharge onset and charge decay from the dielectric layer surface. The research is aimed at better understanding of the back-discharge processes which cause harmful effects in electrostatic precipitators.

Acknowledgement

This research work was supported by the Polish Ministry of Science and Higher Education within the Project No.3520/B/T02/2000/37.

References

[1] Masuda S, Mizuno A, 1977/1978 J. Electrostatics 4 35
[2] Masuda S, Mizuno A, 1978 J. Electrostat. 5 215
[3] Rajch E, Jaworek A, Sobczyk A, Krupa A, 2006 Czech. J. Phys. 56 B803
[4] Czech T, Lackowski M, Krupa A, Sobczyk A, Jaworek A, Rajch E, 2009 European Phys. J. D 54 265
[5] Jaworek A, Krupa A, Czech T, 1996 J. Phys. D Appl. Phys. 29 2439
[6] Krupa A, Lackowski M, Czech T, 2006 Environment Protection Engineering 32 53
[7] Krupa A. 2009 J. Electrostatics 67 291

Carbon structures formation in low current high voltage electrical discharge in hydrocarbon vapours

A T Sobczyk [1], **A Jaworek** [2]

[1)]Institute of Fluid Flow Machinery, Polish Academy of Sciences, Fiszera 14, 80-952 Gdansk, Poland

[2)]Institute of Physics, Pomeranian Academy, Arciszewskiego 22B, 76-200 Slupsk, Poland

E-mail: jaworek@imp.gda.pl

Abstract. The properties of carbon fibers and other carbon structures produced from hydrocarbon vapours decomposed in electrically generated plasma at atmospheric pressure are studied in this paper. The electrical discharge was generated between a stainless steel needle and a plate made of nickel alloy. The carbon fiber has grown at the tip of the needle electrode, while other microflower-like deposits were built at the plate. The physical properties of carbon fibers were investigated by SEM, Raman spectroscopy, XRD, and EDS methods.

1. Introduction

Electrical discharges are used in various industrial processes, for example, for surface modification, sterilization, noxious compound decomposition, particulate matter precipitation, micro- and nanostructures synthesis, plasma etching, or electric-discharge-machining. These processes, operating at pressures ranging from atmospheric to below one pascal, are generated by high voltage of various frequencies: direct-current (DC), alternating-current (AC), radio-frequency (RF), pulsed (PD), or microwaves (μW), supplying electrodes of different geometries.

In industry, carbon fibers are usually produced by thermal pyrolysis of polymer fibers (usually pitch) or by synthesis gaseous light hydrocarbons such as methane, benzene, or acetylene. In our experiments, carbon fibers were synthesized via the pyrolysis of vaporized liquid hydrocarbons in electrically generated plasma in oxygen-free atmosphere.

The aim of the present paper is to investigate the properties of plasma-synthesis carbon fibers, grown in a low-current high-voltage electrical discharge.

2. Experimental

A schematic of experimental set-up is shown in Figure 1. The experiments were carried out at normal pressure, in the atmosphere of argon and cyclohexane as carbon feedstock, in a reactor chamber of 0.6 dm^3, made of plexi glass. The hydrocarbon vapour was obtained by feeding liquid cyclohexane to a heated flask where it evaporated. Next, the vapours were removed to the reactor chamber by flowing argon. The concentration of hydrocarbon was controlled via dosing it by a syringe pump of costant flow rate. The argon-cyclohexane mixture flowed diagonally through the reactor chamber, from the inlet at the ceiling to the outlet in the wall at opposite end of the chamber. The flow rate of the gas mixture was 0.08 dm^3/s. The discharge was generated between a stainless steel needle and a plate made of nickel alloy. The diameter of the needle was 1 mm, and the dimensions of the plate were 67x90 mm. The distance between the electrodes was 15 mm.

The discharge electrodes were supplied from high voltage DC source SPELLMAN HV SL 600W/40kV/PN of positive polarity. The discharge current was stabilized by a series resistance of the value of 5 MΩ. For the discharge current increasing from 1 mA to 3 mA, after the time of synthesis of 30 s, the length of produced carbon fibers changed from 0.5 to 7 mm, and their diameter from about 20 μm to 70 μm [1]. In the following experiments, the concentration of hydrocarbon in argon was kept at 5%, and the current was 2 mA. These conditions were the optimal for carbon fiber synthesis: the growth rate was the fastest (0.25 mm/s), and the surface of carbon fibers was smooth [1].

The as-made carbon fiber grown on the needle tip was carefully removed from the reactor and examined under scanning electron microscope (Zeiss EVO 40) equipped with energy dispersive spectroscope (EDS, Bruker), and next tested by X-ray diffraction method (XRD). The XRD pattern was recorded using Phillips X'Pert system with Cu Kα radiation. Renishaw inVia Raman Microscope with a 100x objective lens and a laser of wavelength of 785 nm of power 1.5 mW was used in order to analyse the Raman spectra.

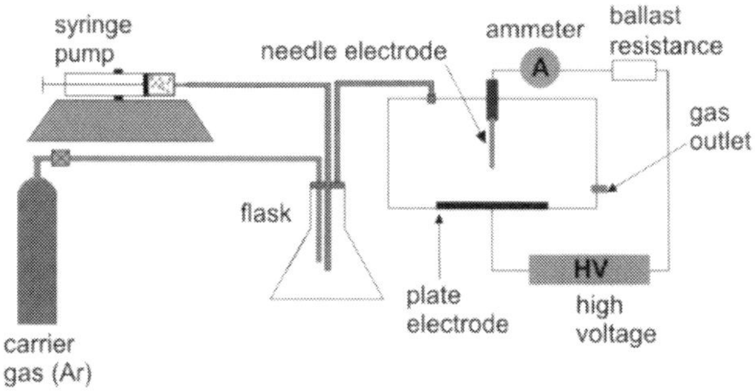

Figure 1. Schematic of experimental set-up.

3. Results

Fig. 2 shows SEM micrographs of a carbon fiber collected from the needle after 30 s of the discharge of the current of 2 mA. In Fig. 2a is presented the tip of carbon fiber, and in Fig. 2b a close-up view of its surface. The surface of the fiber is build from grains of about 1 μm in diameter.

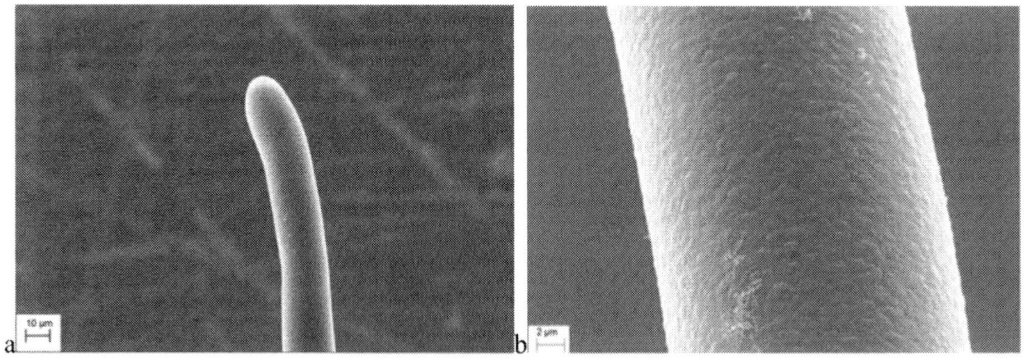

Figure 2. SEM images of carbon fiber obtained in Ar+C_6H_{12} (95:5) mixture, for I=2 mA and U=3.8 kV. a. tip of the carbon fiber, b. surface of the carbon fiber.

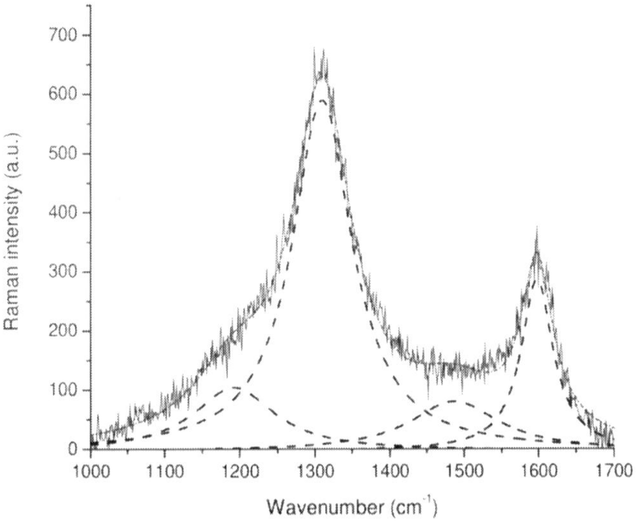

Figure 3. Raman spectrum of produced carbon fiber (I=2 mA).

The Raman spectrum of the as-made carbon fiber, emitted in the range from 1000 cm^{-1} to 1700 cm^{-1} is shown in Figure 3. Deconvolution of the spectra via using the Lorentzian fits (dashed lines in Figure 3) revealed the following peaks: 1192 cm^{-1}, 1309 cm^{-1}, 1486 cm^{-1} and 1598 cm^{-1}. The intense and sharp peaks at 1309 and 1598 cm^{-1} correspond to carbon D- and G-bands, respectively.

The G band is due to the Raman active E_{2g} vibration mode, which is related to the vibration of sp^2 bonded carbon atoms in a hexagonal graphite lattice. The D band is associated with dangling bonds of disordered carbons at the plane boundaries. The G band at 1598 cm^{-1} confirms the presence of graphitic carbon in the synthesized deposit [2]. Strong intensity of the peak at 1309 cm^{-1} implies the existence of large quantity of carbonaceous impurities.

The origin of peaks at 1192 and 1486 cm^{-1} is uncertain. The possible candidates for the 1192 peak are (1) sp^3 C–C stretching vibrations, (2) C–H bonds in the grain boundaries [3], and (3) mixed sp^2-sp^3- bonded carbon structure [4].

X-ray diffraction measurements showed that the carbon fibers obtained during the discharge were partially crystalline and partially amorphous. In Figure 4, the XRD peaks corresponding to 2θ values of 26.5°, 28.5°, 32.3°, 47.5° can be observed. In the case of carbon materials, an XRD peak with 2θ value of about 26.5° corresponds to the 002 plane of graphite. The peaks corresponding to 2θ values of 35.7° and 38.4° correspond to Si (which was the substrate) [5]. The nature of peak with 2θ value of about 32.3° is of unknown origin. Sharp and intense XRD peak at about 2θ=26.5° indicates graphitic nature of the fiber which is in good agreement with Raman spectroscopy investigations.

The elemental analysis was used to identify weight percentage of carbon and hydrogen in synthesized deposit. The analysis confirmed that carbon is the dominant element in the produced fibers. Comparing the concentration of carbon and hydrogen in cyclohexane (C — 85.7 wt.%, H — 14.3 wt.%) it is evident that highly carbonised products (C — 98.7 wt.%, H — 0.9 wt.%, the rest was N and O absorbed from the atmosphere) can be obtained from the decomposition of cyclohexane in electrically generated plasma.

The energy of electrons needed to dehydrogenate or decompose cyclohexane to simpler compounds of carbon and hydrogen should be in the range of 1.6 – 6eV [6]. In an electrical discharge at atmospheric pressure, only the electrons from the tail of energy distribution, which are those flowing near the sharp electrode (needle), could reach sufficiently high energy [7] required for decomposition of hydrocarbons. It can be supposed that carbon atoms can be produced only in the process of hydrocarbon fragmentation in collisions with fast electrons, for example in the following processes:

$$e^- + C_xH_y \rightarrow C_{x-1}H_{y-2} + H_2 + C + e^-$$

or

$$e^- + C_xH_y \rightarrow C_{x-1}H_{y-2} + H_2 + C^-$$

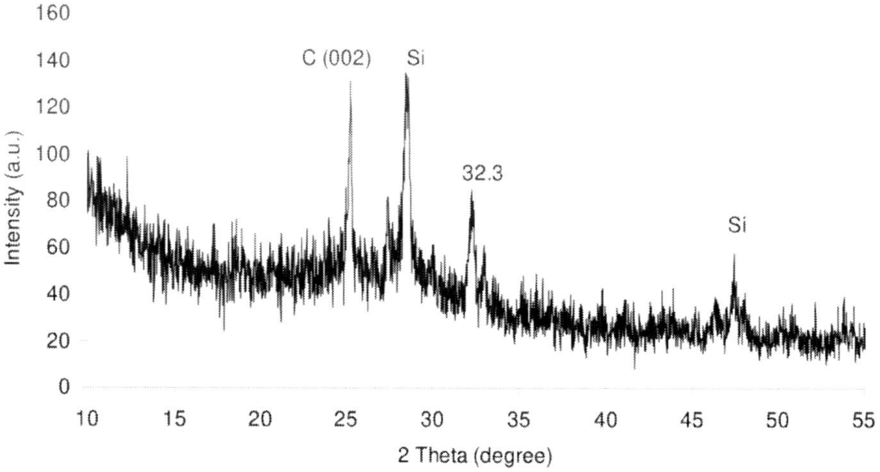

Figure 4. XRD of carbon fiber on silicon substrate.

4. Conclusion

The analysis confirmed that carbon is the dominant element in the produced fibers obtained in the electric-discharge plasma in point-plane geometry, for the discharge current I=2 mA. Raman spectroscopy and X-ray diffraction analysis confirmed the co-existence of crystalline and amorphous structures in the obtained deposit. It is evident that energy of the high-voltage low-current electrical discharge is sufficiently high to decompose cyclohexane to simpler compounds and atomic or molecular carbon from which the carbon fibers are built.

Acknowledgement

This research work was supported by the European Union in the framework of the European Social Fund. The system project of the Pomorskie Voivodeship "InnoDoktorant - Scholarships for PhD students", 2nd edition.

References

[1] Sobczyk A T, Jaworek A, Sozanska M 2009 Electrostatics **67** 275
[2] Zou G, Zhang D, Dong C, Li H, Xiong K, Fei L, Qian Y 2006 Carbon **44** 828
[3] Yang Q, Chen W, Xiao C, Hirose A, Bradley M 2005 Carbon **43** 2635965
[4] Jou S, Huang B-R, Wu M-C 2010 Thin Solid Films **15** 4458
[5] Somani S, Somani P, Noda M, Umeno M 2008 Diamond Relat. Mater. **17** 576
[6] Fridman A, Chirokov A, Gutsol A 2005 J. Phys. D: Appl. Phys. **38** R1
[7] Raizer Y P 1991 *Gas Discharge Physics* Springer, Berlin

Grounding & human health – a review

I.A. Jamieson[1*], S.S. Jamieson[2], H.M. ApSimon[1] and J.N.B. Bell[1]
[1]Centre for Environmental Policy, Faculty of Natural Sciences, Imperial College London, London SW7 2AZ, UK
[2]Department of Epidemiology and Public Health, Faculty of Medicine, Norfolk Place, Imperial College London, St. Mary's Campus, London W2 1PG, UK.

Email: isaac.jamieson@imperial.ac.uk

Abstract. Whilst grounding is often undertaken in industry as a matter of good practice in situations where the risk of excess charge exists, little thought is usually given to the biological effects that such measures may have, or possible benefits that may arise from the more widespread application of electrostatic and other 'electromagnetic hygiene' measures in hospitals and the general built environment. Research, which is still in its infancy, indicates that grounding the human body using suitable methodologies, particularly in low electromagnetic field environments, can significantly enhance biological functioning. It is proposed that there are often a number of electrostatic and 'electromagnetic hygiene' factors that need to be addressed before the beneficial effects of grounding the human body can be fully realised in many everyday environments.

1. Introduction

Though the often dramatic results of electrostatic discharge (ESD) events and ways in which these can be mitigated are well known in industry, the subtler effects that can occur to the human body as a result of the presence of excess charge, and its appropriate mitigation, are seldom documented. Suggestions of possible health benefits that may arise from grounding the body were discussed by White in the late 1920s [1]. This concept was later independently rediscovered by Ober in 1998 [2], with a number of clinical trials being undertaken by others (generally in low field environments) to test his hypothesis on the beneficial effects grounding individuals may have. Whilst the results obtained from such work appear encouraging, and may have far reaching implications related to how environments should be designed to aid performance and wellbeing, there is need for caution.

2. History

White [1] pioneered the scientific concept of grounding humans for health reasons after being informed by some individuals that they could not sleep properly "*unless they were on the ground or connected to the ground in some way*". He undertook sleep studies, using beds that had been earthed using copper wire connected to individual grounding points, and reported that individuals appeared to sleep better under such conditions. This concept was not taken up by mainstream medical science, and remained in obscurity to the general medical profession.

Ober developed the concept of grounding the body independently in 1998, after wondering what the effects of individuals being insulated from the ground might have on their health. After conducting experiments on himself, and noting how his sleep appeared improved when being grounded at night, he conducted similar pilot trials on others [2], and then undertook a study with 60 individuals [3]. Some scientists were then intrigued enough by the results of his published research to conduct studies investigating the possible physiological effects of grounding.

In this paper the term 'biological grounding' is used to denote the deliberate grounding of the body intended to specifically aid biological functioning.

3. Relevant Scientific Studies

3.1 First Published Study

This was a 30-day trial conducted by Ober [3] investigating the possible benefits to subjects with

Published under licence by IOP Publishing Ltd

sleep problems of being 'biologically grounded' (on dissipative carbon-fibre mattress pads connected directly via a ground wire to a ground rod driven into earth) during sleep, as opposed to sleeping on unconnected grounding systems. As with his initial pilot studies, favourable effects were often indicated. In this work, he noted that, in addition to humans often becoming charged with static electricity in standard ungrounded modern environments, they can also be exposed to oscillating electric [and magnetic – *present authors' comment*] fields, which can induce weak currents in their bodies that may cause biological effects.

3.2 Cortisol Study

As a result of the above work, Ghaly & Teplitz [4] undertook a pilot study on test-subjects (n = 12) suffering from sleep dysfunction, pain, and stress problems. This work, undertaken in the test-subjects' own homes, and using a direct grounding system and methods similar to that in the previously mentioned study, investigated how grounding at night might influence the circadian secretion of cortisol, excess production of which is linked with stress, high blood pressure, high blood sugar levels, weakened immune-system functioning and obesity.

Most of the test-group, who had previously exhibited excess night-time cortisol secretions, tended towards idealised circadian secretions of it as a result of grounding. Reduced fatigue too was generally found. Sleeping grounded was also generally indicated as reducing incidence of pain, sleep dysfunction (time to go to sleep, incidence of waking up in the night) and stress. While generally very favourable results were found, in 1 case a worsening of pain was noted. There was also one case where an increase of fatigue was noted and 1 case where increased stress levels were recorded. Though Ghaly & Teplitz [4] observed that both mains frequency fields and radiowaves could influence the results of parameters being assessed, no direct measurements were taken of these as potential confounders that might have influenced test-subjects' results.

3.3 Brainwaves and Muscle Tone

Chevalier et al [5] undertook Electroencephalogram (EEG) readings on 58 healthy adults (including 30 controls) in a double-blind study to record the physiological effects of 'biological grounding' (clinical earthing utilising conductive adhesive patches on each test-subject's feet connected via a ground wire to a ground rod driven into earth). Abrupt changes in brainwave patterns were shown in approximately half of those tested - with grounding occurring covertly halfway through the 56-minute test-period - with the right and left hemispheres of the brain of those subjects appearing to go into increased synchronisation with each other. This effect, which was not found in controls, continued during the remainder of the test periods.

Electromyography (EMG) readings (recordings of electrical activity in muscles) were also undertaken, which revealed increases in the electrical potential of skeletal body muscles (and improvement in muscle tone) of all those tested when grounded. This result is consistent with extra electrons (made available through the grounding) making cell membranes less positive, thereby increasing electrical potential and muscle contraction.

3.4 Thermographic Studies

Electrons can behave as anti-oxidants neutralising positively charged free radicals in the body that can otherwise cause inflammation [6]. It appears that 'biological grounding' can help remedy the body's electron deficit and restrict such damage. Assessments of randomly selected outpatients from a clinical treatment centre were undertaken by Amalu [7], using high-resolution medical infrared imaging equipment. In these tests 'biological grounding' of the test-subjects was undertaken using either clinical earthing or by their sleeping grounded connected to earth via a ground wire and earthed ground rod. The results he obtained further indicate the benefits of using appropriate 'biological grounding' protocols. He stated that it showed great potential for treating both acute and chronic inflammatory medical conditions.

Chronic inflammation is implicated as a potential contributory factor in a number of diseases and ailments including: cardiovascular disease, allergies, Alzheimer's, anaemia, arthritis, asthma, cancers, chronic pain, diabetes, eczema, intestinal disorders, kidney failures, multiple sclerosis and sleeplessness. It is caused, in part, by overproduction of free radicals. When there is a shortage of free electrons, free radicals often take electrons from normal cells & tissues, thereby exacerbating inflammation problems.

3.5 Dark Field Microscopy of Live Blood
As documented by Ober et al. [2], the effects of 'biological grounding' on live blood were assessed by Sinatra (in non peer-reviewed research) using dark field microscopy in 2008 (n = 12). Blood samples were taken both before and 40 minutes after grounding test-subjects, and revealed dramatic differences in their blood viscosity between the two conditions tested, with fewer formations of erythrocytes (red blood cells) associated with clotting and clumping being observed after grounding. Sinatra stated that the results obtained suggest 'biological grounding' could prove particularly beneficial for those with diabetes or cardiovascular problems.

4. Potential Confounders

4.1 Grounding/Earthing Conditions
Whilst grounding via the mains to reduce excess charge is often undertaken to prevent ESD, and is advocated by many companies selling equipment for 'biological grounding', its suitability for the latter has yet to be properly assessed. Grounding in this way may increase exposure to the high frequency transients and mains frequency harmonics sometimes carried on modern mains electricity and created by various electrical items. Indirect evidence of this risk indicates that for some people exposure to such transients when carried on mains electricity may increase incidences of a number of health complaints [8]. In situations where such conditions exist, it is proposed that grounding through the earth connection in the mains might further exacerbate health problems. As mentioned by Chevalier et al [5], the possibility also exists that fields from electrical wiring in rooms could create induced body voltages and that ground wires, such as used in the assessments they undertook, might act as antennae for stray electric fields unless suitable 'electromagnetic hygiene' precautions are taken.

4.2 Electrosmog
As previously mentioned, dark field microscopy demonstrates that grounding the body appears (in certain situations) to reduce blood viscosity, thereby reducing a major risk factor for strokes and heart attacks which are the leading cause of death for both men and women worldwide. However, exposures to 'electrosmog' from different types of commonly encountered electromagnetic field (EMF) regimes may act as potential confounders. Non peer-reviewed research by Havas [9] and others, indicates that whilst exposures to low EMF regimes can also result in reduced blood viscosity, raised field exposures (as can be experienced through exposures from certain types of unshielded electrical items, especially some microwave emitters), may cause clumping of red blood cells similar to that found in cancer patients. It appears that under such circumstances, the cells lose their negative surface charge and become attracted to each other, with this clumping resulting in reduced capillary circulation and oxygen distribution. Whilst the effect of 'biological grounding' under such situations has yet to be assessed, it would appear wise to reduce exposures to 'electrosmog' wherever practical.

4.3 Small Air Ions and Submicron Particles
Both the potential individuals are charged to, and the electromagnetic character of the microenvironments they occupy, can influence their uptake of small air ions (SAI) and charged and charge-neutralised submicron particles when grounded or ungrounded. Andersen [10] indicated that when individuals are at zero-potential in field-free zones, SAI of both polarities can be inhaled in the proportions that occur naturally in the measured air, and that increased DC potentials alter the proportions and concentrations of each polarity inhaled. We hypothesise that a similar situation would appear to be true with regard to charged submicron particles.

Whilst simulations assessing how a subject's potential can influence SAI uptake have been undertaken, similar research does not yet appear to have been undertaken for nanoparticles and other charged and charge-neutralised submicron particles. We propose that it is important to do so, especially for particles in the nanoparticle size range, so that accurate predictions of inhaled and retained particle dosage under different charging and grounding regimes can be made. Such measurements and comparisons should also be undertaken to simulate standard situations where fields exist, as these too can influence local concentrations of SAI and submicron contaminants and their retention when inhaled.

In situations where individuals are grounded and adjacent to active alternating electric field emitters, they will register voltages approaching zero even if they experience induced charges on parts of their bodies (or are exposed to other types of fields that may also be biologically active). Under such conditions, body-voltage may be low but the current flowing through the body in response to the field is likely to be maximised, unless a better pathway to ground is supplied for it, as electricity always flows to the point of lowest potential. The low body voltage recorded in such situations can give a false sense of security. 'Biological grounding' in microenvironments where raised fields exist can cause individuals to experience a concentration of charge on areas of their body, such as their noses, when facing the field source whilst still retaining a net-neutral charge. Such situations can alter the ratio and concentrations of each polarity of SAI and charged submicron particles that they inhale. Additionally, in microenvironments where a high degree of charge exists, subjects will often still be exposed to lower concentrations of SAI and higher concentrations of charged submicron particles (which have been shown to be injurious to health) than would be the case under low field conditions [11]. As a general rule, low field environments should be sought - or created - when seeking to obtain the physiological benefits achievable with this type of grounding.

5. Conclusion

As can be seen by the above, whilst replicating 'biological grounding' as found in Nature holds great promise as a novel means of improving and maintaining human health (and also for helping revolutionise the design of the built environment), there is need for caution.

In instances where the comments of its critics are properly addressed, its proven potential can be refined and cost-effectively applied in many types of environment to provide the opportunity to help increase general wellbeing whilst reducing the likelihood, severity and incidence of a number of serious ailments.

Many of the concepts discussed can now be easily applied in hospital and general building design. 'Biological grounding' measures, alongside ESD prevention and other best practice 'electromagnetic hygiene' measures, have important roles to play in making the built environment safer, more biologically friendly and truly sustainable.

References

[1] White G S 1929 *Finer forces of nature in diagnosis and therapy*, Phillips Printing Company, USA.
[2] Ober C, Sinatra T and Zucker M 2010 *Earthing: The most important health discovery ever?*, Basic Health Publications, Inc., California, 260 pp.
[3] Ober C 2000 *Grounding the human body to neutralize bioelectrical stress from static electricity and EMFs*, ESD Journal, January 2000, http://www.esdjournal.com/articles/cober/ground.htm
[4] Ghaly M and Teplitz D 2004 *The biological effects of grounding the human body during sleep, as measured by cortisol levels and subjective reporting of sleep, pain and stress*, The Journal of Alternative and Complementary Medicine. **10**, pp. 767-776.
[5] Chevalier G, Mori K and Oschman J L 2007 *The effect of earthing (grounding) on human physiology*, European Biology and Bioelectromagnetics, **1**, pp. 600-621.
[6] Oschman J L 2009, *Energy Medicine: Charge transfer in the living matrix*, Journal of Bodywork and Movement Therapies. **13**, pp. 215–228.
[7] Amalu W 2004 *Medical Thermography: Case studies. Clinical earthing application in 20 case studies.* International Academy of Clinical Thermography, Internet article.
[8] Havas M and Olstad A (2008) *Power quality affects teacher wellbeing and student behavior in three Minnesota Schools*, Science of the Total Environment, **402**(2-3), pp.157-162.
[9] Havas M 2009 *Live blood analysis & electrosmog*, http://www.magdahavas.com
[10] Andersen I 1965, *The influence of electric fields on the uptake of light gas ions of a model of man.* International Journal of Biometeorology. **9**, pp. 149–160.
[11] Jamieson I A, Holdstock P, ApSimon H M and Bell J N B 2010, *Building Health: The need for electromagnetic hygiene?,* IOP Conference Series: Earth and Environmental Sciences.

Acknowledgements
Special thanks are given to Dr Jeremy Smallwood and to David Webb for their suggestions and advice given during the preparation of this article.

Health effects of electromagnetic field generated by lightning current pulses near down conductors

Z Á Tamus[1], B Novák[1], L Szűcs[2], I Kiss[1]
[1]Department of Electric Power Engineering, Budapest University of Technology and Economics, 18. Egry J. str., Budapest H-1111, Hungary
[2]Department of Broadband Infocommunications and Electromagnetic Theory, Budapest University of Technology and Economics, 3. Goldmann sqr., Budapest H-1111, Hungary

E-mail: tamus.adam@vet.bme.hu

Abstract. The lightning current generates a time varying magnetic field near down conductors, when lightning strikes the connected Franklin-rod. The down conductors are mounted on the wall of buildings, where residential places can be situated. It is well known that the rapidly changing magnetic fields could generate dangerous eddy currents in the human body. If the duration and the gradient of the magnetic field were high enough, the peripheral nerves are excited. In this study, the authors introduce an improved model of the interaction of electromagnetic fields of lighting current near a down conductor with the human body. The interaction model has two parts: estimation of the magnetic fields surrounding the down conductor and evaluation of health effects of rapid changing magnetic fields on the human body.

1. Introduction

People may suffer electrical lightning injuries in different ways [1, 2]:

- Direct strike: The lightning current flows directly down through the human body.
- Contact injury: The person is contacted to a subject that is connected to the lightning.
- Side flash or splash: The lightning flash jumps to the human from a tree or a building.
- Ground current: The lightning strikes near the person and the finite conductivity of the ground causes potential differences and a dangerous step voltage can appear at the legs of the person.
- Upward streamer or leader: The leader does not attach to the main lightning discharge and a person who is a source of the leader can be shocked.

Besides these electrical injuries of lightning, the authors brought up another deleterious effect of a lightning strike [3]. The lightning current generates time varying magnetic fields near down conductors. The down conductors are mounted on the wall of buildings, where residential places can be situated. It is well known that rapidly changing magnetic fields can generate dangerous eddy currents in the human body and if the duration and the gradient of the magnetic fields are high enough, the peripheral nerves are excited. Gradient magnetic field having higher intensity can cause cardiac stimulation, which may result in fatal cardiac fibrillation.

In this article, an improved model of the interaction of electromagnetic fields of lighting current near down conductor with the human body is introduced. The interaction model has been improved in its two parts: health effects of rapid magnetic fields on human body and estimation of the probability of the dangerous magnetic fields surrounding the down conductor.

2. Excitation of nerves by time varying magnetic fields

Many research groups have addressed interaction between a patient and time varying magnetic fields since the beginning of clinical application of magnetic resonance imaging systems. These studies have resulted in many interaction models, which have been summarized and published elsewhere [4].

The threshold level of excitation of nerves depends on the duration of stimulation, i.e. higher-level stimuli are needed for shorter exciting pulses. If a nerve is excited by a time-varying magnetic field this gradient field induces an electric field in it and this induced field can stimulate the nerve. For long gradient magnetic fields (τ>1 ms) the minimum threshold is expressed by the time derivative of the magnetic flux density (*dB/dt*). If the impulse is shorter than 10 μs, the product of the impulse duration and intensity give the threshold for the nerve stimulation [5]. Between these time intervals, the threshold for nerve excitation depends on the duration of the stimuli. This rule of neurophysiology is summarized in a very simple hyperbolic formula, namely in the Lapicque-formula. At excitation of a tissue by a time-varying magnetic field, the threshold can be expressed as a time derivative of magnetic flux density (*dB/dt*):

$$\frac{dB}{dt} = b\left(1 + \frac{c}{\tau}\right) \qquad (1)$$

where *b* is the rheobase and *c* is the chronaxie, τ is the duration of the stimulus. The rheobase is the asymptotic *dB/dt* for long duration stimuli and the chronaxie is the duration of the stimulus at which *dB/dt* is twice the rheobase (Fig. 1.). This formula can be used when the exciting duration is higher than tens of μs.

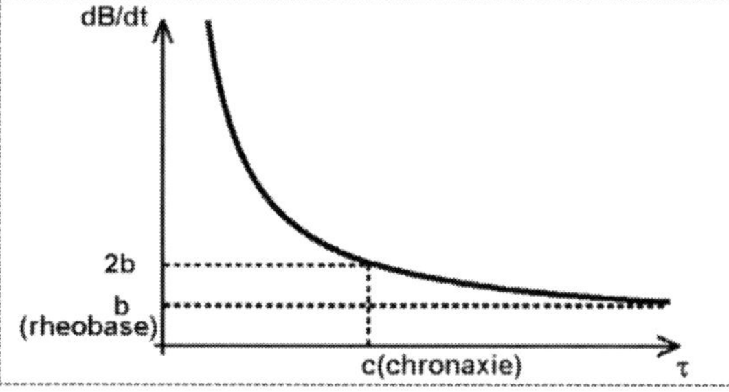

Figure 1. Strength-duration curve of nerve stimulation by time-varying magnetic field

3. The new interaction model

Calculations based on the first interaction model [3] show that the rising edge of the lightning current pulse can cause nerve excitations hence the magnetic field generated by the falling edge was not examined in this case.

In our new model, the probability of nerve excitation, which can lead to harmful heart fibrillation, has been examined as a function of distance from a down conductor, when lightning strikes the connected Franklin-rod. The investigation has two parts regarding the rise time of the lightning current:

- The rise time of the lightning current <10 μs: The product of the strength and the duration (*dB/dt*) has been used for a threshold, which was 0.19 T [4]. The case of positive and negative lightning have been also examined, since both polarities of lightning occurs in this time range.
- The rise time of the lightning >10 μs: the Lapique-formula has been used for the calculations. These calculations have been made for positive lightning since only 5% of negative lightning has higher rise time than 28.6 μs, while the 50% of positive lightning has longer rise time than 35 μs. In this case, two rise times were calculated 35 μs for 50% probability and 250 μs for 5% probability [6]. For cardiac muscle 3 ms and 63 T/s were assumed for chronaxie and rheobase, respectively [7].

4. The magnetic field of lightning current surrounding down conductor

Assuming that the down conductor is filamentary and infinite, the following equation is valid for *dB/dt:*

$$\frac{dB}{dt} = \frac{\mu_0}{2\pi x}\frac{di}{dt} \tag{2}$$

where *di/dt* is the time derivative of the lightning current and x is the distance from down conductor. If the threshold *dB/dt* is given, the corresponding threshold *di/dt* can be calculated at distance x from the down conductor by the arrangement of (2):

$$\frac{di}{dt} = \frac{2\pi x}{\mu_0}\frac{dB}{dt} \tag{3}$$

If the rise time is shorter than 10 μs, the product of *dB/dt* and τ is the threshold for the excitation therefore, Eq. (3) can be simplified:

$$i = \frac{2\pi x}{\mu_0}B, \tag{4}$$

where i is the peak current of lightning, and B is the threshold.

5. Calculations

When lightning strikes the Franklin-rod, the gradient magnetic field surrounding the connected down conductor depends on the peak and the rise time of the lightning current. If the product of *dB/dt* and τ is the threshold for excitation, the current peak determines the adverse health effects. The probability distribution of lightning peak current is described by a lognormal distribution. The parameters of lognormal distribution are in Table 1. according to the IEC 62305-1 standard.

Table 1. Lognormal distribution parameters of lightning current [6]

Polarity	Mean [kA]	σ_{ln}
- < 20 kA	61	1.33
- >20 kA	33.3	0.605
+	33.9	1.21

Table 2. The threshold current and probability of cardiac stimulation in the case of positive and negative lightning. Rise time < 10 μs

Distance [m]	0.1	0.25	0.5	0.75	1
Threshold current peak [kA]	95	237.5	475	712.5	950
Probability (+)	19.72%	5.38%	1.46%	0.59%	0.29%
Probability (-)	4.16%	0.06%	0.00%	0.00%	0.00%

5.1. The probability of excitation, when rise time is lower than 10 μs

The results of the calculations (Table 2, Figure 2) show that the probability of cardiac stimulations by the down conductor was quite high (>0.29%) when positive lightning strikes the Franklin-rod and the distance is less than 1 meter. The probability of this dangerous effect is negligible in the case of negative lightning strike if the distance from the down conductor was greater than 0.5 m. This requirement is generally satisfied in practical conditions.

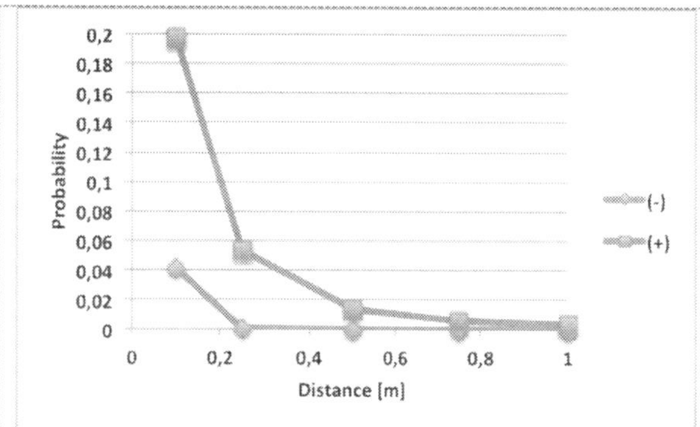

Figure 2. Cardiac stimulation probability as the function of the distance from down conductor.

5.2. The probability of excitation when rise time is higher than 10 μs

In this case, the thresholds (*dB/dt*) are determined by Eq. 1, and from Eq. 3 the threshold currents for stimulations can be calculated (Table 3).

Table 3. The threshold current of cardiac stimulation in the case of positive lightning. Rise time: 35 μs and 250 μs.

Distance [m]	**0.1**	**0.25**	**0.5**	**0.75**	**1**
Threshold current peak (35 μs) [kA]	95.6	239.0	478.0	717.0	956.0
Threshold current peak (250 μs) [kA]	102.4	255.9	511.9	767.8	1023.8

The results of the calculations show that the threshold for cardiac stimulation does not depend on the rise time of the lightning current. This results from the heart muscle chronaxie, which is higher by one order of magnitude than the highest rise time of the lightning current.

6. Conclusion

The harmful effect of time varying magnetic fields generated by lightning current surrounding a down conductor has been investigated. The results show positive lightning strikes can generate cardiac muscle stimulation, which can lead to fatal cardiac fibrillation. Moreover, the analysis of different lightning current parameters bring out that the probability of this life threatening interaction does not depend on the rise time of the lightning current. This results from the excitation characteristic of heart muscle.

References

[1] T. Horváth *Understanding lightning and lightning protection: A multimedia teaching guide*, J. Wiley & Sons, 2006. ISBN-10: 0470030186

[2] M. A. Cooper, R. L. Holle, C. J. Andrews *Distribution of Lightning Injury Mechanisms*, Proc. of the 30th Int. Conf. on Lightning Protection, ICLP2010. Cagliari, Italy, paper 1325.

[3] Z. Á. Tamus, B. Novák, S. Szabó, I. Kiss, I. Berta *Can the Near Field of Lightning Impulse in Down Conductors Cause Harmful Health Effects?* Proc. of the 30th Int. Conf. on Lightning Protection, ICLP2010. Cagliari, Italy, paper 1042.

[4] D. J. Schaefer, J. D. Bourland, J. A. Nyenhuis, 2000 Review of patient safety in time-varying gradient fields, *J. Magn. Reson. Imag.*, vol. 12, pp. 20 – 29.

[5] J. P. Reilly, Peripheral nerve stimulation by induced electric currents: exposure to time-varying magnetic fields *Med Biol Eng Comput.* 1989 Mar, Vol. 27(2) pp. 101-110.

[6] IEC 62305-1 standard: Protection against lightning- Part 1: General Principles

[7] International Commission on Non-Ionizing Radiation Protection, Statement on Medical Magnetic Resonance (MR) Procedures: Protection Of Patients *Health Physics* 87(2): 197-216; 2004.

Aerosol charge state characterisation using an ELPI

J C Matthews[1], M D Wright[1], M F Biddiscombe[2], O S Usmani[2] and D L Henshaw[1]

[1] University of Bristol, UK
[2] NHLI, Imperial College London and Royal Brompton Hospital, UK

E-mail: j.c.matthews@bristol.ac.uk

Abstract. A new technique has been developed to measure the size distribution and charge state of highly charged aerosols using an Electrical Low Pressure Impactor (ELPI). The internal charger was switched alternately on and off and the time between stable charge states found to be ~10 s. The size distribution of aerosols was found when the charger was on, from which the charge distribution can be estimated when the charger is off using the current at each impactor stage. This method was tested in background conditions, when a candle was burning and when a negative air ioniser was used. The ELPI electrometers were not sensitive enough to accurately measure the charge state on background and candle air, but gave a value for air charged by an ioniser. Comparing results from the ELPI with other techniques showed inaccuracies in this method that need to be addressed before further use of this technique.

1. Introduction

The lung deposition probability of inhaled particles is usually described by the ICRP 66 model [1]. In practice, deposition efficiency is below 100%, with a minimum of only 30% at 200 nm, the remaining fraction being exhaled again. The model, however, takes no account of particle charge state and there is evidence that charged particles deposit more efficiently than uncharged or charge-neutralised particles [2]. High voltage power lines are known to emit corona ions [3] which can create a unipolar excess of ions downwind. This can lead to unipolar charging of atmospheric aerosols [4, 5]. It is hypothesised that this increase in charge state could lead to enhanced deposition within the lung [6].

We intend to test whether 100 nm and 200 nm particles with excess unipolar charge have a higher probability of lung deposition than charge-neutralised particles, when inhaled by healthy human volunteers. A key part of the study is characterisation of the charge state and size of inhaled particles.

In this paper we seek to investigate whether the extent of aerosol charging can be determined using an Electrical Low Pressure Impactor (ELPI) (Dekati, Finland). The ELPI is a cascade impactor that uses an internal corona charger to bring particles to a known charge distribution. The current on each impactor stage is measured and the size distribution is obtained in conjunction with the charging probability. Existing research has used the ELPI to measure the total (net) charge present at each stage during a single pharmaceutical inhaler actuation [7]. Here, the charger in the ELPI is alternately switched on and off to allow alternate, on-line measurement of both size and charge distributions (i.e. the probability of a particle of given size carrying a given number of electronic charges).

2. Methods

2.1. Experimental methods
Ambient indoor air was sampled using an ELPI with volumetric flow 9.87 l min^{-1}. When the internal +5 kV charger is on, the particle size distribution can be found from the current values and known particle charging probability at each filter stage [8]. The electrometer measurement range 0 – 100,000 ±15 fA was used. Three different indoor conditions were examined: i) normal background air, ii) with a burning candle, to increase aerosol concentration and iii) with a commercial negative air ioniser, to alter the aerosol charge state. For each run, the ELPI was operated for 30 minutes, with the internal corona charger alternately switched on and off every 30 s. In the candle and ioniser tests, some time was allowed until the readings reached a steady state.

A second test compared aerosol charging results from ionised air using the ELPI to previously used techniques. The bipolar or unipolar electrical environment, when diffusion charging dominates, can be investigated using Aspiration Condenser Ion Mobility Spectrometers (ACIMS) [4] which measure small ion mobility and concentrations, or a 'Dual-Differential Mobility Analyser' (D-DMA) technique [5] which uses size- and mobility-distributions to estimate the ambient charge state of sampled aerosol particles. Both can be used to infer a value for the (steady state) ratio of positive to negative ion concentrations (c_+, c_-) multiplied by their associated mobilities (μ_+, μ_-), referred to as the charge asymmetry ratio, x (1). x can be used to describe the charging probability for aerosols with a Gaussian charge distribution when aerosol concentration is constant, regardless of ion properties [9]. 1 hour of data were collected using the ELPI, ACIMS and D-DMA with the negative room ioniser on.

$$x = \frac{c_+\mu_+}{c_-\mu_-} \tag{1}$$

2.2. Calculation of size distribution, average charge per particle and overall charge distribution
If the size distribution is known, the ELPI can be used to obtain the average charge per particle at each ELPI stage and potentially to estimate the overall charge distribution i.e. the probability that a particle of a given size has n elementary electronic charges (where 1 electronic charge is the charge of an electron). The relationship shown in (2) is used to calculate the mean concentration of particles at each filter stage, C, from the current, I, which enables a size distribution to be estimated. Q is the flow rate of air, e is the charge of an electron and P is the penetration of particles through the corona charger. The quantity Pn_c is a constant for each stage [8].

$$I = CPn_c eQ \tag{2}$$

When the mean concentration for each ELPI stage has been calculated for the times when the internal charger is on, the net average number of elementary charges per particle for each stage can be estimated from the current measured within the ELPI when the internal charger is switched using (2). When the charger and trap voltage are on, P and n_c can be considered theoretically [10] however, the effects when the charger and trap voltage are off are not known. Glover and Chan [7] tested the difference between the total charge measured with the corona charger in place but switched off and with the charger removed from the ELPI, and reported no significant change. Therefore penetration through the charger, P, is assumed to be 1. A simple analytical expression for aerosol charge distributions was put forward by Clement and Harrison [9], which includes an expression (3) for the average number of charges carried by a particle of diameter d, where k is the Boltzmann constant, T is the temperature and ε_0 is the permittivity of free space.

$$n = \frac{2\pi\varepsilon_0 dkT}{e^2}\ln x \tag{3}$$

3. Results

3.1. Characterisation of the instrument response time

Figure 1 shows the current measured on two stages of the ELPI corresponding to average diameters of 120 and 200 nm. A negative room ioniser was activated at the beginning of the measurement. At 180 s the ELPI internal charger was switched on, then turned off and on alternately every 30 s thereafter.

After switching the internal corona charger on or off, all particles undergo a change of charge state, between ambient and corona charged. The current on each filter stage changes until a new stable current is reached corresponding to the concentration at a particular size. Particles larger than 1 μm in diameter also exhibit a transient lasting a few seconds due to image charge effects. The time taken for the electrometers (using the 100,000 fA setting) on all filter stages to settle was ~10 s.

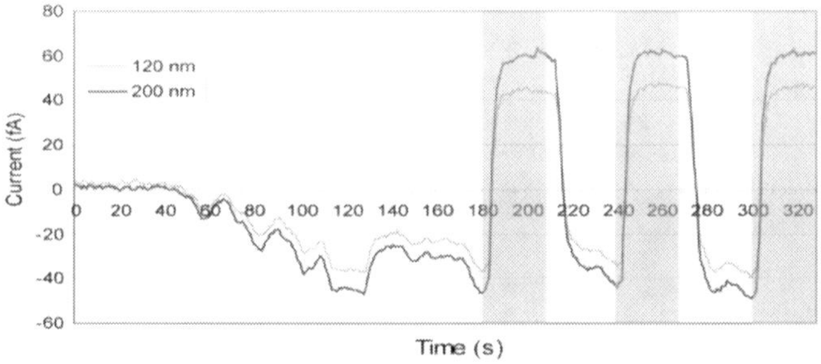

Figure 1. Time series of the current measured in an ELPI in ionised air with the internal corona charger switched on (shaded area) and off.

3.2. Size distributions and average charge of aerosols

The data from the ELPI was separated into times when the internal charger was on and when off. The first 10 s after the charger was switched on or off was discarded so that transitional data was not used in the calculations. Figure 2 shows the size distributions found for background conditions, while an air ioniser was used and when a candle was burning. The size distributions for background and ionised air are broadly similar, while there is a large increase in aerosols smaller than 1 μm when a candle is burnt, consistent with previous measurements of candle burning [11]. The current at each filter stage when the internal corona charger was off, after the first 10 s of settling time was discarded, was used in equation (2) to calculate the average charge per particle for each stage, shown in Figure 3.

3.3. Comparison of x obtained using ELPI, ACIMS and D-DMA

The charge asymmetry ratios (x-value) calculated from the ELPI for each size bin were compared to those obtained from ACIMS and D-DMA results. ACIMS measures ion properties only, while D-DMA infers a value of x based on aerosol charging over a range of diameters (in this case 10-1100 nm), so only a single x value is given in both cases. For ACIMS, $x = 0.032$ while for D-DMA, $x \sim 0.03$ during the period when the room ioniser was switched on, consistent with significant negative unipolar charging. This is in contrast to ELPI results, which were of the order $x \sim 10^{-5}$ to 10^{-7} for the largest particle sizes (3-10 μm, first and second impactor stages) and increased to a maximum of x~0.2 for the smallest sizes (7-30 nm, ELPI filter stage). The very low estimated values of x for the largest sizes (which suggest an unfeasible bias towards negative charging) may result from space charge effects due to the presence of negative ions, enhancing charge deposition on the upper plates. Also, mean particle charge is obtained using currents around the lower resolution limit of the electrometers (several fA).

4. Discussion and conclusions

This study seeks to characterise the charge state of submicron particles for use in a study on the deposition of charged particles in the human lung. For 120 nm and 200 nm particles, the average charge was 4.2 and 8.3 negative charges per particle respectively when the room ioniser was on. The average charge was ~0 for both sizes for background and candle burning, suggesting little deviation from a Boltzmann charge distribution. However in this case the measured currents were of similar order of magnitude to the accuracy limits of the electrometers for each ELPI stage.

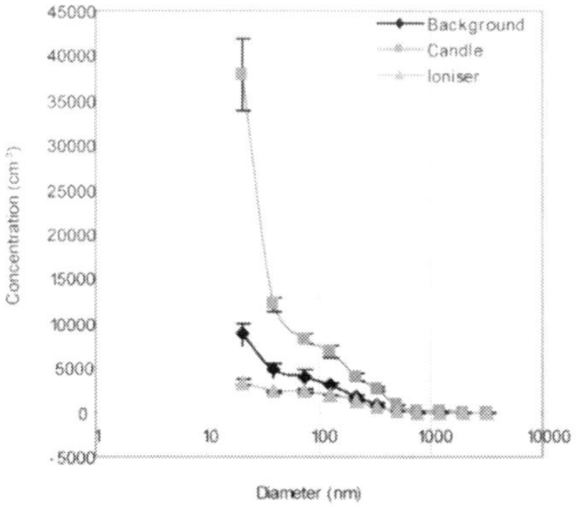

Figure 2. Size distributions (d*N*/dln(*d*)) of aerosols measured by the ELPI.

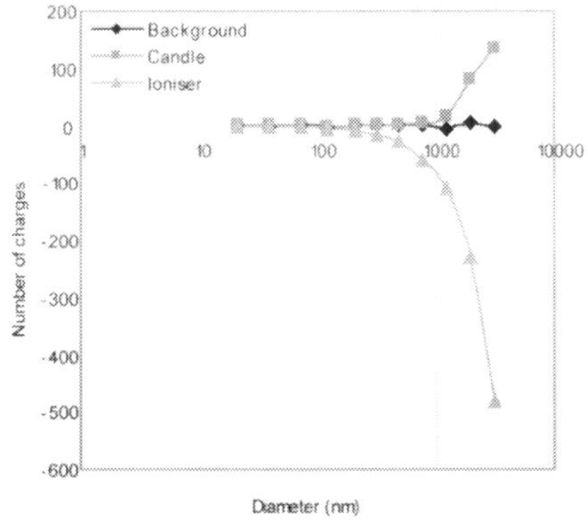

Figure 3. Average particle charge obtained using an ELPI with switched internal charger.

There is good agreement between the overall values of x obtained from ACIMS and D-DMA measurements, and in general poor agreement between these values and those obtained from ELPI. This suggests that, for these charging levels, the present method using the ELPI cannot be used to determine the aerosol charge state. Previous studies have shown that ELPI may not be satisfactory in measuring bipolarly-charged distributions [12] which casts further uncertainty on the results from the background and candle measurements. The assumption given in section 2.2 that $P = 1$ could be found to be incorrect and further work is recommended to assess the effect of space charge and image charge effects of charged particles within the ELPI.

Acknowledgements
This work was funded by Children with Leukaemia, registered charity (UK) number 298405. The authors would like to thank the reviewer for their helpful comments.

References
[1] ICRP. 1994. *Human Respiratory Tract Model for Radiological Protection* ICRP Publication 66 (Oxford: Pergamon)
[2] Melandri C, Tarroni G, Prodi V, De Zaiacomo T, Formignani M and Lombardi C C 1983 *J. Aerosol Sci.* **14** 657–69
[3] Matthews J C, Ward J P, Keitch P A and Henshaw D L 2010 *Atmos. Environ.* **44**(9) 5093–100
[4] Fews A P, Holden N K, Keitch P A and Henshaw D L 2005 *Atmos. Res.* **76** 29–48
[5] Buckley A J, Wright M D and Henshaw D L 2008 *Aerosol Sci. Technol.* **42** 1042–51
[6] Henshaw D L 2002 *Med. Hypotheses* **59** 39–51
[7] Glover W and Chan H K 2004 *J. Aerosol Sci.* **35** 755–64
[8] Marjamaki M, Keskinen J, Chen D R and Pui D Y H 2000 *J.Aerosol Sci.* **31** 249–61
[9] Clement C F and Harrison R G 1992 *J. Aerosol Sci.* **22** 481–504
Harrison R G 1996 *J. Aerosol Sci.* **27** S1 S191-92
[10] Ouf F and Sillon P 2009 *Aerosol Sci. Technol.* **43** 658-698
[11] Wright M D, Fews A P, Keitch P A and Henshaw D L 2007 *Aerosol Sci. Technol.* **41** 475–84
[12] O'Leary M, Balachandran W and Chambers F 2008 in Conf. Rec. IEEE-IAS Annual Meeting, 1–5
Belaca R, Abbod M, Balachandran W and Miller P R 2010 *IEEE Trans. Ind. Appl.* **46** 1181–87

Structure of rat skin after application of electret characterized by DSC

LL Cui[1,3], YY Liang[2,3], FJ Dong[2], L Ma[2], Y Tu[1], HY Liu[1] and J Jiang[2,4]
[1]School of Pharmacy, Second Military Medical University, China
[2]College of Basic Medical Sciences, Second Military Medical University
[3]Both of the authors contributed equally to this work
[4]Author to whom any correspondence should be addressed

E-mail: jjiang@smmu.edu.cn

Abstract. Polypropylene (PP) electrets with surface potential of -500V, -1000V and -2000V were prepared by constant voltage corona charging. The electrets were applied to excised rat skin for 2 hours respectively and then the skin samples were analyzed with the differential scanning calorimetry (DSC) technique to study the alteration of lipid organization of the skin. There were three peaks at 63°C, 82.7°C and 115.1°C in the DSC spectra for rat skin untreated, which have been assigned essentially to lipid, lipid-protein and protein alterations. For -500V electret treated-skin sample, only a single peak appeared at 79.1°C. With the increase of electret surface potential from -500V to -2000V, the transition temperature and peak areas at moderate decreased first and then increased. The negative electret could result in the transition of stratum corneum (SC) lipid from gel to liquid crystal and protein transition from α helix structure to β folding structure. The regulation action of electret to skin mircostructure presented an effect of "potential window".

1. Introduction

An electret is a functional material which can keep space and polar charge for a long time. The external electric field of the electret can result in a series of biological effects, such as improving microcirculation, lowering blood viscosity, enhancing fracture and wound healing, regulating cell growth cycle and apoptosis, bacteriostasis and air cleaning [1-5], etc. With the development of studies on the biological effect of electrets, the regulation effect of electret on skin state and the electret transdermal drug delivery system have become a research focus [6-8], giving a novel formulation and route for clinical application.

Transdermal delivery has a variety of advantages over the oral route as avoidance of the first-pass metabolism, sustained therapeutic action and better patient compliance. However, its prevalent use is restricted due to excellent impervious nature of skin. Therefore, many approaches have been applied to overcome the skin barrier. Since 1980s, we have studied the enhancing effect of electret on transdermal drug permeation. Although our studies and those of Murthy et al. confirmed that an electret could be an excellent physical approach to enhance transdermal drug permeation [8, 9], the mechanism of electret action is still an attractive study area. To well understand the mechanism of electret action, this paper will use differential scanning calorimetry (DSC) and microstructure analysis to study the structure change of rat skin after application of an electret.

Published under licence by IOP Publishing Ltd

2. Materials and Methods

2.1. Electret preparation
The polypropylene (PP) film with a thickness of 13μm was from Toray Fan Co. Japan. All the materials were from the same rolls to keep the same morphometrics. The bare PP film was corona charged at constant voltage and 77% humidity for 5 minutes using a corona charging system (Dalian University of Technology, China) with the point voltages of -20kV and grid voltages of -500V, -1000V and -2000V, respectively. The surface potentials of the samples were measured with a surface potentiometer (ESR102A, Beijing Huajinghui Technology Co. Ltd. China).

2.2. Regeants
Solutions of PBS (0.2M, pH 7.2) and 20% urethane were prepared in our laboratory.

2.3. Animals and grouping
12 Male Sprague-Dawley (SD) rats were randomly divided into groups of control, -500V, -1000V and -2000V electret samples, respectively. The rats were anesthetized by 20% urethane (0.5ml/kg body weight). The abdominal hair of the rat was shaved with a hair clipper. Four test sites were selected and marked, each having an area of 2cm×2cm and an interval of 2cm. The test sites were treated for 2 hours with PP film (0V) and electrets with different surface potentials. Then the rats were killed. The tested skin was cut off quickly, wiped with PBS solution, then the excess fat adhering to the dermis side was removed using cotton, and the skin sample was kept for further study.

2.4. DSC analysis
The endotherm scans of the skin samples were obtained by DSC, using a scanning rate of 10°C/min from 25 to 350°C. A Diamond TG/DTA system (PerkinElmer, USA) was used.

2.5. Microstructure analysis
The skin was dehydrated and paraffin embedded, Hematoxylin and Eosin (HE) stained and sliced for microstructure analysis using the camera microscope (Olympus SF07103, Japan).

3. 3. Results and discussion

3.1. DSC profiles of normal skin
Fig. 1 is the DSC profile of the normal rat skin. It demonstrated three typical endothermic transition temperatures at 63°C, 82.7°C and 115.1°C with the first peak being attributed to either lipid lamella phase transition from a crystalline to a gel like phase or melting of sebaceous lipids, the second peak corresponding to the transformation from a lamellar to disordered state in the lipid structure and protein associated lipid transition from gel to liquid state, respectively, the third peak being associated with the conformation change of protein [10].

3.2. Influence of negative electrets in skin structure
Figure 2 is the DSC profile of rat skin treated by -500V electret. There was only a single peak which appeared at 79.1°C with an increased peak height, decreased peak area and transition temperature. The electrostatic field and microcurrent produced by -500V electret could result in space structure alteration of lamellar lipid from gel to liquid state and resulting in an absence of transition temperature at 63°C. Besides, -500V electret increased the orderliness of SC lipid, reduced the resistance and thermodynamic energy of SC. Thus, the energy needed for lamellar lipid in the same or different SC layer to transform from the gel to liquid state decreased and demonstrated a trend of cooperation, which resulted in shifting of the transition peak to lower temperature and increase of peak height.
Figure 2 further indicated an absence of transition temperature at 115.1°C as compared with control skin, because a large amount of protein in SC denatured from α helix to β fold after action of -500V

electret. Since the secondary and tertiary space structure changes of some protein, both the proportion of protein to absorb heat and the heat absorbed for structure alteration reduced. Therefore, the endothermic transition temperature attributed to conformation change of protein the disappeared after -500V electret treatment.

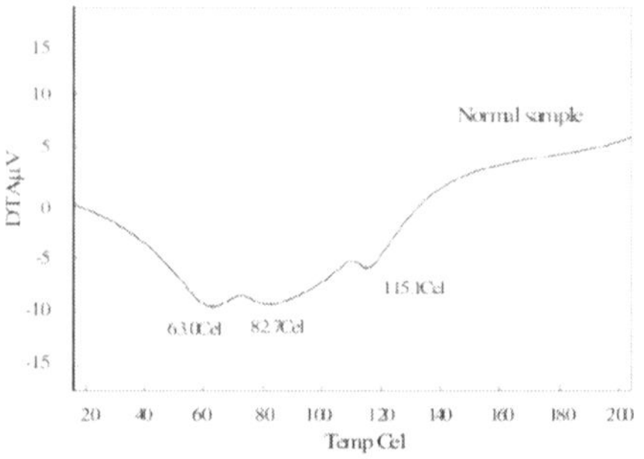

Figure 1 DSC profile of normal rat skin

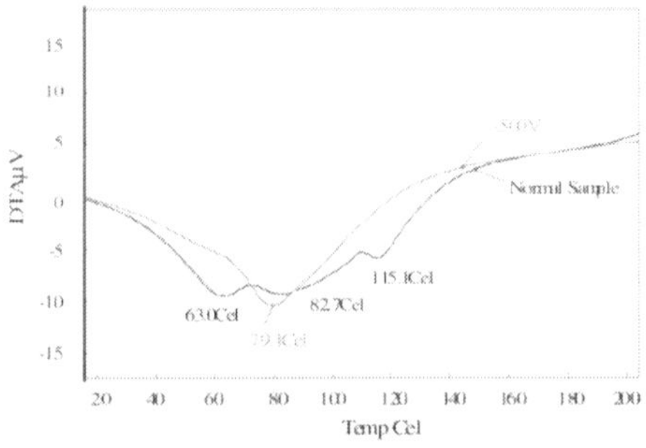

Figure 2 DSC profile of -500V electret treated rat skin.

Compared with the action of -500V electret, the peak area of DSC endotherms for -1000V electret-treated skin sample showed a further decrease (Figure 3). It suggested that with the increase of surface potential of negative electret the orderliness and fluidity of the SC lipid improved, the conformation proportion of liquid state increased, being help for drug transdermal permeation. For skin sample treated by -2000V electret, there was no significant change in peak shape and phase transition temperatures at 78.9°C and 113.7°C when compared to -1000V skin sample except for the obvious increases of peak areas. Besides, the phase transition temperature and peak area between 70°C~200°C were almost the same as that of the control skin (Figure 4), indicating a weaker regulating effect of -2000V electret than -1000V electret on orderliness and fluidity of SC lipid structure and both of the secondary and tertiary structures of the protein. The influence of negative electret in rat skin structure presented an effect of a "potential window" with the -1000V electret having a better regulating effect on skin structure.

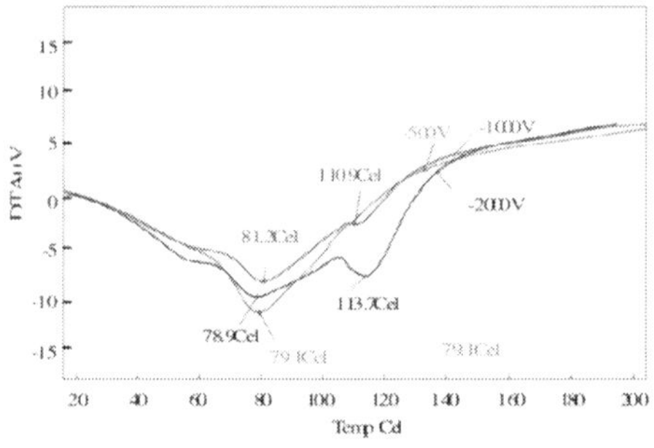

Figure 3 DSC profile of rat skin treated by different potential electrets.

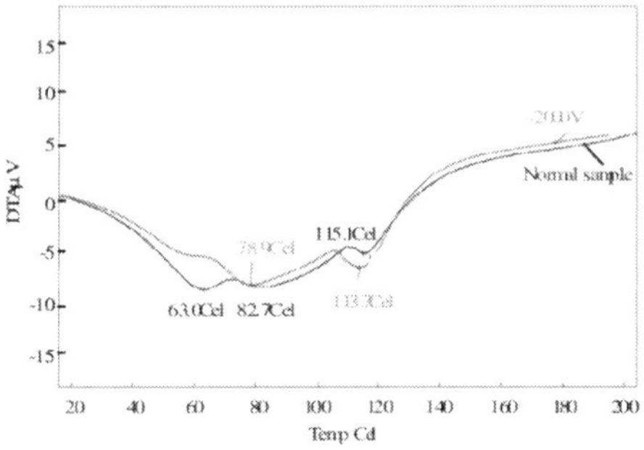

Figure 4 DSC profile of -2000V electret treated rat skin.

3.3. Microstructure analysis

The regulating effect of -1000V electret on skin structure was further confirmed by microstructure analysis. Figure 5 and Figure 6 are the microphotographs of normal rat skin and -1000V electret-treated rat skin, respectively. A considerable change could be observed for the -1000V electret treated-skin (figure 6). The structure of the rat skin changed obviously with a loosening and thickening of the SC, disengagement of SC lipid cells, and separation of part of the SC from the skin epidermals when compared to normal skin. Therefore, -1000V electret could result in alteration of the arrangement of SC lamellar lipid, reducing the barrier function of the skin.

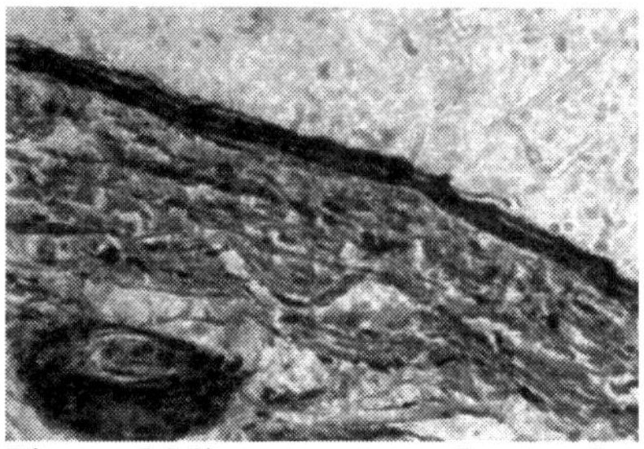

Figure 5 Microstructure of normal rat skin

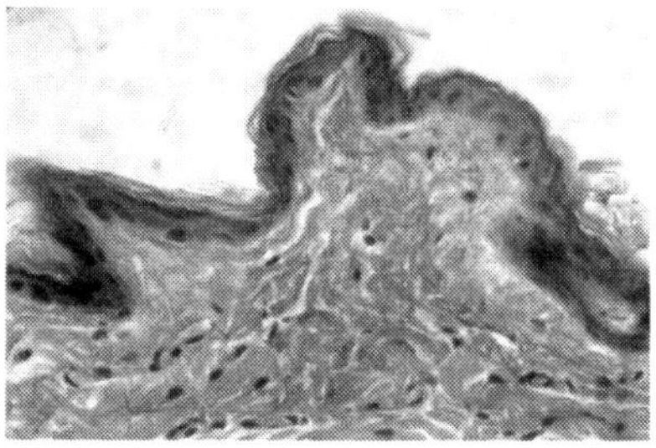

Figure 6 Microstructure of -1000V electret treated rat skin

4. Conclusion

Negatively charged electret can regulate the skin structure by alteration of the SC lipid organization and α helix to β fold of the protein, regulating the lipid fluidity by transformation of gel to liquid state of the lamellar lipid. The effect of negative electret on rat skin structure presented an effect of a "potential window". Over the potential range from -500V to -2000V, -1000V the electret showed a stronger skin structure regulation effect. Electrets have an excellent effect on enhancing transdermal drug delivery by skin structure regulation.

Acknowledgments

The authors thank the National Nature Science Foundation of China (Grant No. 50977089) and Natural Science Foundation of Shanghai (Grant No. 10ZR1437600) for the financial support provided.

References

[1] Jiang J, Song M H, Cui L L, Wang X P, Fang Y, Li L and Ye X T 2001 Effects of negative PTFE electrets on the apoptosis of fibroblast cells *Academic J Second Military Medical University* 22 409-411

[2] Jiang J, Cui L L, Song C R, Wang X P, Song M A, Fang Y and Sun Q W 2001 Effect of negative electret on surface charge and free calcium in 3T3 cells *Proc. 4th Intern Conf. on Applied Electrostatics* (China: Dalian) 558-560

[3] Jiang J, Cui L L, Xia Z F, Wang X P and Song C R 2002 Effect of electret on surface charge of red blood cell in scalded rats *Proc. Intern Symp. on Electret* (Australia: Melbourne) 294-297

[4] Wang Z Z and Jiang J 1995 The study on the treatment of fracture of god's femur by FEP electrets and on the mechanism of action *Chinese J. Med. Phys.* **12** 81-85

[5] Tian T, Hao L M, Wu J H, Lin S and Yang J Q 2010 Study on antibacterial activity of polypropylene fiber electret filter medium *Chinese J. Disinfection* **27** 268-272

[6] Kong Y X, Cheng L, Xiao Y H, Wang D, Jiang J and Cui L L 2009 Enhancing effects of electret on transdermal delivery of lidocaine patches in vitro *Academic J Second Military Medical University* 30 469-472

[7] Jiang J, Liang Y Y, Cui L L, Hou X M, Tang Y, Ye X T, Yang Y J and Song M H 2008 Influence of porous PTFE/LDPE/PP composite electret in skin ultrastructure *J Phys.: Con. Series* 012050

[8] Cui L L, Hou X M, Jiang J, Li G D, Liang Y Y and Xin X 2008 Comparative enhancing effects of electret with chemical enhancers on transdermal delivery of meloxicam in vitro *J Phys.: Con. Series* 012015

[9] Murthy N S, Boguda V A and Payasada K 2008 Electret enhances transdermal drug permeation *Biol Pharm Bull* 31 99-102

[10] Golden G M, Guzek D B, Harris R R, Mckie J E and Potts R O. 1986 Lipid Thermotropic Transitions in Human Stratum-Corneum *J Invest Dermatol.* 86 255–259

Electrohydrodynamic pumping in microsystems

Antonio Ramos

Deptartamento de Electronica y Electromagnetismo, Universidad de Sevilla, Avenida Reina Mercedes s/n. 41012-Sevilla, Spain

E-mail: ramos@us.es

Abstract. The physical principles behind the electrohydrodynamic (EHD) actuation in microsystems is presented by reviewing five different EHD micropumps. These are classified into two groups: micropumps that exert electric forces in the liquid bulk and micropumps that exert forces in the diffuse double layer. This review of five EHD micropumps allows us to analyse the EHD actuation ranging from very insulating liquids to electrolytic solutions.

1. Introduction

Among the most promising techniques to handle small liquid volumes in microdevices are those that employ electrical forces, which have the advantages of voltage-based control and dominance over other forces. Microfluidic transport is sometimes achieved by using macroscale pumps such as syringe pumps or pressure/vacuum chambers and valves. However, many applications would benefit from an on-chip active micropump: dispensing therapeutic agents into the body, cooling of microelectronic devices, dispensing liquids in miniature systems for chemical and biological analysis, and more. Here, we are going to deal with dynamic micropumps that exert electric forces on liquids, called electrohydrodynamic (EHD) micropumps. EHD micropumps can work as on-chip active micropumps with no moving parts.

EHD dynamic micropumps may be classified into: a) pumps that exert electric forces in the liquid bulk: the EHD injection, conduction and induction pumps; and b) those that exert electric forces in the diffuse double layer: the electroosmotic and AC induced-charge electroosmotic pumps. The main goal of the present work is to describe the physical principles behind the EHD actuation in microsystems.

2. Common features of EHD Micropumps

The electrical current density in the liquid bulk can be written as the sum of electromigration, diffusion and convection:

$$\mathbf{j} = \sum_i q_i(n_i\mu_i\mathbf{E} - D_i\nabla n_i + n_i\mathbf{u}) \tag{1}$$

where D_i and μ_i are the diffusion and mobility coefficients of an ion, q_i its charge, n_i the ion density, and $\mathbf{u}$ the velocity of the fluid. Species conservation equations for ions and molecules must be included to complete the description of the system [1]:

$$\frac{\partial n_i}{\partial t} + \nabla\cdot(n_i\mu_i\mathbf{E} - D_i\nabla n_i + n_i\mathbf{u}) = r_i\,. \tag{2}$$

Here r_i is the rate of production due to chemical reactions ($\mu_i = 0$ for neutral species).

Working with fluids with conductivities smaller than 10 S/m, with applied ac signals of frequencies up to 1 GHz, the electric fields in microsystems hold the quasi-electrostatic limit of Maxwell equations [1; 2]:

$$\nabla \cdot (\varepsilon \mathbf{E}) = \rho; \quad \nabla \times \mathbf{E} = 0; \quad \nabla \cdot \mathbf{j} + \frac{\partial \rho}{\partial t} = 0. \tag{3}$$

The liquid motion is governed by the Navier-Stokes equations which, under steady-state conditions in microfluidics and in the limit of low Reynolds number, can be written:

$$\nabla \cdot \mathbf{u} = 0; \qquad 0 = -\nabla p + \eta \nabla^2 \mathbf{u} + \mathbf{f}. \tag{4}$$

Joule heating is an important consideration in EHD systems. In many situations in microsystems, heat convection is small compared to heat diffusion and the temperature equation in the steady-state reduces to

$$0 = k\nabla^2 T + \mathbf{E} \cdot \mathbf{j}' \tag{5}$$

where $\mathbf{j}'$ is the current density measured in the rest frame of the fluid element.

We are interested in cases where the main body force density is of electrical origin. The electric volume force density is [1]

$$\mathbf{f}_E = \rho \mathbf{E} - \frac{1}{2} E^2 \nabla \varepsilon + \frac{1}{2} \nabla \left[E^2 \rho_m \left(\frac{\partial \varepsilon}{\partial \rho_m} \right)_T \right] \tag{6}$$

where the first term is the Coulomb force, the second term is the dielectric force, and the third is the electrostriction, which can be incorporated into a redefined pressure for an incompressible fluid. The Coulomb force is the term primarily responsible for generating pumping in Electrohydrodynamics.

3. Micropumps that exert forces in the liquid bulk

For the following EHD pumps we will analyse the geometry depicted in figure 1: Two perforated electrodes separated a distance L, subjected to a potential difference V, are inside a pipe of diameter D filled with the working liquid. Using a one-dimensional analysis, the total electric force per unit area exerted on the liquid (the generated pressure) is

$$\Delta p_g = \int_0^L \left(\rho E - \frac{E^2}{2} \frac{\partial \varepsilon}{\partial x} \right) dx = \int_0^L \frac{1}{2} \frac{\partial \varepsilon E^2}{\partial x} \, dx = \left. \frac{\varepsilon E^2}{2} \right|_{x=L} - \left. \frac{\varepsilon E^2}{2} \right|_{x=0} \tag{7}$$

where we have used that $\rho = \partial(\varepsilon E)/\partial x$. The expression says that the generated pressure is equal to the jump in electric energy density from $x = 0$ to $x = L$.

3.1. Injection pump

The injection pump is also known as ion-drag pump. Ions emitted from one electrode (the emitter) are collected in the other electrode (the collector). Since ions do not accelerate but move at the terminal velocity $\mathbf{v} = \mu \mathbf{E}$ (ion inertia is negligible) the electric force acting on them is transmitted directly to the liquid by viscous friction. Therefore, in the bulk we have $\mathbf{f} = \rho \mathbf{E}$. The typical working liquid is a very insulating liquid. If the liquid has some residual conductivity σ, we expect that EHD operation is satisfactory if the time required from an ion to go from emitter to collector is shorter than the charge relaxation time, i.e. $L/\mu E < \varepsilon/\sigma$. To ensure that the ions are injected from the emitter rather than from the collector, the local electric field at the emitter is increased due to presence of sharp edges or points.

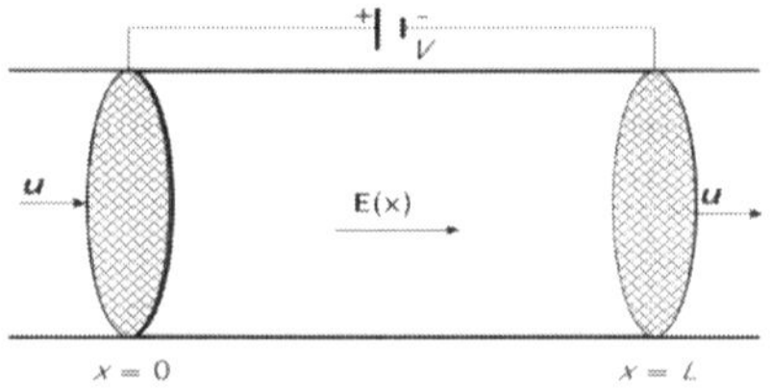

Figure 1. Geometry for injection, conduction, and induction pump analysis.

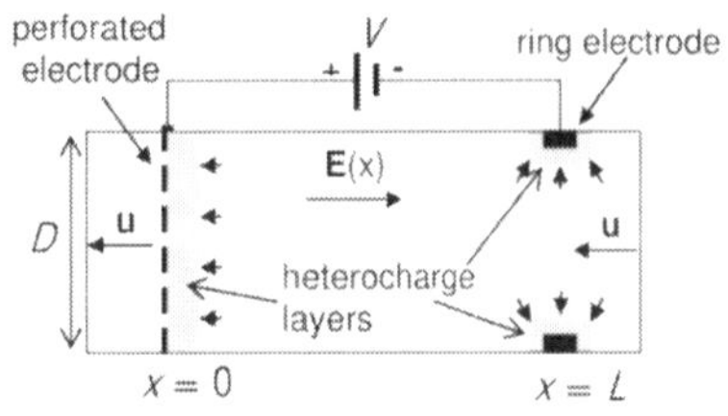

Figure 2. Scheme of conduction pump.

For dielectric liquids the ratio of diffusion to electromigration current is negligible to a very good approximation. In microsystems, fluid velocities are commonly on the order of 1 mm/s or less and this implies that convection of charge is also negligible compared to electromigration. Under these approximations and considering unipolar injection, the generated pressure in the regime of space-charge-limited emission (SCLE) is $\Delta p_g = (9/8)\varepsilon V^2/L^2$ [2].

In microsystems, Richter and Sandmier [3] fabricated ion-drag micropumps made of two opposite metallic grids through which the pumped fluid moves. Ion-drag micropumps made of co-planar microelectrodes have been studied by Ahn and Kim [4] and Darabi and Wang [5]. Kazemi et al [6] studied an injection micropump made of 3D microelectrodes (metallic pillars) and HFE-7100 as working fluid. A maximum pressure of 2240 Pa was generated at an applied voltage of 900 V.

3.2. Conduction pump

This kind of EHD actuation is based upon the Coulomb force on non-equilibrium charged layers adjacent to electrodes that appear in the conduction regime. The representative working fluid is a semi-insulating liquid. For electric fields less than 10^7 V/m, ions generated in the bulk from dissociation of neutral species move by electromigration towards the electrodes where they discharge. Near the electrodes, charged layers appear of opposite sign to that of the adjacent electrode and are called heterocharge layers [7].

A good picture is to consider a bipolar conduction model where a neutral species n_0 dissociates reversibly (at a rate $k_d n_0$) into univalent positive n_+ and negative n_- ions (that recombine at a rate $k_r n_+ n_-$) [7; 8]. Counter-ions arrive at the electrode and discharge on it but co-ions cannot come from the electrode below the electric field threshold for charge emission. This implies that near the positive (negative) electrode the majority of ions are negative (positive), and that there is an electric force directed towards each electrode. The thickness of this charge layer, λ, can be estimated for the positive electrode as follows. At $x = 0$ the flux of positive ions is zero ($\mu_+ n_+ E = 0$) and at $x = \lambda$ it reaches the value in the bulk ($\mu_+ n^{eq} E_{bulk}$). This jump in positive ion flux should be on the order of the ions produced per unit of time in that distance ($k_d n_0 \lambda$).

Each electrode is subjected to an electric pressure from the bulk side. The total generated pressure for a geometry like the one depicted in figure 1 is $\Delta p_g = (\varepsilon E_b^2/2)((\mu_-/\mu_+) - (\mu_+/\mu_-))$ [2]. In order to have a pressure difference even in the case of equal mobilities, the electrodes are made nonsymmetric [7; 8] (see an example in figure 2).

EHD conduction pumping has been mainly reported in the millimeter scale [7; 9]. Pearson and Seyed-Yagoobi [10] reported recently a micro-scale EHD conduction pump. Working with refrigerant HCFC-123 at 4 kV they obtained $\Delta p_g \sim 500$ Pa.

In order to increase the generated pressure, we can increase the voltage or reduce the dimensions, however, we can then reach either the saturation current regimen or the charge injection regime. In some configurations, the appearance of charge injection is manifested by the liquid changing direction.

3.3. Induction pump

Induction pumping is based on the forces upon charges induced in a liquid by the electric field due to a gradient or discontinuity of the electric conductivity. Gradients in conductivity are usually obtained by imposing a gradient of temperature. The common induction pump consists of an array of electrodes that generate a traveling-wave potential together with a perpendicular gradient of temperature [11]. The travelling-wave electric field attracts or repels the induced charges, leading to fluid flow. The temperature gradient can be imposed externally or it can be obtained by Joule heating generated by the imposed electric field. In microsystems, the EHD induction micropumps fabricated by Fuhr and co-workers [12; 13] employed Joule heating to establish the temperature gradient. Water saline solutions were used as working fluids, which are very difficult (if not impossible) to be pumped using EHD induction at the macroscale. Unidirectional motion can also be obtained with electrodes subjected to a single phase ac signal with an imposed longitudinal temperature gradient [14].

An induction pump can be fabricated with two parallel perforated electrodes subjected to both a potential difference and a temperature difference (geometry like in figure 1 where we impose a temperature T_0 at $x = 0$ and T_L at $x = L$). The charge induced is obtained from $\rho = \nabla \cdot (\varepsilon \mathbf{E}) = \nabla \cdot (\varepsilon \mathbf{j}/\sigma) = \sigma \mathbf{E} \cdot \nabla (\varepsilon/\sigma)$. In an induction pump like this, the Coulomb force is directed from high to low conductivity. It can be shown that the generated pressure for small increments of temperature is $\Delta p_g = (\varepsilon V_0^2/2L^2)(2\alpha - \beta)(T_0 - T_L)$ [2], where $\alpha = (1/\sigma)(d\sigma/dT)$ and $\beta = (1/\varepsilon)(d\varepsilon/dT)$. For water saline solutions, $\alpha = 0.02\,\mathrm{K}^{-1}$ and $\beta = -0.004\,\mathrm{K}^{-1}$ at room temperature, and the force goes from high to low temperature.

Stubbe et al. [14] fabricated an induction micropump using a single phase ac signal in the frequency range 0.3-50 MHz and voltages in the range 0-30 V. They employed aqueous solutions of conductivity in the range $\sigma = 0.1$-1.5 S/m and observed velocities up to 120 μm/s.

4. Micropumps that exert forces in the double layer

Electrical double layers are formed at solid/liquid interfaces spontaneously because of differences in electrochemical potentials between the solid and liquid phases [15]. The solid surface becomes charged and ions in solution of opposite charge (counterions) are attracted until they screen the electric field created by the surface charge. At equilibrium, the liquid is electroneutral except for a charged layer adjacent to the solid surface, with characteristic thickness given by the Debye length, λ_D [15]. We can distinguish to ways of double layer generation at the solid/liquid interface: (a) a chemical mechanism such as ionization of bound surface groups or ion adsorption; (b) an electrostatic mechanism in which a solid metal surface becomes charged because is subjected to a potential difference with respect to the liquid. We can identify two kinds of pumps based on these charging mechanisms.

4.1. Electroosmotic pump

An example of a spontaneous double layer is the case of silica-based ceramics, like glass, with Si–OH groups at the surface. In contact with water a fraction of the Si–OH bonds change into Si–O– releasing H^+. Therefore, we have a negative surface charge balanced by positive counter-ions. When an electric field is applied along the surface, the Coulomb force causes the motion of the mobile ions of the double layer dragging the fluid with them. This is called electroosmosis (EO) and has been known for 200 years (F.F. Reuss discovered electroosmosis in 1809). The EO velocity generated in the double layer is seen as a superficial or slip velocity since the thickness of the Debye layer is very small (1-100 nm in water). It is given by the Helmholtz-Smoluchowski expression [15]: $v_{eo} = \varepsilon \zeta E/\eta$, where ζ is the potential difference between the plane of non-slip and the bulk. The pressure difference that can be generated in a capillary of diameter D and length L for $\lambda_D \ll D$ is $\Delta p_g = 3\varepsilon \zeta EL/2D^2$ [2]. In miniaturized systems for chemistry and life

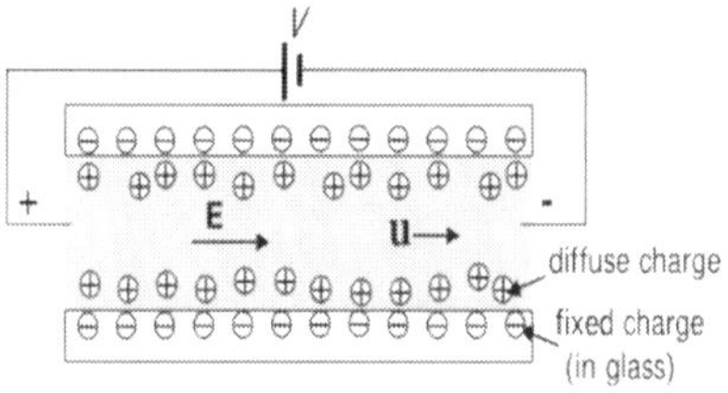

Figure 3. Scheme of electroosmotic pump.

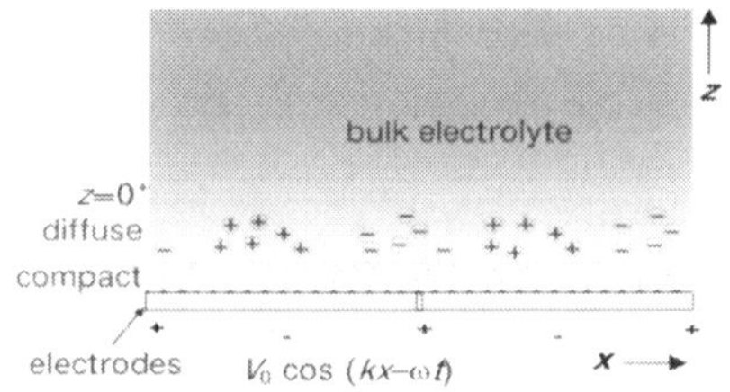

Figure 4. Scheme of TWEO pump.

sciences, a number of important techniques make use of EO pumping: flow injection analysis [16], on-chip electrophoretic separation [17] and on-chip liquid chromatography [18].

In practice, EO pumps that generate high pressure make use of porous structures so that each pore acts as a tortuous capillary of small diameter [19; 20]. The pressure is high because D is small and the flow-rate is high because there are many pores per cross-section. Wang et al [20] employed a porous structure made of silica-based monoliths inside a capillary with 100 μm diameter (0.5 mm^3 micropump). They conducted experiments with deionized water as working fluid at 6 kV and obtained a maximum pressure around 3×10^5 Pa.

4.2. AC induced charge electroosmotic pump

The alternating current electroosmotic micropump, also known as induced charge electroosmotic micropump, actuates upon the diffuse charge in the double layer that is induced between an electrolyte in contact with an electrode that is subjected to a potential [21]. Typically, unidirectional flow is obtained by arrays of asymmetric pairs of electrodes subjected to a single phase ac signal, known as ACEO pump [22], or by arrays of symmetric electrodes subjected to a travelling-wave signal, known as TWEO pump [23].

The mechanism responsible for the fluid pumping at low voltages generated in a travelling-wave electrode array is as follows (see figure 4). When a travelling-wave potential is applied to the electrodes, counter-ions accumulate in the electrical double layer in response to the applied signal. The typical time for charging the double layer is given by the RC time of the equivalent circuit, where C is the capacitance of the double layer and R is the typical bulk resistance. Due to this double-layer RC charging time, a delay exists between the time of maximum induced charge and maximum applied signal. The electric field acts on the charge pulling the fluid in the direction of the traveling wave. The fluid velocity is maximum for angular frequencies of the order of $\omega = 1/(RC) \sim (\sigma\lambda_D)/(\varepsilon\ell)$, with ℓ a typical distance. The maximum generated pressure in a microchannel of rectangular cross-section is $\Delta p_g = \Lambda 3\varepsilon k V_0^2 L/2h^2$ [2], where k is the wavenumber of the travelling wave, L is the length of the array of electrodes, h is the height of the channel, and Λ is a factor less than one that accounts for compact layer effects.

Yang et al [24] performed experiments on TWEO and ACEO pumping using the same microelectrode array by changing the electrical connections of gold microelectrodes. The ACEO velocity was around five times smaller than the TWEO velocity for the same applied voltage. The TWEO slip velocity was around 480 μm/s for a TW signal of amplitude 1.5 volts peak to peak and frequency 1 kHz, using a saline solution of 100μM KCl in water. Huang et al [25] fabricated an ACEO micropump made of long serpentine array of 3D asymmetric stepped microelectrodes. Using an ac signal of amplitude 1 Volt rms at a frequency of 1 kHz and distilled water as working fluid, the pump was capable of generating a maximum pressure of 1.4 kPa (much higher than previous ACEO pumps) and an effective slip velocity of 1.3 mm/s.

To increase the generated pressure, we could increase the applied voltage, however, this cannot be increased beyond some volts because it eventually leads to Faradaic reactions, which limit the use of the polarisation mechanism. Even before this threshold voltage is reached, fundamental limitations such as ion crowding play an important role [26]. Liquids that have been pumped using AC induced charge electroosmosis had conductivities below 0.1 S/m.

5. Conclusions

In this work we have reviewed the basic principles of five different EHD micropumps: from those that exert forces in the liquid bulk to those that exert forces in the double layer. Very insulating liquids can be actuated by the injection pump. Semi-insulating liquids can be pumped by the injection, conduction and induction pumps. Aqueous solutions can be pumped in microsystems by the induction, electroosmotic and ac electroosmotic pumps. Electrolytes with conductivities much greater than 1 S/m are very difficult, if not impossible, to pump by using electrical forces in the bulk because of Joule heating leading to boiling. For $\sigma \geq 1$ S/m, the temperature rise can be on the order of 100 K for a voltage amplitude of 10 V [27].

Acknowledgments

This work has been supported by the Spanish Government Ministry MEC (contract FIS2006-03645) and the Regional Government Junta de Andalucía (contract P09-FQM-4584).

References

[1] Castellanos A 1998 *Electrohydrodynamics* ed Castellanos A (Springer-Verlag)
[2] Ramos A 2007 *Microfluidic Technologies for Minituarized Analysis Systems* ed Hardt S and Schönfeld F (Springer)
[3] Richter A and Sandmaier H 1990 *IEEE Proceedings of Micro Electro Mechanical Systems, 1990* pp 99–104
[4] Ahn S and Kim Y 1998 *Sensors and Actuators A: Physical* **70** 1–5
[5] Darabi J and Wang H 2005 *J. Microelectromech. Syst.* **14** 747–755
[6] Kazemi P, Selvaganapathy P and Ching C 2009 *IEEE Trans. Dielectr. Electr. Insul.* **16** 483–488
[7] Atten P and Seyed-Yagoobi J 2003 *IEEE Trans. Dielectr. Electr. Insul.* **10** 27–36
[8] Feng Y and Seyed-Yagoobi J 2004 *Phys. Fluids* **16** 2432–2441
[9] Jeong S and Seyed-Yagoobi J 2004 *IEEE Trans. Ind. Appl.* **40** 900–904
[10] Pearson M and Seyed-Yagoobi J 2009 *2009 Electrostatics Joint Conference*
[11] Melcher J and Firebaugh M 1967 *Phys. Fluids* **10** 1178–1185
[12] Fuhr G, Hagedorn R, Müller T, Benecke W and Wagner B 1992 *J. Microelectromech. Syst.* **1** 27–36
[13] Fuhr G, Schnelle T and Wagner B 1994 *J. Micromech. Microeng.* **4** 217–226
[14] Stubbe M, Holtappels M and Gimsa J 2007 *J. Phys. D: Appl. Phys.* **40** 6850–6856
[15] Hunter R 1981 *Zeta Potential in Colloid Science* (Academic Press)
[16] Dasgupta P and Liu S 1994 *Anal. Chem.* **66** 1792–1798
[17] Manz A, Effenhauser C, Burggraf N, Harrison D, Seiler K and Fluri K 1994 *J. Micromech. Microeng.* **4** 257–265
[18] Jacobson S, Hergenroder R, Koutny L and Ramsey J 1994 *Anal. Chem.* **66** 2369–2373
[19] Yao S and Santiago J 2003 *J. Colloid Interface Sci.* **268** 133–142
[20] Wang P, Chen Z and Chang H 2006 *Sensors Actuators B* **113** 500–509
[21] Ramos A, Morgan H, Green N and Castellanos A 1999 *J. Colloid and Interface Science* **217** 420–422
[22] Brown A, Smith C and Rennie A 2001 *Phys. Rev. E* **63** 016305
[23] Ramos A, Morgan H, Green N, González A and Castellanos A 2005 *J. Appl. Phys.* **97** 084906
[24] Yang H, Jiang H, Ramos A and García-Sánchez P 2009 *Microfluid Nanofluid* **7** 767–772
[25] Huang C, Bazant M and Thorsen T 2010 *Lab Chip* **10** 80–85
[26] Bazant M, Kilic M, Storey B and Ajdari A 2009 *Adv Colloid Interface Sci* **152** 48–88
[27] Ramos A, Morgan H, Green N and Castellanos A 1998 *J. Phys. D: Appl. Phys.* **31** 2338–2353

Characterisation of pharmaceutical materials produced by electrospraying

M Nyström*, M Murtomaa and J Salonen

Department of Physics and Astronomy, University of Turku, FI-20014 Turku, Finland

*Corresponding author. E-mail address: mhnyst@utu.fi

Abstract. Electrospraying of solutions or suspensions provides a method for production of fine particles, in certain conditions even down to nanometer size. In the present study, electrospraying was used in drug particle production. The most important properties of the produced particles in a pharmaceutical sense are size distribution, degree of crystallinity and porosity. The size of the produced particles can easily be tailored by varying the concentration of the dissolved material, electric field strength or liquid flow rate. Most experiments on electrospraying are performed at atmospheric pressure. In the present study, the produced particles were dried under atmospheric or reduced pressure, the latter in order to improve the drying process. The effect of pressure reduction on the degree of crystallinity of the produced drug particles was studied by XRD and DSC. Also the porosity of the particles depends mainly on the drying conditions. The porosity and size of the produced particles was studied by SEM imaging.

1. Introduction

In the present study, electrospraying was used in drug particle production. In this process, a drug powder is dissolved in a convenient solvent. The solution is atomised using electrostatic forces. The solvent is evaporated from the formed droplets in a drying medium, and a dense cluster of the dissolved drug remains.

Most experiments on electrospraying are performed at atmospheric pressure [1]. Nevertheless, there are several studies of electrospraying into vacuum mainly, but not exclusively considering electrical colloidal thrusters [1-5]. In the present study, the produced drug particles were dried both under atmospheric and reduced pressure, the latter in order to improve the drying process.

It is estimated that more than 95 % of new drug candidates suffer from limited bioavailability [6]. For poorly soluble and highly permeable drugs, the rate of oral absorption is often controlled by the dissolution rate in the gastrointestinal tract [7]. Therefore together with the permeability, the solubility and dissolution behaviour of a drug are key determinants of its oral bioavailability. A commonly used process in modifying bioavailability of poorly soluble drugs is particle size reduction, since it increases the relative surface area of the particles. Another commonly used technique is the formation of a high-energy amorphous form. The solubility and dissolution rate can be enhanced since the physical links are weaker in the amorphous structure [8]. A porous particle has a very large surface area and small density compared to a dense particle. This has a remarkable influence on the functional properties of a drug. Hence the most important properties of the electrosprayed particles in pharmaceutical sense are size distribution, degree of crystallinity and porosity.

In the present study, budesonide was dissolved in chloroform and the solution was electrosprayed. Budesonide is known to be poorly soluble in water [8]. By electrospraying, the size of the drug particles can be reduced down to the micrometer range. The size and the size distribution of the particles can easily be tailored by varying the concentration of the dissolved material, electric field strength or liquid flow rate. The size distribution of the produced particles was determined by SEM.

The porosity of the particles produced by electrospraying depends mainly on the drying conditions. The drying can be done quickly using heat. This tends to lead to the formation of porous or hollow particles because the drug does not have time to diffuse to empty places inside the cluster [9]. However, there are several pharmaceutical materials which are temperature sensitive or highly resistant to migration of moisture from within the solid. The structures of these materials may be damaged at increased temperature, so drying without addition of heat may be required. In the present study, the drying was done both under atmospheric and reduced pressure at room temperature. The effect of pressure reduction on the formation of pores was studied by SEM.

In addition, the pressure reduction during drying may increase the proportion of the amorphous form in the product. Due to the fast solidification, the molecules do not have enough time to arrange in the crystal lattice, which causes the formation of amorphous material. The effect of pressure reduction on the amount of amorphous content in the electrosprayed product was previously studied by DSC and XRD. The results of this research are presented in a separate paper [10]. The results obtained on budesonide are discussed briefly here.

2. Materials and methods

2.1. Materials

Budesonide is an anti-inflammatory corticosteroid. Budesonide was dissolved into chloroform at room temperature using drug material concentration of 15 g/dm^3. In electrospraying the most important properties of the solution are its conductivity, surface tension and viscosity. The solvent was chosen optimising these properties and taking into account that the drug material has to be soluble, but does not degrade in the chosen solvent. Chloroform has also been found to be volatile enough in the previous studies [10]. The prepared solution was stored at room temperature and protected from light. Micronised budesonide was acquired from Orion Pharma (Finland).

2.2. Particle fabrication

In the electrospraying studies, stainless steel capillaries (manufactured by EFD, USA) with inner diameter of 0.33 mm were used. Electrostatic atomisation was carried out using Alpha Series II high voltage source (Brandenburg, UK) set to positive potential and connected to the stainless steel capillary. A grounded plate electrode with a hole in the centre was placed beneath it. The distance between these electrodes was set to approximately 1 cm. A circular metal plate was attached to HV-conductor above the capillary. This was to make the electric field near the capillary tip more uniform, and to prevent some external electric disturbances. The electrospraying equipment is presented in more detail in a previous paper [11].

In the present study, atomisation voltages ranged between 3.0 – 3.6 kV. All samples were fabricated using the stable cone-jet mode. Highly charged droplets were neutralised in order to avoid adhesion to grounded surfaces. This was done using a corona neutraliser, which was connected to E.H.T. unit Type 532/D voltage source (The Isotope Developments, UK) and set to negative potential. Liquid flow to the capillary was controlled using a syringe pump (NE-500, New Era Pump Systems, USA). Liquid flow rates ranged from 1.5 to 2.0 ml/h. Atomised droplets were spray dried at room temperature at a pressure of 0.5 – 1 atm. When atomisation was done under reduced pressure, the solution was set in a vacuum prior to electrospraying. This was done in order to prevent formation of bubbles in the capillary while electrospraying.

2.3. Particle characterisation

Produced particles were studied using SEM (Cambridge S200, UK). SEM samples were prepared by collecting particles on a nylon filter. The samples were coated with a 20-30 nm layer of gold to improve their conductivity. Samples were stored at room temperature in a desiccator containing silica gel for two days before the imaging.

Calorimetric measurements were carried out using differential scanning calorimeter (DSC, PerkinElmer Instruments, USA). The apparatus was controlled with a computer using Pyris DSC Software -program (PerkinElmer Instruments, USA). The equipment was cooled with Intracooler 2D (PerkinElmer Instruments, USA). Nitrogen was used as a scavenging gas using mass flow rate controller (B.V. 5850S, Brooks Instruments, USA), which was monitored using *0152* -control unit (Brooks Instruments, USA). Data was collected and analysed using Pyris Software Version 4.02 -software.

The crystallinity of the samples was measured using x-ray diffraction using a PW 1830 generator, PW 1820 goniometer and PW 1710 diffractometer controller (Philips Electronics, Netherlands). Data was analysed with X'Pert HighScore Version 1.0 -program. The samples were measured between 3 - 45 ° values of 2θ with scanning speed of 0.02 °/s.

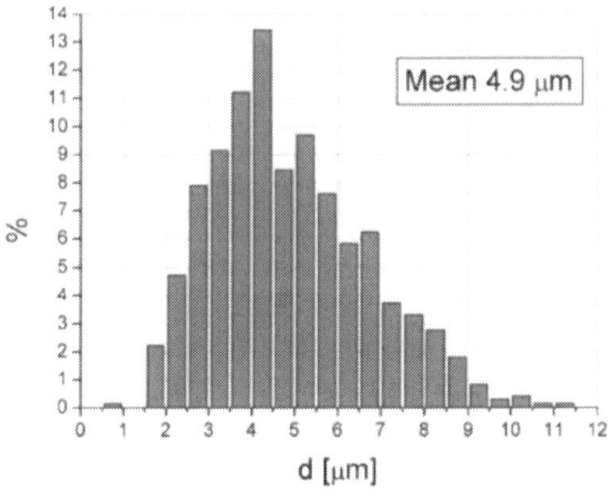

Figure 1. The size distribution of budesonide particles fabricated by electrospraying.

3. Results

3.1. Particle size

The fabricated particles appeared to be spherical and quite regularly shaped (figure 2). Based on the SEM images, particle size distribution was determined using Image-Pro Plus (Version 1.3) - program. More than 700 particles were analysed. The particle size distribution was found to be quite narrow with the mean particle diameter of 4.9 μm. The size distribution is presented in figure 1.

3.2. Porosity

The formation of pores on the produced budesonide particles was also studied with SEM. The fabrication of the samples was as described in the previous paragraph. That is, the values of the atomisation parameters were the same for the two samples, except for the drying pressure. The drying was done under atmospheric and reduced pressure, 1.0 and 0.7 atm, respectively. SEM images of the produced particles are presented in figure 2.

Figure 2. Budesonide particles electrosprayed in 1.0 atm (upper row) and in 0.7 atm (lower row).

Based on the SEM images, budesonide particles fabricated under atmospheric pressure seem to have a smooth surface and very regular spherical form. The particles dried under reduced pressure seem more porous and collapsed. The size of the pores can be approximated to be around one micrometer. The pressure reduction presumably increases the evaporation rate of the solvent from the particles, which in turn seems to increase the formation of pores. The structure on the background of the particles in figure 2 is the nylon filter on which the particles were collected.

3.3. Degree of crystallinity

Another important property of a drug material on which the pressure reduction seems to have an effect is the degree of crystallinity (DOC). In previous studies it was noted that for chloroform solutions, the amount of amorphous material in the product increased when the drying pressure was decreased [10]. This was studied by DSC (differential scanning calorimeter) and XRD (x-ray diffraction).

For budesonide electrosprayed at 0.5 atm, a peak most likely caused by crystallisation of the amorphous content was detected in the DSC scan. This peak did not occur on the sample dried at atmospheric pressure.

On the diffractograms of the electrosprayed budesonide samples, a clear decrease in the peak intensities was detected compared to the peaks of a crystalline reference sample. Both the samples prepared at 1.0 and 0.5 atm included a notable proportion of amorphous material, 80 and 90 %, respectively. Based on the XRD measurements on budesonide, it was also calculated that the crystallinity obtained in 0.5 atm was 56 % of the crystallinity in 1.0 atm.

4. Conclusions and Discussion

Production of microparticles of pharmaceutical materials by electrospraying was studied. The most important properties of the fabricated particles in the pharmaceutical sense are size distribution, porosity and degree of crystallinity. The bioavailability of a poorly soluble drug can be improved by modifying these properties. For particles produced by electrospraying, the particle size distribution is quite narrow and tailoring the size of the particles is relatively simple. By electrospraying into a reduced pressure, the degree of crystallinity of the product can be decreased to some extent. It also seems that porosity of the electrosprayed drug particles depends on the drying pressure. However, electrospraying into a reduced pressure is not entirely free of complications. Bubbles tend to form in the capillary when liquids with low boiling points are electrosprayed. This causes instability to the jet. Also, maintaining a stable corona discharge in the neutralisator is somewhat complex.

Nevertheless, electrospraying into reduced pressure might offer a useful method for improving the bioavailability of poorly soluble drugs by modifying the three mentioned quantities simultaneously. In the future, the porosity of the produced drug particles will be studied in more detail and the dissolution behaviour will be tested *in vitro*.

References

[1] Ku B K and Kim S S, *J. Electrostat.* **57** (2003) 109-128
[2] Bailey A G and Borzabadi E, *IEEE T. Ind. Appl.* **14** (1978) 162-167
[3] Kidd P W, *J. Spacecraft Rockets* **5** (1968) 1034
[4] Gamero-Castano M, *J. Fluid. Mech.* **604** (2008) 339-368
[5] Swarbrick J C, Taylor J B and O'Shea J N, *Appl. Surf. Sci.* **252** (2006) 5622-5626
[6] Brayden D J, *Drug Discov. Today* **8** (2003) 976–978
[7] Shokri J, Hanaee J, Barzegar-Jalali M, Changizi R, Rahbar M and Nokhodchi A, J. *Drug. Deliv. Sci. Tec.* **16** (2006) 203-209
[8] Dudognon E, Willart J F, Caron V, Capet F, Larsson T and Descamps M, *Solid State Commun.* **138** (2006) 68-71
[9] Ciach T, *Int. J. Pharm.* **324** (2006) 51-55
[10] Nyström M, Murtomaa M and Salonen J, *J. Electrostat.* (2011) (Submitted)
[11] Nyström M, Murtomaa M and Salonen J, *J. Electrostat.* **68** (2010) 42-48

Acknowledgements: Financial support from the Academy of Finland is acknowledged.

Post-dispersion electrification of droplets in a system with pneumatic atomization

R Kacprzyk, M Lewandowski

Institute of Electrical Engineering Fundamentals, Wroclaw University of Technology, 50-370 Wroclaw, Wyb. Wyspianskiego 27, Poland

E-mail: ryszard.kacprzyk@pwr.wroc.pl

Abstract. The process of electrifying aerosol particles dispersed by a pneumatic sprayer with supersonic air flow is presented. To avoid the electric-field shielding effect, confirmed by investigations of induction charging of aerosol droplets with application of a concentric induction electrode placed in the region of liquid dispersion, the droplet electrification process following the dispersion of liquid was proposed. The supersonic atomizing head was equipped with an external high voltage contact electrode placed concentrically and perpendicularly to the droplets stream and closely to the atomizing head. Experiments were conducted in air, at ambient conditions (T=18±2 °C, RH= 55±3%), for standard air feeding rate (0.5 m^3/min, 0.4 MPa) and regulated dispersed liquid rate (0.1 – 0.55 l/min). Results of the applied electrification process, characterized by a (Q/m) factor measured as a function of liquid feed rate, have shown that the (Q/m) values achieved for post-dispersion electrification are comparable to the values obtained for typical induction electrification with application of a concentric electrode.

1. Introduction

Investigations of the induction charging process in relation to electrostatic spraying were undertaken in the early 1960's [1,2]. Induction or conduction (contact) charging of aerosol droplets were usually realized with the application of a concentric charging electrode placed in the region of liquid dispersion. The electrode was usually placed at the tip of the liquid feeding capillary and the charging process occurred during the removal of a particular droplet from the continuous liquid stream (being at the earth or high voltage potential). It was generally assumed that the charge-to-mass (Q/m) parameter could be treated as a droplet electrification effectiveness measure.

Previous research of the electrification process for aerosol droplets carried out on various types of pneumatic spraying heads with de Laval nozzles has shown that using the induction method obtained (Q/m) parameter of 0.1 - 2.3 mC/kg. It was found that the (Q/m) value was closely related with both the geometry of the electrostatic system of the atomizing head as well as the media feeding conditions (air pressure and liquid feed rate). The mentioned values for the (Q/m) parameter were obtained for the electric field intensity (existing in the electrification region – capillary tip) being close to the electric strength of air. The (Q/m) values obtained in the case of pneumatic heads with supersonic gas flow were much lower in comparison to the values predicted by generally known limits (Rayleigh's, Paschen's, air electrical strength limits or Maxwell time constant limit) [3]. Experimental investigations of the induction charging process of aerosol droplets of pneumatically atomized liquid with using a concentric induction electrode (placed in the region of liquid dispersion at the tip of liquid feeding capillary) seemed to confirm the appearance of a strong electric-field shielding (field compensation) effect [4] limiting the value of the (*Q/m*) parameter. To avoid the shielding effect the droplet electrification process following the dispersion of liquid was proposed. The corona discharge method is usually applied in the case of post-dispersion electrification of particles of different nature. The corona discharge method requires a relatively long time (or relatively large space with strong electric field) for efficient electrification of particles. Alternatively the post-dispersion conductive electrification method was proposed to avoid the shielding effect in case of pneumatic liquid dispersion.

2. Atomizing head and electrification system

The post-dispersion electrification of droplets was conducted using typical supersonic atomizing nozzle head (Telesto®, EFEN 110,5) producing droplets within the diameter range 10-15 μm. The head was equipped with an external high voltage contact electrode in the form of a metal ring of 70 mm diameter. The electrode cross-section was filled with a metal mesh (3 x 3 mm, ϕ=0.3 mm wire diameter) in order to increase contact probability with aerosol droplets. The electrode was placed concentrically and perpendicularly to the droplet stream along its axis. The distance separating the atomizing head and conduction electrode was adjustable in a range of 20 – 100 mm and kept in position by two insulating rods. Experiments were conducted in air at ambient conditions (T=18±2 °C, RH= 55±3%), for a standard air feeding rate (0.5 m^3/min, 400 kPa) and a regulated dispersed liquid rate (0.1 – 0.55 l/min). In all the experiments tap water with electric conductivity 0.40±5 mS/cm was used.

3. Measurement circuit

The efficiency of the aerosol electrification process was characterized by determining the (*Q/m*) parameter value using the measuring system shown in figure 1.

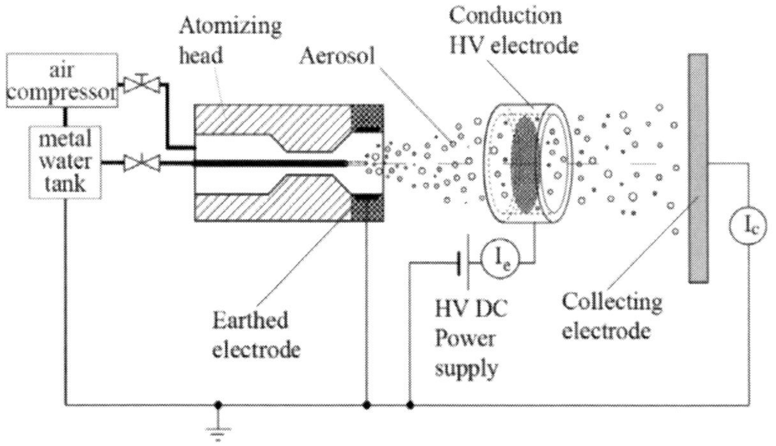

Figure 1. Sketch of the measurement set-up for spray electrification.

The measuring system included air and liquid media conditioning stations (equipped with a metering circuitry) supplying the atomizing head which was equipped with a conduction electrode. The conduction electrode was supplied from a dc stabilized high voltage power supply (Glassman EW40/50 type) and could be regulated in the range 0 ±50(40) kV of both polarities. The cloud of electrified particles was collected on the surface of the collecting electrode system containing three metal electrodes (1.30 m × 1.30 m) equipped with droplet velocity damping screens (metal mesh screens 1.00 m × 1.00 m kept at the collecting electrode potential). The collection electrode system was isolated from the earth and its leakage current I_c was measured by an analogue pico-ammeter, PA100. The value of the charge to mass ratio (Q/m) parameter was determined from the relation

$$(Q/m) = \frac{I_c t_c}{\gamma V_c} \tag{1}$$

where I_c = the steady state current value determined for the collected clouds of electrified droplets, γ = mass density of the dispersed liquid, V_c = the volume of liquid collected on the collector electrodes within the time t_c.

4. Results and discussion

The performance of the conduction electrode system was determined by plotting the conduction electrode charging current vs. applied voltage and (Q/m) vs. applied voltage characteristics as shown in figures 2 and 3 respectively, for a fixed head-to-conduction electrode separation of 75 mm.

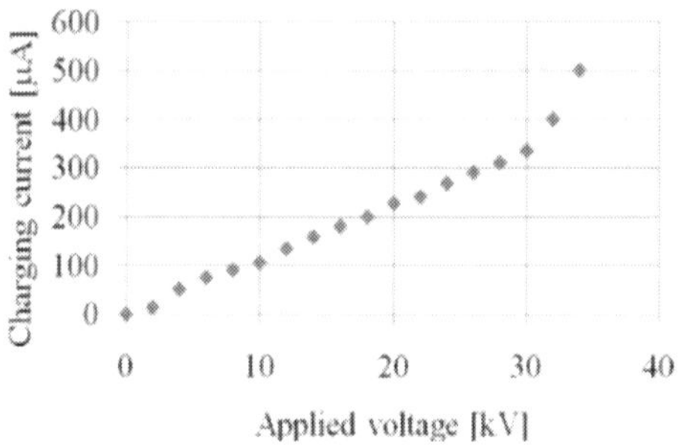

Figure 2 Dependence of the conduction electrode charging current on the electrode supply voltage.

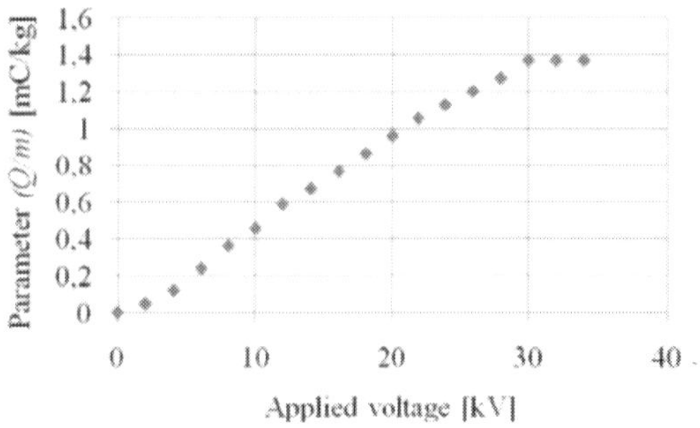

Figure 3. Dependence of the *(Q/m)* parameter on the electrode supply voltage.

Results show that a saturation of the (Q/m) parameter is closely related with the observed sudden increase of the charging electrode supply current. The corona discharges appearing for applied voltages above ca. 30 kV are probably responsible for both of the observed effects. The influence of the head-to-conduction electrode separation distance on the (Q/m) parameter is shown in figure 4. The results were obtained for an average electric field intensity $E = 400 \pm 5$ kV/m = constant.

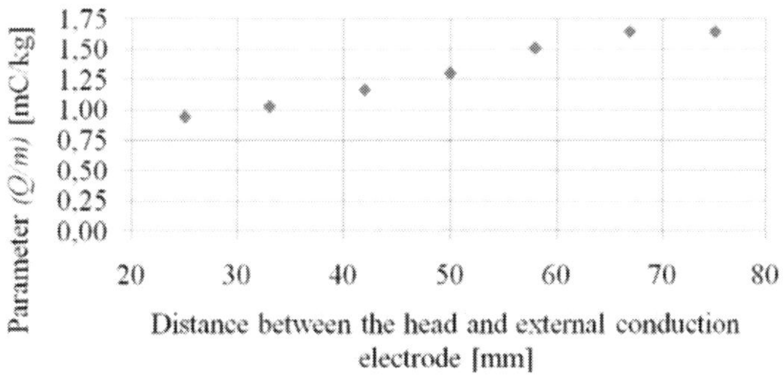

Figure 4. Dependence of the (Q/m) parameter on the atomizing head-to-conduction electrode separation distance (E=const=400 ±5 kV/m).

The (*Q/m*) parameter dependence (figure 4), may be explained assuming dependence of the droplet-mesh contact probability on the head-to-charging electrode separation distance. The distance increase leads to a decrease in the droplet numerical density in the electrode region as well as to an average contact surface increase (of the mesh of conduction electrode) determined per one droplet. A simplified model [5] assuming proportionality of the droplet – electrode contact probability to the average contact surface (per one droplet) and constant droplets diameter (not affected by liquid feed rate) leads to the expression

$$(Q/m) = C\frac{Ex^2}{\gamma W_w} \tag{2}$$

where C = constant, E = electric field intensity in the electrifying system, x = atomizing head-to-conduction electrode separation distance, γ and W_w = liquid mass density and feed rate, respectively. The results shown in figures 3 and 4 support the validity of relation (2) from the point of view of the expected dependences on field intensity E and the distance x. The expression (2) suggests additionally a hyperbolic dependence of the (*Q/m*) parameter on the liquid feed rate. Results of appropriate investigations shown in figure 5 confirm a power-type dependence of the (*Q/m*) parameter value on the liquid feed rate W_w but the curve slope n = -1.44 was found to be higher than the expected one (n = -1.0). The observed discrepancy is probably due to an influence of the liquid feed rate on the droplet diameter. Decreasing the liquid feed rate usually leads to a decrease in droplet diameter and an increase in the (*Q/m*) parameter.

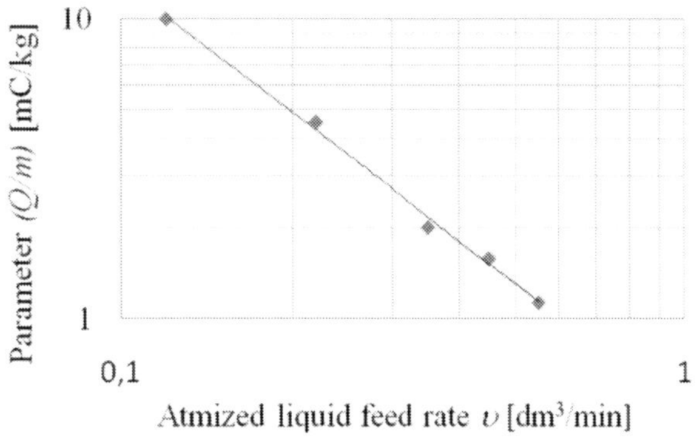

Figure 5. Dependence of the (Q/m) parameter on the liquid feed rate. Measurements conditions: temperature *T*=18±2 °C, relative humidity *RH*= 55±3%, *E*=400±5 kV/m, dispersed liquid – tap water conductivity (0.40 mS/cm).

The obtained results have shown that conduction electrification using an external mesh electrode allows the (*Q/m*) parameter to reach values similar to those obtained for induction electrification (with an internal, ring electrode) [4].

References

[1] Law S E Agricultural electrostatic spray application: a review of significant research and development during the 20th century *Journal of Electrostatics* 51 – 52 (2001), **25 – 42**

[2] Lifschitz M N, Moisiejev V M, Elektriceskije javlenia v aerozoliach i ich primienenije, *Izd. Energija*, Moskva, (1965) (in Russian)

[3] Bailey A G, Electrostatic Spraying of Liquids, *EEE Res. Stud. Electrostatics and Electrostatic Applications Series*, RSP Ltd. (1988)

[4] Kacprzyk R, Zylka P, Electrostatic Shielding Effect in Atomizing Heads with Supersonic Gas Flow, *7th Conference of the French Society of Electrostatics SFE 2010*, Montpellier, France 30/08/2010 - 01/09/2010, in printing (IEEE TDEI, 2011)

[5] Kacprzyk R, Lewndowski M, Contact electrification of droplets using an external electrode (to be published)

Journal of Physics: Conference Series **301** (2011) 012031 doi:10.1088/1742-6596/301/1/012031

Electrohydrodynamic actuation of co-flowing liquids by means of microelectrode arrays

Pablo García-Sánchez, Mathieu Ferney, Antonio Ramos
Depto. de Electrónica y Electromagnetismo. University of Sevilla, Spain
E-mail: pablogarcia@us.es

Abstract. Electric fields induce forces at the interface between liquids with different electrical properties (conductivity and/or permittivity). We explore how to use these forces for manipulating two coflowing streams of liquids in a microchannel. A microelectrode array is fabricated at the bottom of the channel and one of the two liquids is labelled with a fluorescent dye for observing the phenomenon. The diffuse interface between the two liquids is deflected depending on the ac signal and conductivity (or permittivity) ratio between the liquids. Only a few volts are needed for observing the interface destabilization, in contrast with other electrode configurations where hundreds of volts are applied.

1. Introduction

The integration of microelectrode arrays in microchannels represent an opportunity for direct actuation either of the liquid or suspended particles [1]. Thus, electrohydrodynamics (EHD) in microsystems is of central interest in current *microfluidics* research and *Lab-on-a-Chip* technologies [2, 3].

Manipulation of two co-flowing fluids is a common operation in Lab-on-a-Chip devices. In this work, we present an experimental characterization of the actuation of coflowing fluids with different electrical properties (conductivity and/or permittivity) by an array of microelectrodes. Figure 1a) shows a sketch of the microelectrode array and the microchannels. Figure 1b) shows a picture of the actual device: platinum microelectrodes are fabricated on glass and a polymeric microfluidic channel (500 μm high and 100 μm wide) is built on it. Each liquid is inserted on one side and they flow together down the main channel, where the electrode array is placed. The microelectrode array consists of 100 parallel electrodes oblique to the channel. Electrodes are 10 μm wide separated by 10 μm gaps and they are energized with an ac signal of amplitude between 0-10 V and in the frequency range 1 kHz-1 MHz. Both fluids are driven by hydrostatic heads and the pressures are adjusted to have the same flow rate of both liquids prior to the application of the electric field. Working liquids are aqueous solutions where KCl is added to control the conductivity. In some experiments, ethanol with dissolved salts and water were used in order to have liquids with the same conductivity but different permittivity (for example: water with KCl as liquid 1; and ethanol with KCl as liquid 2). As demonstrated in [4], the boundary between the two liquids is deflected upon application of ac signals of only a few volts of amplitude. Figure 2 shows an example for two aqueous solutions. We use fluorescence microscopy and the liquid with higher conductivity is labelled with a fluorescent dye (fluorescein), while the less conducting liquid appears dark. When the voltage is on, the whole channel appears fluorescent,

indicating the presence of the more conducting liquid.
The electrical force in an incompressible fluid is given by $\mathbf{f}_E = \rho\mathbf{E} - (1/2)|\mathbf{E}|^2\nabla\varepsilon$, where ρ is the free electrical charge density and ε the dielectric constant. The first term corresponds to the Coulomb force, while the second accounts for polarization forces. When we use water solutions, the polarization force vanishes since the permittivity is the same for both streams. However, as a consequence of the difference in conductivities, net free charge is induced at the interface between the two streams. It can be shown that an ac electric field of frequency ω acting on a sharp interface between two media of equal permittivities and conductivities σ_1 and σ_2, respectively, induces a surface charge given by $q_s = \varepsilon[(\sigma_1 - \sigma_2)/(\sigma_2 + i\omega\varepsilon)]\mathbf{E}_1 \cdot \mathbf{n}$ where $\mathbf{E}_1$ is the electric field in medium 1 and $\mathbf{n}$ a vector normal to the interface. The charge, and therefore the electrical force, vanishes if the two liquids have the same conductivity and/or the applied frequency is high enough.

In the next section we present an experimental study of the interface deflection as a function of frequency, voltage amplitude and conductivity ratio between the two liquid streams. We also present results using ethanol and water as working liquids with the same conductivity but different permitivitties. It demonstrates that the strength of polarization forces is also sufficient for producing an important deflection.

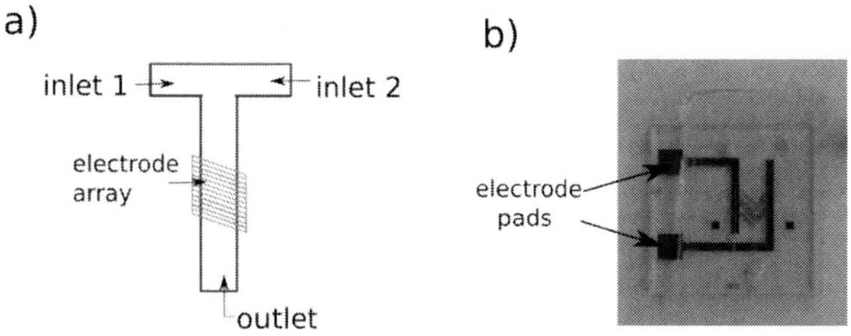

Figure 1. a) Sketch of the microfluidic device showing the inlets for the two fluids and the outlet. An array of interdigitated microelectrode is placed in the middle of the main channel. b) Photograph of the device. Microelectrodes are 10 μm wide and the channel is 500 μm wide and 100 μm high. The pads are for electrical connections.

2. Experimental Results

The experimental procedure relies on quantitative measurements of the fluorescence intensity in a given position of the channel. Grayscale images of two coflowing streams are captured to a PC and, from these images, we extract the intensity profiles along a straight line transverse to the channel (see yellow lines in figure 2). This intensity results from the light coming from all fluorescent molecules within the depth of the channel. In other words, we do not just collect the light coming from the focal plane (as in confocal microscopy) but integrate the light over the thickness of the fluid layer. We checked this by taking images while focusing at different planes and we obtained the same intensity profiles. Note that, with this measurement, we obtain how much of the fluorescent liquid moved to the region of lower conductivity and vice versa but we do not obtain a full 3D picture of the phenomenon. Therefore, we have also used fluorescent beads suspended in the more conducting liquid. Focusing the beads allows us to distinguish the region occupied by each liquid (these beads diffuse slowly for the typical time of the experiment). We have checked that, after application of the electric field, the more conducting liquid occupies the region near the electrode array, i.e. the region where the electric field is more intense. Because

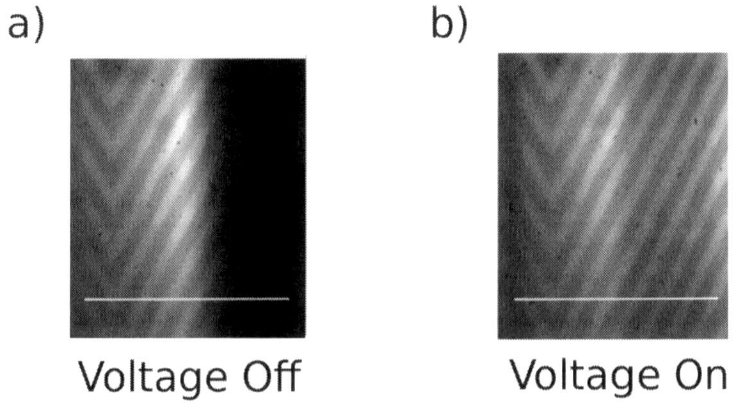

Figure 2. a) When no voltage is applied the two liquids flow together down the channel. The liquid with a higher conductivity bears a fluorescent dye, while the lower conductivity liquid appears dark. b) Upon application of an ac signal, the whole channel is fluorescent indicating that the more conducting liquid is occupying the whole width of it. The horizontal lines indicate the lines along which intensity profiles were obtained

of incompressibility, the upper part is occupied by the less conducting liquid.

For analyzing the effect of frequency, voltage and conductivity ratio, we fix our attention on a given point in the channel. We chose a position on the right half of figure 2, that is, a region occupied by the less conducting liquid in the absence of applied field. Therefore, the fluorescence is zero when no voltage is applied and relative changes in the fluorescence intensity indicate the presence of the more conducting liquid. Figure 3 shows the fluorescence intensity as a function of the signal amplitude for two different conductivity combinations. Figure 4 shows the fluorescence intensity as a function of frequency. The results are normalized so 0 corresponds to no fluorescence at all and 1 to maximum intensity (that is, fluorescence intensity given by the channel filled with the more conducting liquid). As expected, the phenomenon disappears for frequencies higher than the reciprocal of the charge relaxation time and, interestingly, it also vanishes for low frequencies. Figure 5 shows the impedance of the complete array with the liquids of conductivity $0.75\,\mathrm{mS/m}$ and $1.5\,\mathrm{mS/m}$ on it. The impedance of the system increases for frequencies below tens of Hertz. This is due to polarization of the electrodes, i.e. counterions accumulate at the electrode-electrolyte interface and screen the electric field. This is the reason that the phenomenon also disappears at low frequencies.

Finally, figure 6 shows results for water and ethanol with the same conductivity. Dielectric forces are also capable of producing an important interface deflection. In this case, the phenomenon is almost independent of frequency.

3. Conclusions

Using fluorescent dyes is a practical way of characterizing the behaviour of the two coflowing liquids under ac fields. We found that ac signals of a few volts are enough to produce the deflection of the diffuse interface between two water streams with different conductivities over a wide frequency range. The more conducting liquid occupies the region near the electrodes, where the electric field is higher. At low frequencies the phenomenon disappears because of the electrode polarization (the electric field in the bulk vanishes) in accordance with impedance measurements. Experiments with water and ethanol show also that dielectric forces are capable of producing a considerable interface deflection, the liquid with higher permittivity occupies the

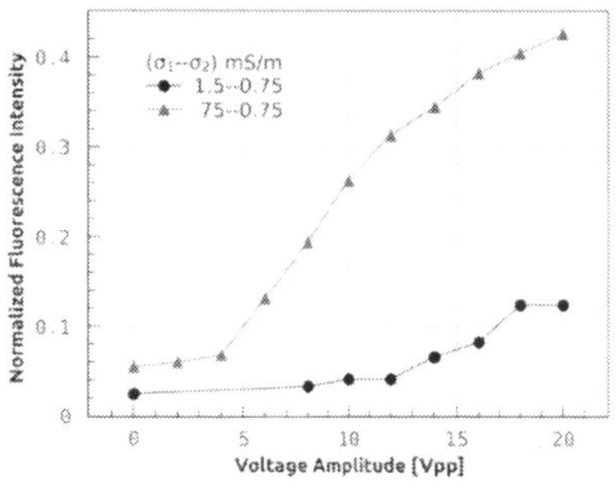

Figure 3. Fluorescence intensity vs voltage at a position in the channel where, with no applied field, there would only be non-fluorescing liquid.

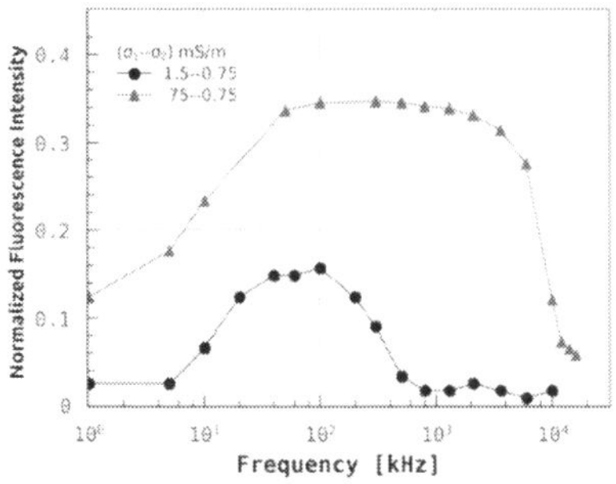

Figure 4. Fluorescence intensity vs frequency, the phenomenon disappears at both high and low frequencies.

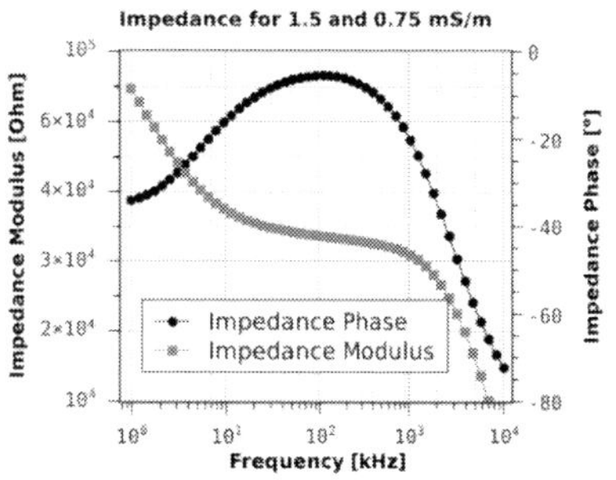

Figure 5. Impedance measurements of the system vs frequency. The increase of the impedance at low frequencies is due to electrode polarization.

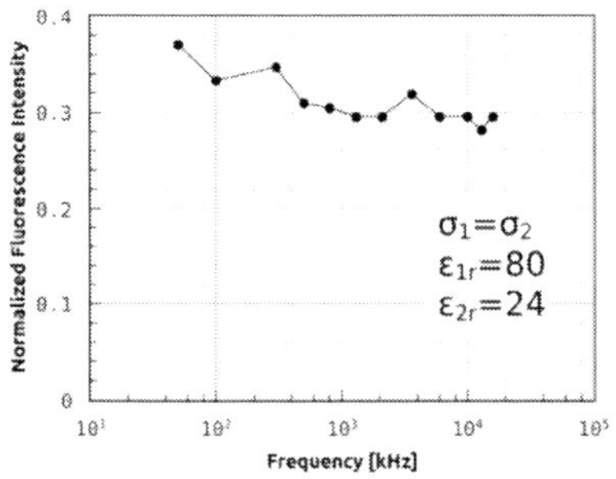

Figure 6. Fluorescence intensity vs frequency for water and ethanol with the same conductivities. Intensity 1 corresponds to the channel filled with water.

region of more intense electric field.

Acknowledgments
Authors are thankful to Dr. N.G. Green and Prof. H. Morgan from Southampton University for providing the microelectrode arrays. Financial support from the Spanish Government Ministry MEC and Regional Government Junta de Andalucía under Contracts FIS2006-03645 and P09-FQM-4584, respectively, are acknowledged. M.F. acknowledges financial support from the Rhône-Alpes region (France).

4. References
[1] Morgan, H.; Green, ,N.G. *AC electrokinetics: colloids and nanoparticles* **2003** "Research Studies Press Ltd.".
[2] Squires, T. M.; Quake, S. R. *Review of Modern Physics* **2005**, *77*, 977–1026.
[3] Ramos, A. *Electrohydrodynamics and Magnetohydrodynamics Micropumps.* In *Microfluidics technologies for miniaturized analysis systems*; *Springer*, **2008**.
[4] Morgan, H.; Green, N. G.; Ramos, A.; Garcia-Sanchez, P. *Applied Physics Letters* **2007**, *91*, 254107.

Electro-hydrodynamic instability patterning of polymers

S. Cargill[1], M.P.Y. Desmulliez[1], W.Yu[2]

[1] Heriot-Watt University, MIcroSystems Engineering Centre (MISEC), Institute of Integrated Systems, Edinburgh, EH14 4AS, United Kingdom

[2] Changchun Institute of Optics, Chinese Academy of Sciences, No 3888, Southeast Lake Road, Changchun City, Jilin Province 130033, P. R. China

Email: sc277@hw.ac.uk

Abstract. Electro-HydroDynamic Instability Patterning, EHDIP, is a novel micro-manufacturing process that makes use of the instability of viscous polymeric thin films when exposed to electrostatic fields. By using non uniform electrostatic fields, it is possible to shape the polymer into defined meso- and micro-scale structures which are subsequently cured to defined 2D and 3D microstructures. The relatively rapid process time, the one-step manufacturing approach, as well as the ability to produce hitherto unrealised topographies - such as continuous profile structures, makes EHDIP an attractive manufacturing process.

1. Introduction

The last decade has witnessed an expanding literature examining the effects of electrostatic fields upon the fluid-fluid interface of polarisable leaky dielectrics [1][2]. These studies have demonstrated the formation of structures under uniform and non uniform E-fields. The phenomenon, called here electro-hydrodynamic instability patterning or EHDIP, offers several advantages over traditional photolithographic methods. Firstly, as it is a maskless method, no expensive optical equipment or photomask are needed. Secondly, a large range of polymer materials are, in principle, suitable for this technique, allowing thereby the possibility to use cheaper resists for the patterning of thick structures. Thirdly, there is no further polymer development process needed due to the one-step pattern formation after annealing of the polymer material. This article describes the progress made towards the modeling and experimental characterization of this process.

2. Theory and modeling

The combination of the Navier-Stokes equation [1],

$$\rho(\partial_t v + (v.\nabla)v) = -\nabla p + \eta\Delta \mathrm{v} + \mathrm{f}\ (1)$$

And the body force term, f, coupled to the electromagnetic forces through the Maxwell stress tensor as derived by Landau [1],

$$f = QE - \frac{1}{2}E^2\nabla\varepsilon + \frac{1}{2}\nabla\left[\rho\frac{dz}{d\rho}\right]_\theta E^2 \quad (2)$$

Published under licence by IOP Publishing Ltd

Constitute the basis of the modelling of the EHDIP process. In the first equation, the fluid stress is the sum of a diffusing viscous term, proportional to the gradient of the fluid velocity, *v*, and a pressure term. In the second equation, the first term on the R.H.S. is the Coulomb force, the second term is the dielectric force and the last term is the electrostrictive force.

The pressure of the electrostatic force at the interface between air and the polymer causes a deformation in the polymer surface. This deformation is more pronounced and quicker in areas where the field is stronger. A patterned metal mask subjected to an electric potential will create a strong non-uniform field that will redistribute the volume of the polymer into lateral mesas with the same dimensions as the mask, effectively transferring the mask pattern to the target polymer [3].

2.1 Schwarz-Christoffel mapping and COMSOL™ simulation

The complex electrostatic field distribution at the interface was computationally modeled using the multiphysics modelling software package COMSOL™. Simulation results are provided in fig.1 and shows strong non-uniformity of the electrostatic field near the surface of metal mask. In the simulation, the mask is taken a binary grating; the redistribution of volume of the polymer is schematically described by fig.2.

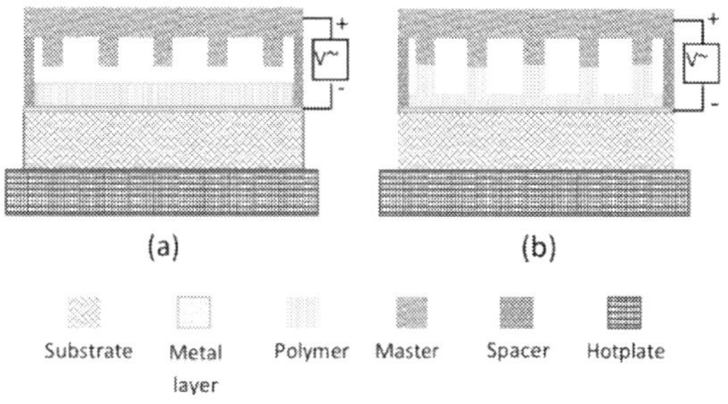

Figure 2: Schematic of the EHDIP process with basic experimental set up

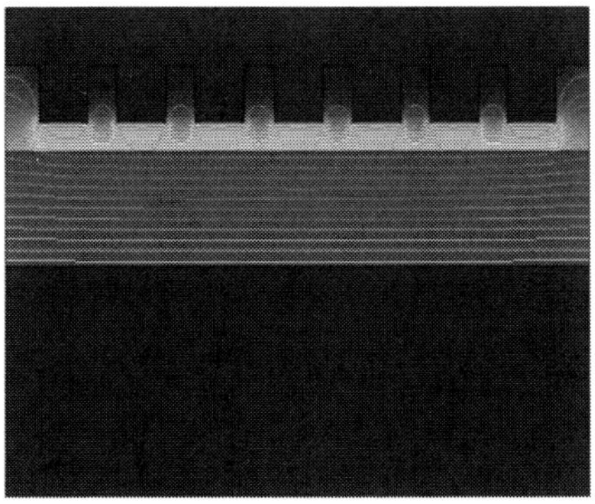

Figure 1: COMSOL modelling of the electrostatic field strength

An analytical solution of the electrostatic field would be desirable such that it can into the previous two equations. In that end, the Schwarz-Christoffel (SC) conformal mapping has been used to map the binary structure onto a specific complex plane in order to solve the electrostatic potential problem more easily [5]. This method allows for the binary structure described above to be mapped from the spatial domain (z) to an auxiliary plane (w). The complex co-ordinates (x, y) in the z–domain are mapped to the axes (u, v) of the w–domain. The basic SC equation is as follows:

$$z(w) = \int \prod (w - a_j)^{\alpha_j} + k \text{ (3}$$

Where a_j are the pre-vertices and α_j is the turning angle of the boundary in the z-domain. The two constant parameters, C and K, affect the scaling and the positioning of the conformal map, respectively. By selecting the correct pre-vertices it is possible to construct a conformal map of a periodic binary pattern like that seen in the experimental set up. For every period, one vertex is set to infinity creating thereby a map of periodic trigons, where the two top vertices have 90° turning angles. This allows us to construct the map function:

$$z(w) = \frac{l}{\pi} \int \sqrt{\frac{\cos\left(\frac{\pi w}{l}\right) + \cos\left(\frac{\pi}{l}\right)}{\cos\left(\frac{\pi w}{l}\right) + 1}}\, dw \text{ (4)}$$

The electric field strength and electrostatic pressure at the fluid interface can be computed and compared to the COMSOL™ simulation results as shown in (fig.3) where the continuous lines show the COMSOL modeling and the points show the SC mapping, it was observed that there is a very close correlation in the two methods at the points of interest, where the field is highest underneath the mask protrusions:

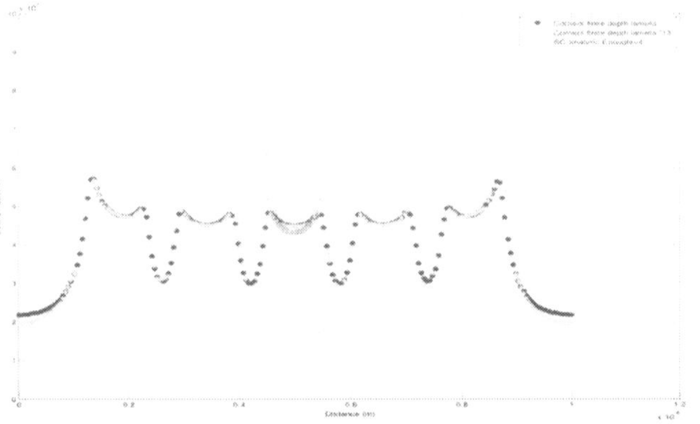

Figure 3: SC vs. COMSOL™ electrostatic field modelling results

3. Experimental set up and real time sensing

In order to properly characterise the EHDIP process, a mechanical rig was built to allow for six degrees of freedom of movement. Linear Variable Differential Transformers (LVDT) and a three-axis accelerometer were configured for static angle readings to control the orientation of the rigs stages and the separation of the mask from the polymer surface. A vacuum chuck was used to suspend the mask above the surface as shown in fig.2. The rig accommodates a hot plate upon which rests a wafer with the spun-on polymer. The temperature is raised beyond the glass transition temperature of the polymer, to allow the redistribution of the polymer volume. The electric field is then applied to the mask through the conductive vacuum chuck.

In order to examine the real time effects of the polymer growth in the micron sized gap between the mask and target polymer, we manufactured a capacitive sensing system that measures the change in polymer volume as a function of time. This allowed us to measure the instantaneous change in the target material as shown in fig.5.

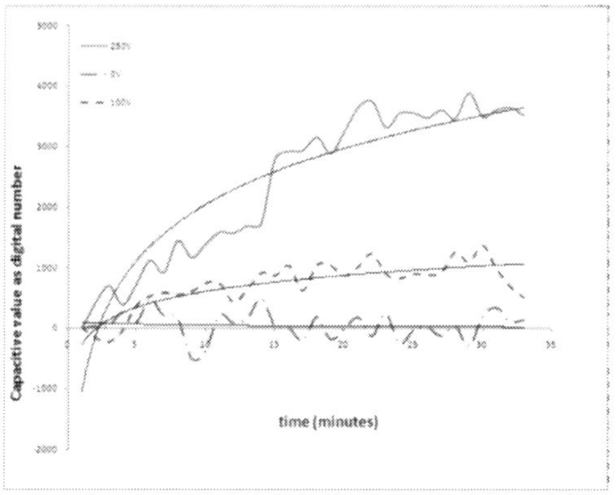

Figure 4: Time evolution of the polymer volume

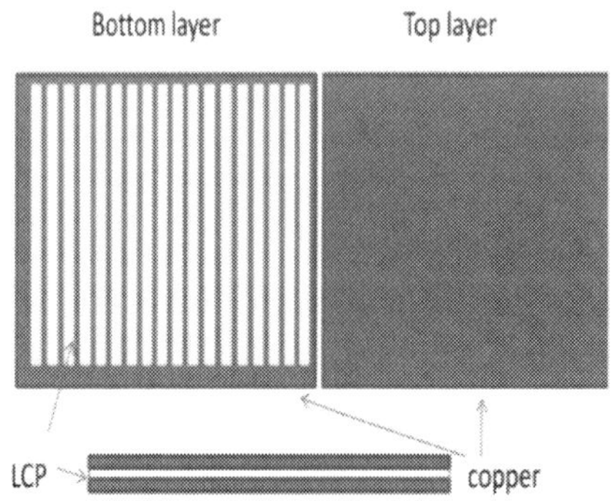

Figure 5: Example of LCP mask

We were able to characterise the change of polymer volume over time and observe the effect of different magnitudes of electrostatic fields on the interface. We observed that the majority of polymer displacement occurred early in the process time.

4. Mask design

EHDIP is wholly dependent upon the use of the non uniform electrostatic fields generated by the binary gratings, used here as the master electrode. The electrode was manufactured using the LIGA process [6]. This method involved the photolithographic development of a pattern into a polymer photoresist spun on a wafer with a suitable seed layer for the subsequent electroforming process. An alternative, and somewhat simpler process, was also implemented using copper clad Liquid Crystal Polymer (LCP) as shown in fig.6.LCP is a dielectric material chosen for its large breakdown potential, and its ability to be fabricated in very thin thicknesses.LCP is usually clad on both sides by a copper sheet. One side of the copper is patterned using well proven PCB development techniques, leaving exposed LCP patterns in the underside copper. A large electric potential is applied to the undeveloped copper side, and a much lower electric potential is applied to the developed side. This gives rise to a field effect where depletion regions build up in the areas adjacent to the developed copper areas. The electric field developed by the large potential is effectively shaped by these regions and concentrates in the exposed LCP areas. By choosing appropriate patterns for development, it is possible to shape the E-field in the same dimensions as that achieved using traditional LIGA techniques in a much shorter time frame and with less demanding mechanical properties.

5. Experimental results and discussion

A full factorial experiment was carried out in order to identify the most important factors promoting the polymer growth. Our investigation looked at the applied potential, the temperature applied to the polymer, the film thickness of the target polymer, and the separation (air gap) between the mask and the polymer. The properties examined were the average growth rates at the corner and middle of the structures as well as the uniformity of the structure dimensions over the patterned area.

The most important factor influencing both growth and uniformity is the film thickness. A greater thickness causes greater undulations in the target material fluid interface. A thicker film allows for more available material for redistribution than thinner films, thereby increasing uniformity. We observed that greater temperatures caused the solvent in thinner films to dissolve quicker, retarding growth rates, while larger air gaps induced a larger thermal flow, the effect of which is to increase the surface pressure on some parts of the film but not others, negatively influencing uniformity. EHDIP is usable with a broad class of materials; any polarisable polymer fluid can be used in the process. EHDIP also has several advantages with respect to traditional polymer machining techniques. It is a non contact process, making it suitable for constructing devices for bio-MEMS fields where contamination is a risk; the non contact also prolongs the useful life of the mask as well. It is a one step process, capable of completing a useful microstructure in two minutes, after curing. We also observed the formation of continuous curved surfaces, opening the possibility of creating new micro devices for optical and microfluidic applications, such as hydro dynamically efficient micro channels, diffraction gratings and lenses for short wavelengths.

References

[1], Xiaopeng C, Jiusheng C & Xiezhen Y, *Advances and applications of electrohydrodynamics Chinese Science Bulletin* 2003 Vol. **48** No.11 1055 1063

[2] Yu W, Cargill S, Leonard M, Desmulliez M P Y, *Micro-fabrication on 3D Surfaces by Electrostatic Induced Lithography, ESTC proceedings* 2008,111-116, ISBN 978 1 4244 2813 7

[3] Schäffer E, Thurn-Albrecht T, Russell T P and Steiner U *Electrically induced structure formation and pattern transfer, letters to nature* **403**, 24 February 2000

[4] Chou S, Zhuang L, and Guo L, *Lithographically induced self-construction of polymer microstructures for resistless patterning, Applied Physics Letters*, Vol **75**, No 7

[5]Driscoll T, *Schwarz-Christoffel mapping* 2002, ISBN 0 521 80726 3

[6] Khan C, Saile V, *Applications of LIGA technology to precision manufacturing of high-aspect-ratio micro-components and -systems: a review MicroElectronics Journal* Vol **35**, No 2

Statistical analysis of fires and explosions attributed to static electricity over the last 50 years in Japanese industry

A. Ohsawa

Electrical Safety Research Group, National Institute of Occupational Safety and Health, 1-4-6 Umezono, Kiyose, Tokyo 204-0024, Japan

E-mail: ohsawa@s.jniosh.go.jp

Abstract. This paper presents a statistical analysis of 153 accidents attributable to static electricity in Japanese industry over the last 50 years. A more thorough understanding of their causes could help prevent similar incidents and identify hazards that could assist in the task of risk assessment. Most of the incidents occurred during operations performed by workers. In addition, more than 70% of the flammable atmospheres resulted from the presence of vapours. A noteworthy finding is that at least 70% of the ignitions were caused by isolated conductors including operators' bodies leading to spark discharges, which could have easily been prevented with earthing. These tendencies indicate that, when operators handle flammable liquids with any conductors, the ignition risk is significantly high. A serious lack of information regarding fundamental countermeasures for static electricity seems to be the main cause of such hazards. Only organised management, including education and risk communication, would prevent them.

1. Introduction

Accidents involving fires and explosions attributable to static electricity still occur in industry even though controls are in place. Some accidents have been minor; however, others have been very serious sometimes unfortunately involving loss of life and considerable damage to property.

Accidents, however, offer an opportunity for learning that may prevent similar incidents as far as accident analysis permits proper cause identification and application of risk reduction measures to prevent future accidents. In addition, their results should lead to new knowledge, rules and practices with the goal of higher reliability and safety. Since the analysis of a single case can give helpful information, comprehensive statistic analysis of a number of incidents would be even more useful. Accident trends could be identified to assist in the task of risk assessment, in particular, providing important insight into hazard identification.

In this paper, 153 accidents attributable to static electricity in Japanese industry, in particular, the petroleum and chemical industries, over the last 50 years (1960–2010) are investigated. Although data from 310 accidents was collected, 157 cases were excluded because of a lack of information or a strong suspicion that ignition might have been caused by sources other than static electricity.

It is noteworthy that some of these accidents were similar to those reported in other countries [1, 2, 3], which indicates that similar types of accidents occur repeatedly.

2. General scenario of static ignitions

All the cases investigated in this paper fit the process (scenario) in a similar manner. A prerequisite for this process is the formation of a flammable atmosphere. If the rate of charge relaxation is lower than that of charge generation caused by charge separation, charge accumulation occurs. An electrostatic discharge may occur if the electric field generated by the charge is sufficient to cause an electrical breakdown of the air under the boundary conditions of the ground provided in the surrounding area. In addition, isolated conductors in the electric field generated by the accumulated charge may lead to spark discharges caused by induction. Ignition occurs when the released energy by the discharge is higher than the ignition energy of the flammable atmosphere.

A fault tree of static ignition according to the scenario is shown in figure 1. It indicates that, if the fault events shown by thick circles, which correspond to failures in fundamental measures against electrostatic hazards, could be corrected, the top event, static ignition, would never occur. The events shown below the numbered triangles indicate examples of processes and operations that caused charge separation.

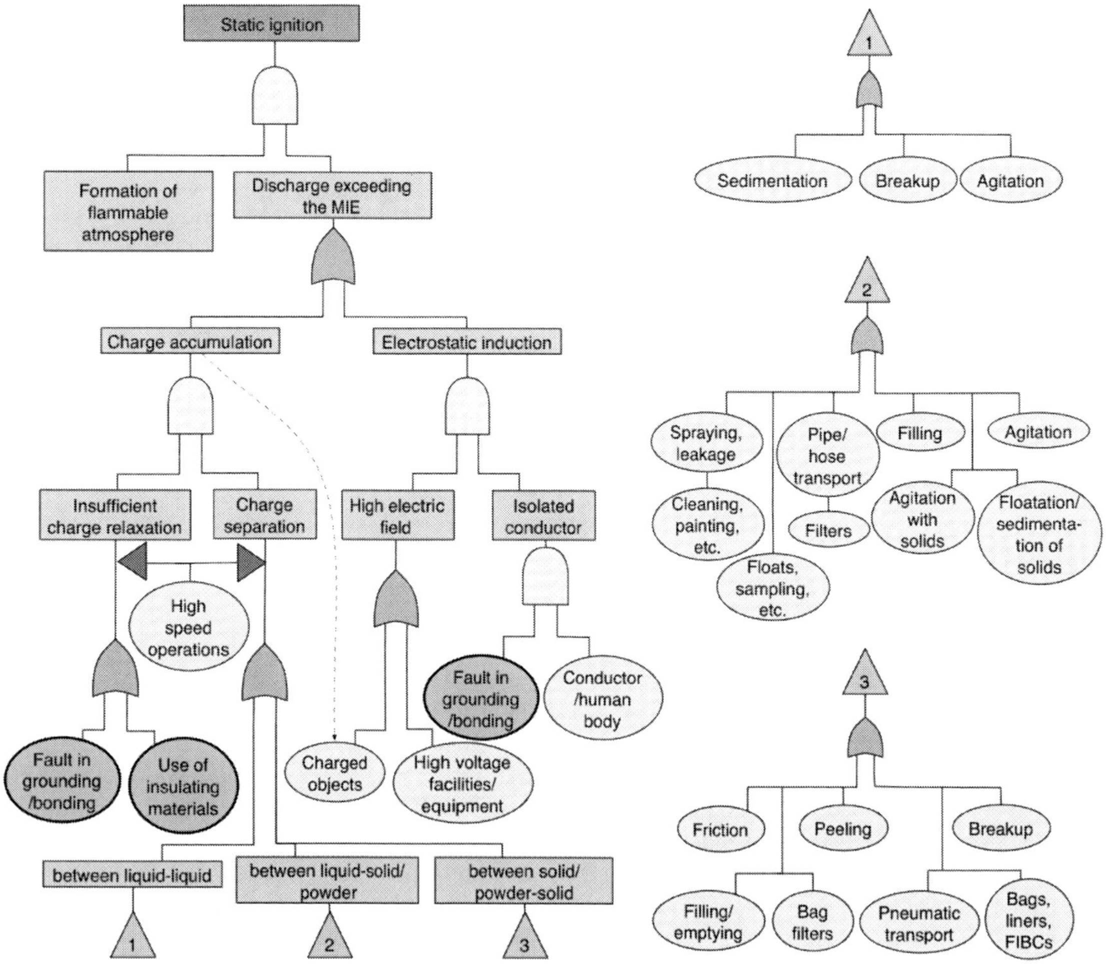

Figure 1. Fault tree of static ignition

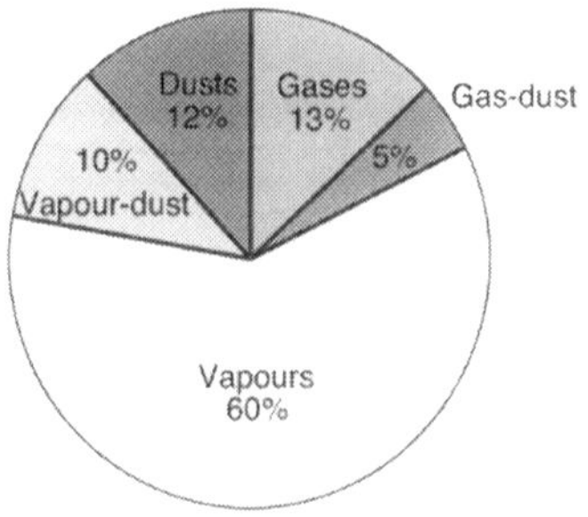

Figure 2. Flammable atmospheres

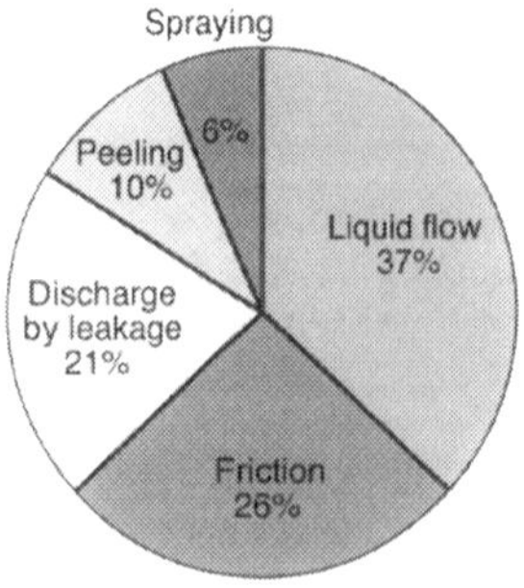

Figure 3. Causes of charging

3. Formation of flammable atmospheres

As shown in figure 2, most flammable atmospheres investigated were the results of the presence of vapours.

In 13.1% of the cases investigated, flammable atmospheres were considered to be the result of the presence of flammable gases. Among them, as many as 14 cases (~70% of flammable gas atmospheres) involved hydrogen, which is highly susceptible to ignition. In some cases, hydrogen was directly involved, while in others, hydrogen was generated by the reactions of hydrochloric acid, hydrogen peroxide, or hydrogen sulphide. Other gases causing the formation of flammable atmospheres were ethylene (4 cases) and LP gas (2 cases). In addition, most of the flammable gas atmospheres were formed due to gas leakage accompanied with liquid discharge, which can generate highly electrostatic charges.

In 60.1% of the cases investigated, flammable atmospheres were determined to have been created by the presence of flammable vapours. With the addition of hybrid mixtures containing flammable dusts, vapours formed 70.6% of the flammable atmospheres in those cases. The vapours include toluene (20 cases), gasoline (11), hexane (9), benzene (7), ethyl acetate (6), thinner (4), ethanol (4), xylene (3), methanol (3), acetone (2), naphtha (2), styrene (2), light oil (2 switch loadings), heavy oil, dioxane, and waste oil. The hazard identification of these vapours in terms of their potential to form flammable atmosphere can be estimated by the relationship between their flash points and temperatures.

In 11.8% of the cases investigated, flammable atmospheres were created by the presence of flammable dusts. The dusts were aluminium (3 cases), polyethylene (2 cases), poval, decaborane, para-octylphenol, bisphenol, anthracene, sulphur, and powders used in coating operations. In all cases, ignition was caused either by sparks by isolated conductors or by propagating brush discharge that might have occurred in flexible intermediate bulk containers (FIBCs) and insulating pipes.

In 15.0% of the cases, flammable atmospheres were estimated to be created by the presence of a hybrid flammable mixture of gases or vapours with dusts. The flammable gas-dust hybrid mixtures (4.6%) include ethylene-polyethylene mixtures (4 cases) and propylene-polypropylene mixtures (3 cases), which were formed due to insufficient degassing in a drying process. On the other hand, the flammable vapour-dust hybrid mixtures (10.5%) were created by solvents or materials such as styrene, heptane, alcohols, and ketone, which were used during the process of adding powder to a flammable liquid, filling powder into a container cleaned with solvent, or cleaning containers with solvent near the powder-filling site. Flammable dusts contained in those mixtures were polypropylene, polyethylene, polyvinyl alcohol, maleic anhydride, aluminium isopropoxide, rubber powders, hydrazine, and pharmaceutical agents.

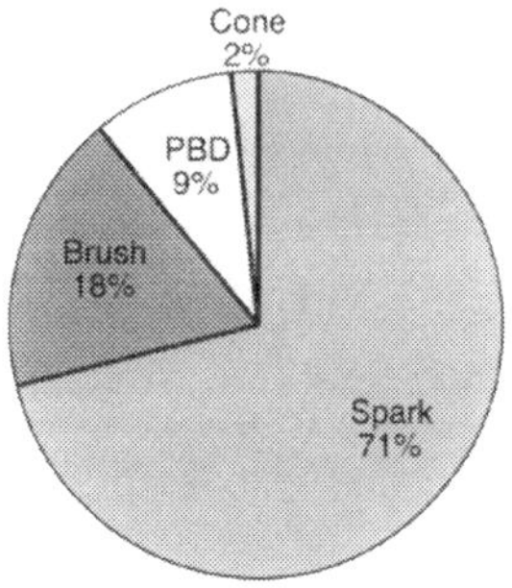

Figure 4. Types of discharges

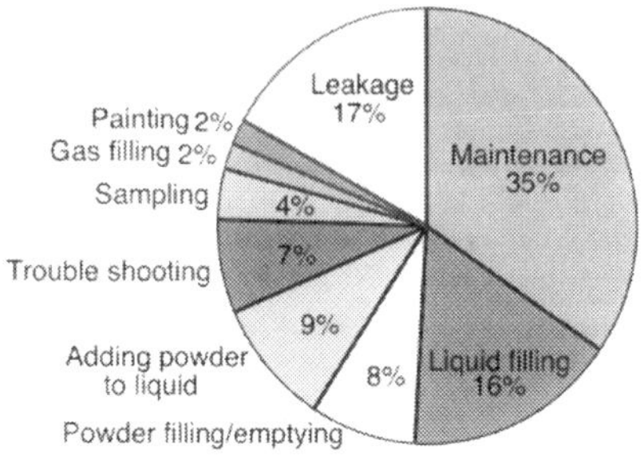

Figure 5. Operations

4. Causes of charging

According to the cases investigated, the main factors responsible for the charging are charge separation at the liquid-solid/powder interface resulting from the flowing, agitating (specially, with powder), spraying, and accidental discharging of liquids; charge separation at the solid/powder-solid interface resulting from the friction and collision caused by powders; or friction and peeling between adjacent solid materials. These findings are summarised in figure 3, which shows that the leading cause of the charging is the flowing of liquids (36.8%), followed by friction and collision mainly involving powders (26.5%), liquid discharge by leakage (20.6%), peeling (9.7%), and spraying (6.5%). In addition, the use of insulating materials in hoses/pipes and containers/bags increased the charging in many of the cases due to the loss of the charge leakage path to earth. Furthermore, these charges accumulated often caused induction in the presence of an isolated conductor, resulting in many spark ignitions.

The charging of liquids occurred during any processes involving liquid flow, such as filling, transferring liquid via pipes/hoses, draining liquid from a tank or a pump, sampling liquid from a valve, and agitation. The use of a high flow rate, splash loading, filters (in particular at the proximity of receiving containers), and two-phase liquids (liquid-immiscible liquid or liquid-solid particles mixtures) was found to accelerate the build-up of static charges.

Powder charging was caused by the charge separation by friction and collision between powder particles and a solid object during the process of transporting powder (including pneumatic transport) and filling/emptying. Other causes of charging by friction were wiping and scraping actions during cleaning.

Accidental discharges by leakage could also cause the charging of liquid, in which solid particles contained in discharged liquid accelerated the charge. Peeling was also responsible for the charging, such as pulling liner bags out of drums and transferring insulating film with rollers. In addition, charging hazards by spraying, accompanied with the processes of solvent spraying, steam spraying, spray painting, electrostatic powder coating, and tank cleaning could be identified.

5. Types and causes of electrostatic discharges

Ignitions were caused by four types of electrostatic discharge: spark, brush, propagating brush, and cone discharges. As shown in figure 4, sparks are responsible for the majority of static ignition sources. The spark discharges were caused by the presence of an isolated conductor, which accounts for 71.1% of the cases investigated. The following isolated conductors were identified: metal parts of cleaning tools, spatulas, small containers, hoppers, weighting scales, funnels, metal part of bag filters, sieves, rods, scoops, shovels, stool (resulting also in the isolation of a human body on it), plates for dividing pipelines, metal parts on insulating pipes/hoses, hose nozzles, platform trucks, a partially conductive FIBC, human bodies with insulating footwear/floor, highly conductive liquids in insulating containers, and a liquid level float (a design error). Most of the items listed above are either handled (e.g., treated with insulating gloves) or movable conductors or human bodies with insulating footwear/floor; therefore, earthing is frequently overlooked. It is noteworthy that over 70% of the accidents could have been prevented by earthing and bonding, which is the basic measure against electrostatic hazards.

In the cases investigated, brush discharges occurred when a grounded conductor approached charged insulating objects (powders, liquids, and solid items) or when insulating objects in contact with a grounded conductor separated from one another (such as liner bags in grounded metal drums). The following brush discharges could be identified: discharges between insulating hoses or funnels and a grounded conductor (a container or a finger tip) during liquid filling, a discharge at a liquid surface due to improper insertion of a grounded dip pipe, and those regarding insulating bags or bag filters for powders in hybrid flammable atmospheres.

Propagating brush discharges (PBDs) estimated as an ignition source were mostly caused by charging of insulating bags for powders, which occurred when charged powder was discharged from an insulating bag/FIBC or peeled from a bag filter. In addition, a PBD was generated in a PVC exhaust pipe used for a powder coating process. The use of insulating bags and pipes (including insulating coats) that can create a potential hazard of PBD should be avoided using appropriate bags and pipes.

Three cases of cone discharge were identified, in which ignition occurred during the pneumatic filling of polymer powder into a silo in an atmosphere of a hybrid flammable gas-dust mixture caused by insufficient gas removal from the powder.

6. Operations

This section is a description of the types of operations that were carried out when accidents happened as summarised in figure 5. In 94.1% of cases investigated, ignition occurred during operations performed directly by workers. Other cases of ignition occurred due to accidental liquid discharges, an abnormal spraying by pump cavitation, electrical isolation of a metal part of a bag filter and a liquid level float due to insufficient maintenance. In addition, it is noteworthy that over half of incidents occurred in except manufacturing operations.

Maintenance, such as cleaning, is more susceptible to accidents than normal manufacturing operations. Cleaning operations resulted in 37 accidents, which included cleaning filters, containers, pipes, paint nozzles, and metal hand tools with water or a solvent that was sometimes pressurised. In the operations, the ignition sources were mainly spark discharge due to isolated conductors. In the cleaning of paint nozzles, the bottom of a container for solvent circulation was coated by the paint, resulting in its isolation even though it was placed on an earthed metal plate. In two cases of tank cleaning, a sensitive atmosphere was created due to insufficient removal of hydrogen before cleaning and charge accumulation was enhanced by the cleaning of a Teflon-lined container, or the use of an insulating hose. Draining or sampling liquids from a valve, accompanied with the cleaning of tanks, pipes and pumps (17 cases), resulted in ignitions due to spark discharges from isolated conductors (mainly receiving containers). Regarding the maintenance of bag filters, 7 accidents occurred while removing powder. The incidents were caused by a poking of a bag filter with a metal rod in one case and by poor maintenance resulted in ungrounded filters in others. In such cases, hybrid flammable gases/vapours-dusts mixtures were often created. In addition, many cases of leakage occurred at the flanges or valves on pipes; such cases were observed during maintenance or in trial operation and start-ups following maintenance.

On liquid filling, ignitions were caused by sparks from isolated conductors including conductive liquids and human bodies or by brush discharges between an insulating material and an earthed conductor. Such ignitions mainly occurred with small containers. Other cases regarding liquid filling took place during two switch loadings of light oil to road tankers, which had previously contained gasoline, and a ship tanker loading with a defect level float.

The majority of powder-filling cases were performed by pneumatic transport. In about half of these cases, a flammable hybrid mixture was created due to insufficient degassing, in which cone discharges also caused the ignitions. Other cases included emptying powder into an FIBC to clean a silo, emptying an FIBC to fill a hopper, manual-filling of drums with a PP liner bag, and filling a movable container (electrically isolated). In the cases involving FIBCs, the ignition source of PBD was estimated. Accidents related to the PP liners occurred in a hybrid flammable vapour-dust mixture which possibly led to ignition by brush discharge, in which some accidents occurred when the liner was extracted from a drum.

In the operations of adding powder to flammable liquid, powder poured into a container with entrained air potentially forms a hybrid flammable vapour-dust mixture. Therefore, ignitions were identified to occur not only by sparks from isolated conductors (a hopper on a glass lining container and human bodies) but also as a result of brush discharges (insulating bags and insulating liners of paper bags; in one case, a liner was pulled out of a paper bag with powder into a container).

Troubleshooting was also susceptible to accidents. For example, some cases involved steam spraying against leakage, and, in another, leakage occurred as a valve was opened to release an accidental rise in pressure. In addition, liquid sampling from a manhole or a valve led to accidents, in which ignitions were caused by a spark from an isolated metal container for sampling. Gas filling operations also led to accidents, in which an LP gas leakage due to improper inspection of hose connections and a hydrogen leakage due to a high-pressure hose being severed by a truck occurred. In one case of gas lighter filling, the ignition source was a spark discharge from an isolated operator. In powder coating processes, the isolation of a metal hook resulted in the isolation of the object to be coated, which caused a spark. Painting using a spray can while wearing insulating shoes led to a spark. A metal scoop treated with insulating gloves in powder bagging led to a spark. In addition, discharges of liquids or gases with liquids by leakage caused by pipe corrosion over time, maintenance errors of flanges, pump cavitation, pressure increase due to abnormal reaction and decomposition, and valve mishandling were also susceptible to accidents, in which most ignition sources were sparks from isolated conductors.

7. Concluding remarks

More than 70% of the ignitions was caused by sparks from an isolated conductor or operator by induction; furthermore, well over 70% of the flammable atmospheres examined were formed by vapours. In addition, over 90% of all accidents occurred during work performed by operators. These findings suggest that any work performed by operators using any conductors or with insulating shoes/floor in areas where flammable liquids are handled can be extremely hazardous. However, thorough understanding and implementation of the principles of anti-static measures [4] based on proper risk assessment under organised management would have certainly prevented all these accidents from happening.

Acknowledgments

This work is a part of a study supported by Health and Labour Science Research Grants from Ministry of Health, Labour and Welfare of Japan.

References

[1] Lüttgens G, Wilson N 1997 *Electrostatic hazards* (Butterworth-Heinemann, Oxford)
[2] Pratt T H 1997 *Electrostatic ignitions of fires and explosions* (Burgoyne Incorporated Consulting Scientists & Engineers, Marietta)
[3] Center for Chemical Process Safety of the American Institute of Chemical Engineers 2005 *Guidelines for safe handling of powders and bulk solids* (AIChE, New York)
[4] For example, CLC/TR 50404 (2003), IEC 60079-32 (to be published), JNIOSH TR-42 (2007), NFPA 77 (2007)

Journal of Physics: Conference Series **301** (2011) 012034 doi:10.1088/1742-6596/301/1/012034

The effect of reduced oxygen levels on the electrostatic ignition sensitivity of dusts

Graham Ackroyd, Mike Bailey, Robert Mullins

Syngenta Process Hazards Section, Huddersfield, HD2 2GZ, UK

email address

Abstract. When handling powders with low values of minimum ignition energy (MIE), it is often necessary to employ additional protective measures such as explosion venting, suppression, containment or inerting. Inerting generally involves reducing the oxygen concentration to around 5% v/v; however, it has been shown with gases that more modest reductions in the oxygen content can still have a significant effect on the MIE. Therefore, a test program was carried out to assess the impact of reduced oxygen levels on the MIE of a series of sensitive powders. In addition, this work was also used to investigate whether testing of such sensitive materials in the standard equipment but with reduced oxygen levels could enable the prediction of MIEs <1 mJ at standard atmospheric oxygen levels.

1. Introduction

The sensitivity of flammable dust clouds to ignition by electrostatic discharges is characterised by the minimum ignition energy (MIE), commonly determined according to EN 13821 [1]. The lower the MIE, the more potential electrostatic ignition hazards that require addressing and, for powders that have very low MIEs, it may not be possible to control all potential electrostatic ignition sources reliably. For example, even if all fixed metallic parts are earthed and insulating plastics are excluded, small pieces of tramp metal (eg bolts) carried in an insulating powder may still be capable of producing spark discharges that can ignite such sensitive powders [5]. As a consequence, when handling low MIE powders, CLC/TR 50404 [2] advises that it may be necessary to implement explosion protection measures such as venting, suppression or containment. In situations where explosion protection measures are not feasible, either due to operational or financial constraints, inerting is often employed to prevent the formation of a flammable atmosphere.

According to CEN/TR 15281 [3], inerting generally requires the oxygen concentration to be reduced to <60% of the Limiting Oxygen Concentration (LOC), which in most cases results in an operating level of around 5 - 6% v/v. For some equipment and operations, it may not be possible to achieve such low oxygen levels. However, previous work [6] has shown that more modest reductions in the oxygen content can still have a significant effect on the MIE of flammable gas / air atmospheres. Assuming that a similar effect occurs with dust / air atmospheres, if the MIE can be raised above the igniting power of the largest electrostatic ignition source present, then this may present a possible basis for safe operation.

2. Initial Work

The original phase of work was designed to assess the electrostatic hazards of a particular organic powder (referred to as "Powder T") in a specific handing / processing operation. Using the commercially available Kühner MIKE3 apparatus and the standard test method [1], the MIE of the powder in air was already known to be <1 mJ. At this level, the reliable control of electrostatic ignition sources in the system was not considered possible. The powder also had an unusually low limiting oxygen concentration (LOC) [3] of 7 - 9% v/v, which meant that basing safety on inerting was also not possible due to constraints within the system. As a result, explosion protection measures had been put in place previously. However, in order to reduce the probability of the protective systems being called upon, it was decided to investigate whether a slight reduction in the oxygen concentration, or partial inerting [6], would have a beneficial effect upon the dust cloud MIE, and what oxygen level might be required.

The test program was carried out using the standard test procedure of EN 13821 [1] and the Kühner MIKE3 equipment. In order to determine the MIE an oxygen depleted atmosphere, the equipment was simply fed from a cylinder of compressed gas with the required nitrogen / oxygen composition, rather than the normal air compressor. The flushing facility around the electrodes was use to establish the required oxygen concentration in the tube before dispersion, which enabled the atmosphere to be purged without disturbing the powder in the dispersion cup. Oxygen measurements were carried out prior to each series of tests and showed that the concentration in the equipment was consistently within 0.5 % v/v of the nominal concentration of the cylinder.

The results on Powder T showed that ignitions still occurred at 1 mJ with 14% v/v oxygen in the atmosphere. Below this level the change in MIE was rather rapid, increasing to >1000 mJ (the maximum range of the equipment) with only a further 4% change in oxygen concentration. Published work with propane / oxygen / nitrogen mixtures [6] indicates that reducing the oxygen concentration from 21% to 14% v/v increases the MIE by an order of magnitude. If a similar effect were occurring with Powder T, then an MIE <1 mJ at 14% oxygen suggests an atmospheric MIE of around 0.1 mJ, which would clearly present an unusual degree of hazard.

3. Additional Work

These results highlight a limitation of the standard MIE equipment, that differentiation of MIEs <1 mJ is not possible. In the past this was not a major concern as few materials had MIEs <1 mJ, and it was generally assumed that they would only be slightly below 1 mJ. However, results <1 mJ are becoming more common, and different hazards / risks may be presented by a material with an MIE just below 1 mJ and one at 0.1 mJ. Therefore, the work was extended to see if more precise MIE values could be determined from values measured in the standard equipment with a reduced oxygen atmosphere. Additional samples of powders with MIEs <1 mJ were obtained, most being the same as tested in [5] & [9], obtained directly from the relevant workers to minimize any differences. The results of the tests are given in Table 1, together with those for Powder T.

Oxygen (% v/v)	21	18	16	14	12	10	8	6	LOC
Ti Grade E	0.012					<1	3-10	>1000	4-7%
Sulphur	0.043				10-30	>1000			9%
SIBS-K32	0.1	3-10	300-1000						16.0-16.5%
ZiH Grade S	0.13				3-10	100-300	>1000		8%
TiH Grade VM	0.19				3-10	>1000			10%
Ti Grade S	0.36						3-10		4-7%
CARO 03	0.54		100-300	300-1000					12.6-14.9%
Powder T				<1	30-100	>1000			7-9%

Table 1 - Measured MIEs with Reduced Oxygen, with Atmospheric MIEs from [5] & [9], and LOC data

The results clearly show a general trend of increasing MIE with reducing oxygen, though with some apparent inconsistencies. For example, the two Ti grades have very different MIEs at 21% oxygen, but at 8% oxygen they are the same; and Sulphur has the second lowest MIE at 21% oxygen, but increases rapidly to >1000 mJ at 10%. To determine whether differences in the LOCs might be responsible, measurements were carried out according to EN 14034-4 [4], and these did appear to account for some of the apparent anomalies. This agrees with [10], which suggested the general relationship shown in Figure 1. This is simply an exponential curve, with asymptotes at the LOC and MIE in pure oxygen. However, the precise form of the curve warrants further discussion.

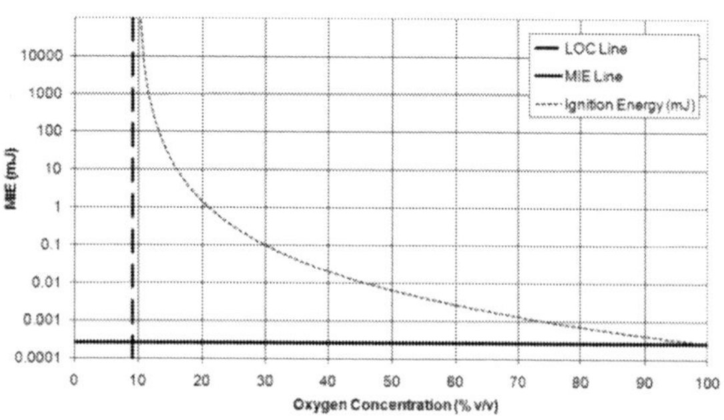

Figure 1 - General form of Relationship between Oxygen Concentration and MIE as given in [10]

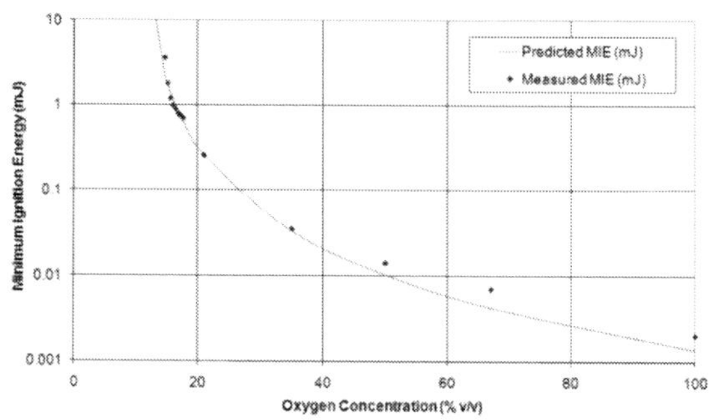

Figure 2 - MIE of Propane in Oxygen / Nitrogen Atmospheres

Journal of Physics: Conference Series **301** (2011) 012034 doi:10.1088/1742-6596/301/1/012034

4. Data Analysis

In [10], where testing was carried out on a number of powders with both reduced and elevated oxygen levels, the form of the equation was proposed as being $IE = MIE * e^{23.2-(23.2-21)*O2}$. This was reported to give a good fit at reduced oxygen levels for some of the dust examined in that paper, though not with others, and also no agreement was found with the LOC or with elevated oxygen concentrations. Some of the issues may have occurred as not all of the measurements were carried out using the standard MIE apparatus; at higher ignition energies a 20 litre sphere was used with both spark (5 to 50 J) and chemical igniters (100 J to 10 kJ), and a 1 m^3 vessel was also used with 100 J to 10 kJ chemical igniters.

In [11], where testing was carried out with hydrocarbon gases at elevated oxygen levels, a different form of equation was proposed, with the ignition energy being proportional to the oxygen concentration raised to a powder of 4 or 5 (depending whether conditions were quiescent or flowing respectively). The LOC was not used as a factor in that study, presumably as only elevated oxygen levels were of interest. However, incorporating the LOC into the equation (ie using $[O_2 - LOC]^x$ rather than $[O_2]^x$) results in this type of correlation giving a good fit to the data on propane in reduced and enriched oxygen atmospheres presented in [7] & [12]/[12] respectively, as shown in Figure 2, though with a modified exponent of 2.5. In view of these factors, the following correlation was proposed for the relationship between the MIE of the dusts tested and the oxygen concentration in the atmosphere:

$$MIE_{[O2]} = MIE_{21} * ((21 - LOC)/([O_2] - LOC))^{3.75} \qquad \{1\}$$

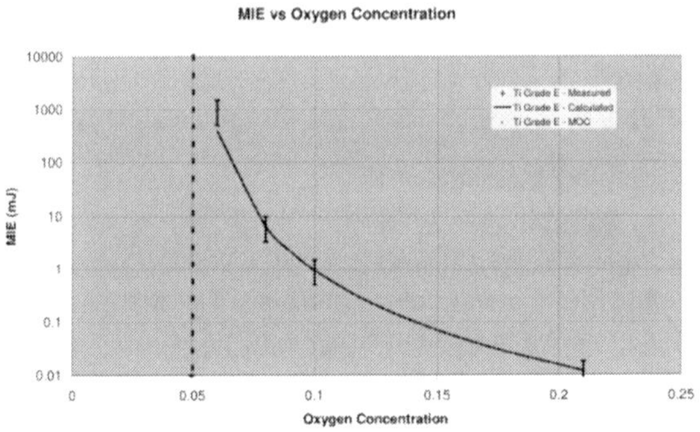

Figure 3 - MIE vs O_2 Concentration for Ti Grade E

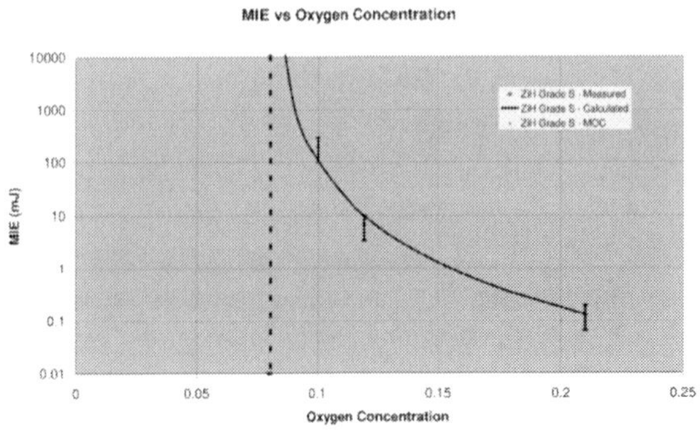

Figure 6 - MIE vs O_2 Concentration for ZiH Grade S

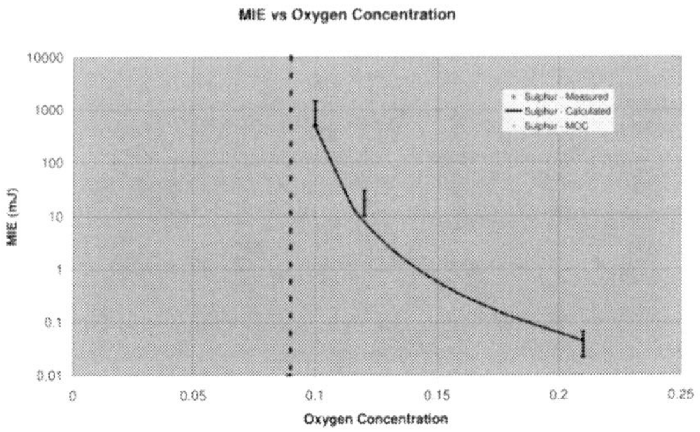

Figure 4 - MIE vs O_2 Concentration for Sulphur

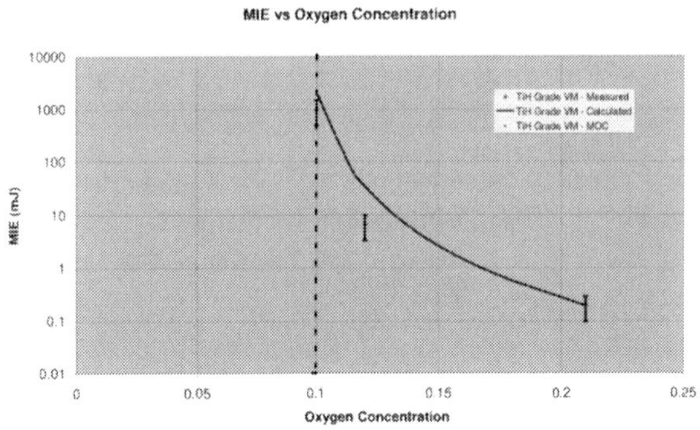

Figure 7 - MIE vs O_2 Concentration for TiH Grade VM

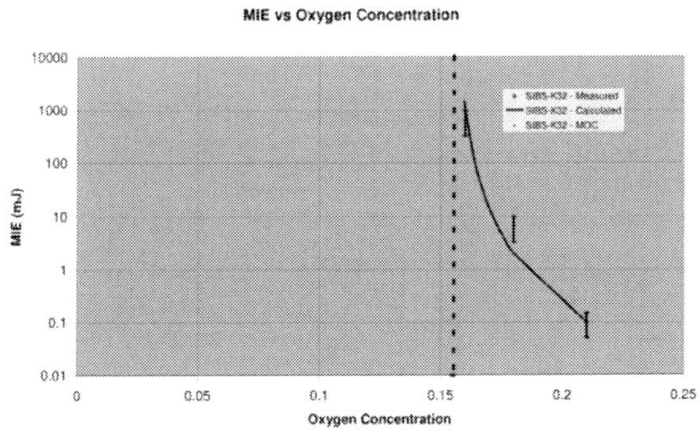

Figure 5 - MIE vs O_2 Concentration for SIBS-K32

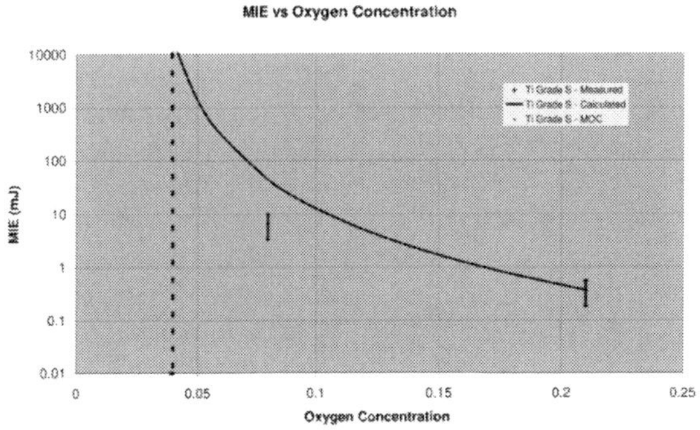

Figure 8 - MIE vs O_2 Concentration for Ti Grade S

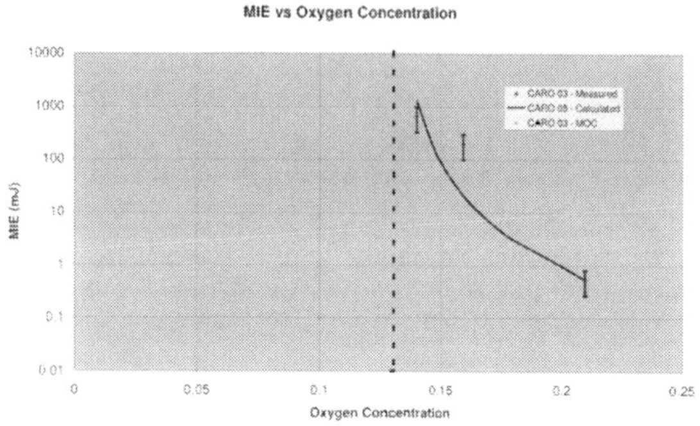

Figure 9 - MIE vs O_2 Concentration for CARO 03

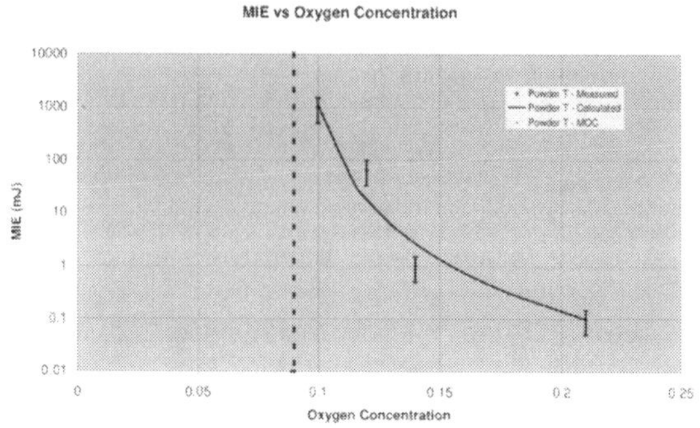

Figure 10 - MIE vs O_2 Concentration for Powder T

The experimental and predicted data are shown in Figures 3 to 10 and illustrate that, although there is some scatter, the equation does provide a good correlation with the MIE and LOC values measured. Furthermore, it does appear to confirm that the original organic powder under investigation, Powder T, may indeed have an MIE in the region of 0.1 mJ under normal atmospheric oxygen conditions, and so presents an unusual degree of hazard.

5. Conclusions

The experimental technique devised enabled the reliable determination of the MIE of dust clouds at reduced oxygen concentrations, with only minor changes to the standard method and equipment. The results of such tests may enable partial inerting to be applied with more confidence in situations where either standard inerting or control of ignition sources if not feasible or, as in the case initially investigated, to reduce the possible demands placed on explosion protection systems.

In addition, a correlation was devised to describe the relationship between the MIE of a dust cloud and the oxygen concentration of the atmosphere. This is based on an exponential relationship between two asymptotes, one at the LOC and the other at the MIE in pure oxygen. Finally, the work also confirmed that at atmospheric oxygen conditions, the organic powder originally examined may have an MIE in the region of 0.1 mJ. This clearly means that it may present increased hazards and risks compared to powders with MIEs only slightly below 1 mJ.

In view of the increased frequency at which measurements of <1 mJ are being obtained for dust clouds, the ability to obtain a more precise MIE value for sensitive powders may become increasingly important. In particular, it may be relevant when evaluating the relative merits of different bases of safety, as lower values of MIE are likely to mean an increased ignition risk, which may result in a preventative measure, such as inerting, being preferred to protective measures, such as venting or suppression.

References

[1] EN 13821 (2002): "Potentially Explosive Atmospheres - Explosion Prevention and Protection - Determination of Minimum Ignition Energy of Dust/Air Mixtures"

[2] CENELEC CLC/TR 50404 (2003): Electrostatics - Code of Practice for the Avoidance of Hazards due to Static Electricity

[3] CEN/TR 15281 (2006): Guidance on Inerting for the Prevention of Explosions

[4] EN 14034-4 (2005): Determination of the limiting oxygen concentration LOC of dust clouds

[5] Randeberg & Eckhoff, "Electrostatic Spark Ignition of Sensitive Dust Clouds of MIE <1 mJ", Journal of Loss Prevention in the Process Industries, Volume 20, Issues 4-6, Jul-Nov 2007, p396-401

[6] Eckhoff, "Partial Inerting - an Additional Degree of Freedom in Dust Explosion Protection", Journal of Loss Prevention in the Process Industries, Volume 17, 2004, p187-193

[7] Glor, "Ignition of Gas/Air Mixtures by Discharges between Electrostatically Charged Plastic Surfaces and Metallic Electrodes", Journal of Electrostatics, Volume 10, 1981, p327-332

[8] Prevention of Fires and Explosions in Dryers, J Abbott, IChemE, 1990, 2nd Edition

[9] Randeberg & Eckhoff "Measurement of Minimum Ignition Energies of Dust Clouds in the <1 mJ Region", Journal of Hazardous Materials, Volume 140, Issues 1-2, 9 February 2007, p237-244

[10] Schwenzfeuer, Glor & Gitzi, "Relation between Ignition Energy and Limiting Oxygen Concentration for Powders", 10th EFCE Symposium on Loss Prevention, Stockholm, 2001, p909

[11] Chin, "Analysis of the Effect of Oxygen Addition on Minimum Ignition Energy", Journal of Energy, Nov-Dec 1983, Volume 7, Number 6, p710-715

[12] Britton, "Avoiding Static Ignition Hazards in Chemical Operations", AIChemE / CCPS, 1999, p91

[13] Lewis & von Elbe, "Combustion, Flames and Explosion of Gases", 2nd Edition, 1961

Journal of Physics: Conference Series **301** (2011) 012035 doi:10.1088/1742-6596/301/1/012035

Simultaneous measurement of ignition energy and current signature for brush discharges

Lars Fast[1], Birgitta Andersson[2], Jeremy Smallwood[3], Paul Holdstock[4], Jaakko Paasi[5]

[1]SP Technical Research Institute of Sweden, Sweden. [2] Örebro University, Sweden. [3]Electrostatic Solutions Ltd, UK. [4]Holdstock Technical Services, UK. [5] VTT Technical Research Centre of Finland, Finland.

jeremys@static-sol.com

Abstract. Accurate prediction of the probability of ignition arising from charged insulators is a crucial element of risk assessment in process industry. Incendiary brush discharges can occur when a large or grounded conductor approaches a charged insulator in the presence of a flammable atmosphere. This paper describes ignition tests based on an IEC standard method and simultaneously recorded temporal distribution of current released in the discharges, using a discharge probe integrated with the ignition probe. Ignition and non-ignition results are compared with peak discharge current and charge transferred in the discharge. No clear ignition threshold was found for either of these parameters. No major differences were found between igniting and non-igniting waveforms.

1. Introduction

Accurate prediction of the probability of ignition arising from charged insulators is a crucial element of risk assessment in process industry. Incendiary brush discharges can occur when a large or grounded conductor approaches a charged insulator in the presence of a flammable atmosphere and have long been studied by various authors, for example [1], [2], [3], [4]. It is not possible to calculate the brush discharge energy in the way that can be done for sparks. It is customary to evaluate brush discharges in terms of an "equivalent energy" given by the Minimum Ignition Energy (MIE) of the flammable atmosphere which they will ignite. It is known that for brush discharges it is not just the total released energy that is of importance in determining whether or not an ignition will occur. Even if the total energy released in the discharge exceeds the MIE, there may not be an ignition [4]. There has been much discussion about the parameters that are important, such as polarity of the source, peak discharge current and waveform, and charge transferred in the discharge. The latter has been used in standards for evaluation of ignition risk from ESD from insulating surfaces, with limits of 60nC for Group I or IIA, 30nC for Group IIB and 10nC for Group IIC apparatus [5], [6]. A common method for measurement of charge transferred in the discharge is to collect it in a capacitor and measure the voltage produced. This method, however, does not give the ESD waveform and fails when an ignition occurs, as the flame plasma provides a conduction path for the charge to ground [3].

The probe design and electrode shapes and sizes, and the gap across which the discharge occurs, also have an effect [7]. For these reasons, in recent research the current waveform of the discharge has often been measured, and von Pidoll et al. have investigated the use of the magnitude of the charge transferred in the discharge for incendivity evaluation [8]. We are not aware, however, of any published measurements of the electrostatic discharge current waveform simultaneously with an ignition test.

In this paper we describe ignition tests based on an IEC standard method [9] and simultaneously recorded temporal distribution of current released in the discharges, measured using a discharge probe integrated with the ignition probe. While it is known that shielded probe designs capture the discharge current more efficiently [10], we have chosen to use an unshielded design as this enabled use of a standard probe and is most commonly used by other workers.

2. Methodology

The ignition tests with simultaneous discharge current measurements were performed with a modified ignition probe designed according to IEC 61340-4-4 [9]. The probe design allowed an ethylene - air mixture to flow through the probe inside a containing wall, past a grounded spherical electrode. The gas mixture used was as specified in [9] with volume concentration of 5.4% ethylene and MIE of 0.14 mJ. When a discharge occurs to the electrode at the centre of the gas flow, and its energy content is high enough, an ignition of the gas mixture can occur.

With the standard probe it was not possible to measure the electrostatic discharge current occurring to the grounded probe at the time of the ignition. In the modified ignition probe used in this work, the ground connection to the probe was cut. A resistance consisting of 8 parallel 6.81Ω resistors was inserted in-between the discharge electrode and ground giving a resistance to ground of 0.85Ω in series with the ESD current, in a manner that minimized the high frequency impedance in order to maintain the waveform shape.

An oscilloscope used to detect the discharge current was connected to the ignition probe by a 50Ω coaxial cable in parallel with the resistors. As the ESD current flowed through the parallel resistors to ground the voltage developed across these resistors was recorded by the oscilloscope. Waveform parameters such as peak current could be directly measured from the digitized current waveform. Charge transferred in the discharge was calculated by integration of the discharge current over the full duration of the waveform, i.e. to the point at which the current flow became insignificant.

The test object was a highly insulating polypropylene tray, charged by rubbing with a polyester fabric. The charging level of the test object was not controlled or measured during the tests, but all charging of the test object and tests were done by the same person and in a similar way.

Each ignition test was performed in a manner similar to [9] except that the ignition probe was kept in a fixed position during the entire experiment, while the charged brush discharge source (a charged plastic tray) was moved towards the probe.

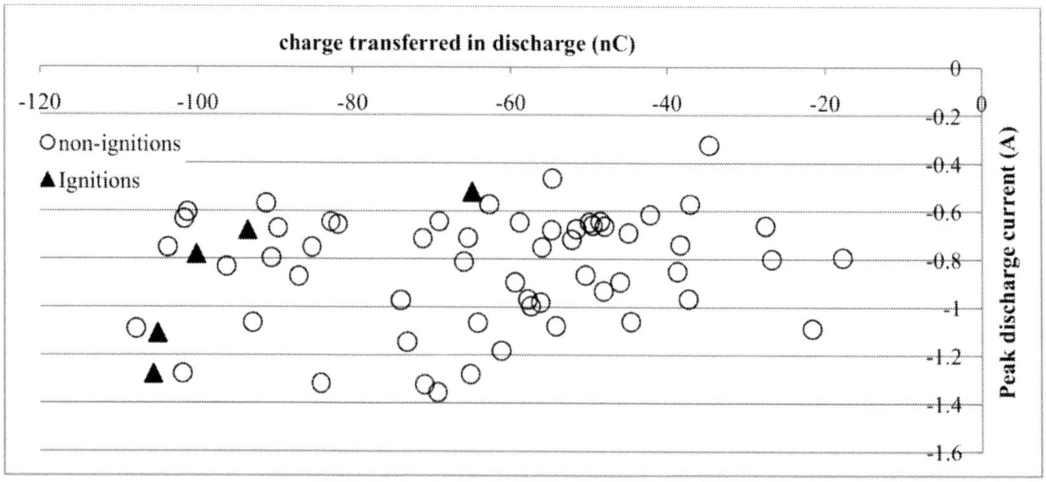

Figure 1: Peak current as function of the total charge transfer for brush discharge ignition trials.

3. Results and discussion

Figure 1 shows the peak current and charge transferred in all our igniting and non-igniting test results. The charge transfer given is that of the first single brush discharge occurring. If more than one brush discharge occurred in the test, those after the first were not recorded. Ignitions are represented as solid triangles on the graph. Non-ignitions are shown as hollow circles.

Peak discharge current varied between about -0.3A and -1.4 A, and charge transferred in the discharge varied from about -20nC to -110 nC. There did not seem to be any clear relationship between peak current and charge transferred in a discharge.

There were no ignitions below -60nC charge transferred. This result is in line with generally accepted charge transfer threshold values for discharge incendivity. Most ignitions occurred between about -90nC and -110 nC. However there were only 5 ignitions in total out of 62 tests. It may be necessary to do many more trials before any threshold for ignition could be reliably established. The situation is similar with regard to ignitions and peak current flowing in the discharge, with ignitions occurring between about 0.5 A and 1.3A peak current.

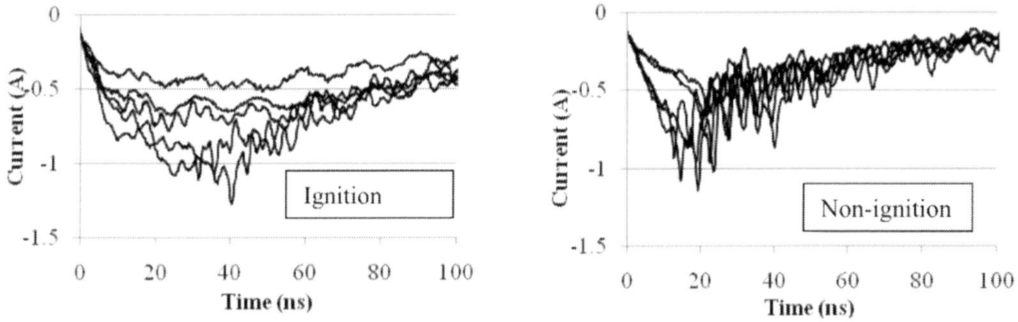

Figure 2. Five igniting waveforms (left) compared with five non-ignition waveforms (right)

Typical waveforms were unidirectional and very similar in shape with duration around 70-100ns and risetimes around 10-40 ns (Figure 2). There is some indication that there may be small differences between igniting and non-igniting waveforms, with some of the latter having shorter risetime and duration and a slightly less rounded shape. Further work with a greater number of tests will be required to clarify these apparent differences have significance.

No clear threshold between igniting and non-igniting regimes was found for charge transfer or for peak discharge current. Only a small number of ignitions were obtained (5 in 62 tests). This is consistent with ignition by electrostatic brush discharges having a large probabilistic element and that a large number of tests would be required to adequately demonstrate an ignition threshold, or determine factors leading to ignition with reasonable certainty. The 61340-4-4 ignition test method requires no ignitions to be found with at least 200 tests, presumably with this in mind. Gibson and Harper [3] proposed that a probability of less than 1 in 1000 discharges under laboratory test conditions may be acceptably safe for use in most industrial situations. They commented that in specialised conditions such as low humidity, continuous charge generation and flammable atmosphere continuously present, this may not be valid. They suggested that using a normal distribution function, 0.001 probability could be predicted by extrapolating from data from 0.01 probability and above. Furthermore, data produced using clean dry material surfaces may not be representative where surface contamination or other factors may change the characteristics and incendivity of discharges from the surface. Buhler et. al. [11] have found that with some materials ignitions occurred under 50% humidity conditions whereas at 20% humidity there were none.

It should also be noted that in these experiments, if more than one brush discharge occurred, only the first would be recorded. If further discharges occurred within a short time period they could, however, be expected to contribute to an ignition.

These considerations make the study of simultaneous measurement of electrostatic discharge waveforms and ignition tests of great interest and it is hoped that further work will explore some of these issues in more detail.

4. Conclusions

Simultaneous ignition tests with measurement of the electrostatic brush discharge waveforms have been demonstrated. Igniting and non-igniting discharge waveforms were found to be very similar with unidirectional shape, risetimes around 10-40ns and duration around 70-100ns. Small differences between igniting and non-igniting waveforms have been noticed but their significance is undetermined.

No clear threshold between igniting and non-igniting regimes was found for charge transfer or for peak discharge current. Only a small number of ignitions were obtained (5 in 62 tests). This is consistent with ignition by electrostatic brush discharges having a large probabilistic element and that a large number of tests would be required to adequately demonstrate an ignition threshold, or determine factors leading to ignition with reasonable certainty.

Acknowledgement
The authors would like to thank Dr John Chubb for some useful comments on this paper.

References

[1] Gibson N and Lloyd FC. 1965 Incendivity of discharges from electrostatically charged plastics. *Brit. J. App. Phys.* V16
[2] Gibson J and Harper D J. 1981Evaluation of electrostatic hazards associated with non-conducting materials. *J. Electrostat.* 11 27-41
[3] Gibson J and Harper D J. 1988 Parameters for assessing the electrostatic risk from non-conductors – a discussion. *J. Electrostat.* 21 27-36
[4] Glor M and Schwenzfeuer K, 2005 Direct ignition tests with brush discharges, *J. Electrostat.* 63 463-468
[5] British Standards Institute. Electrical equipment for explosive atmospheres – Part 0: General requirements. BS EN 60079-0:2004
[6] British Standards Institute. Non-electrical equipment for potentially explosive atmospheres – Part 1: Basic method and requirements. BS EN 13463-1:2001
[7] Salmela H, Paasi J, Kalliohaka T and Fast L, 2005Measurement of air discharge from insulating electrostatic dissipative and conducting materials with different ESD probes *J. Electrostat.* 63 539
[8] von Pidoll U, Brzostek E and Froechtenigt H-R. 2004 Determining the incendivity of electrostatic discharges without explosive gas mixtures. *IEEE Trans. Ind. App.* V40, 6 1467-75
[9] IEC 61340-4-4 Standard test methods for specific applications - Electrostatic classification of flexible intermediate bulk containers (FIBC)
[10] Chubb J N. 2006. Measurement of charge transfer in electrostatic discharges. *J. Electrostat.* 64 (5) 301-5
[11] Buhler C, Calle C, Clements S, Ritz M and Starnes J. 2006. Test methodology to evaluate the safety of materials using spark incendivity. *J. Electrostat.* 64 744-51

Comparison of current signatures for brush discharges using different resistance values in the discharge probes

Lars Fast[1], Birgitta Andersson[2], Jeremy Smallwood[3], Paul Holdstock[4], Jaakko Paasi[5]

[1]SP Technical Research Institute of Sweden, Sweden, [2] Örebro University, Sweden, [3]Electrostatic Solutions Ltd, UK, [4]Holdstock Technical Services, UK, [5] VTT Technical Research Centre of Finland, Finland

E-mail: lars.fast@sp.se

Abstract. Incendiary brush discharges can occur when a large or grounded conductor approaches a charged insulator in the presence of flammable atmosphere. The probability of ignition of these discharges is essential to risk assessment in process industry. It is known that even if the total energy released in the discharge exceeds the minimum ignition energy (MIE), there may not be an ignition [1]. In a companion paper in this conference, we have reported simultaneous measurements of ignition and discharge current waveforms for brush discharges in an ethylene-air mixture in ignition tests based on an IEC standard test method [2]. In this paper we show that the resistance of the electrostatic discharge measurement system can have an effect on the peak discharge current signatures and charge transferred in the brush discharge from an insulating surface. The resistance of the discharge probe seems to affect the peak current value, but also to lesser extent the amount of charge transferred in the discharge.

1. Introduction
A brush discharge occurring in the presence of flammable atmosphere may or may not be incendiary, even if the total energy release in the discharge exceeds the minimum ignitions energy (MIE) there might not be an ignition [1]. A risk assessment that reflects the actual risk of ignition is therefore of interest for the whole community. An electrostatic discharge (ESD) is initiated when the local electric field exceeds the electrical breakdown strength of the surrounding atmosphere. This is affected by various factors like amount and polarity of charge on the charged object, electrical properties and geometry of the discharging object, and air gap length of the discharge channel [3]. In this paper we will present some results from ignition tests done with a modified ignition probe, where the current signature has been recorded for two different resistance values. In addition we have made some discharge measurements in a controlled environment using a stripped down discharge probe with similar properties as the modified ignition probe, but without the flowing gas apparatus. The recorded discharges are taken from an insulating surface charged by triboelectrification.

2. Experimental setup
The experiments were performed at SP (address 1) at three different occasions. The first two of the experiments were done using the modified ignition probe in a heated outdoor facility designed for tests including flammable atmospheres. The relative humidity was around 60%, the temperature 20° C and

a resistance of 7 Ω in the ignition probe was used for the first experiment. During the second experiment the relative humidity (RH) was around 30% and the resistance of the ignition probe was 1 Ω. At the last experiment the discharges were made in 12 % RH and 23° C at a SP indoor facility with an altered ignition probe.

The discharge probe used was made according to the details presented reference [2]. Essentially it consists of a spherical ball electrode made of stainless steel connected to ground through a wire. The gas mixture is fed through a plastic container filled with glass beads, which are prevented from falling out by a metallic mesh. After the gas mixture has passed the metallic mesh it is confined within a plastic cone with an opening at its top where the gas passes the ball electrode and out in the air. This implies that the gas mixture is well defined is a small volume around the ball electrode under the assumption that no other objects are present. A schematic picture of the experimental set up is indicated in figure 1. For details about design and test procedure see reference [2].

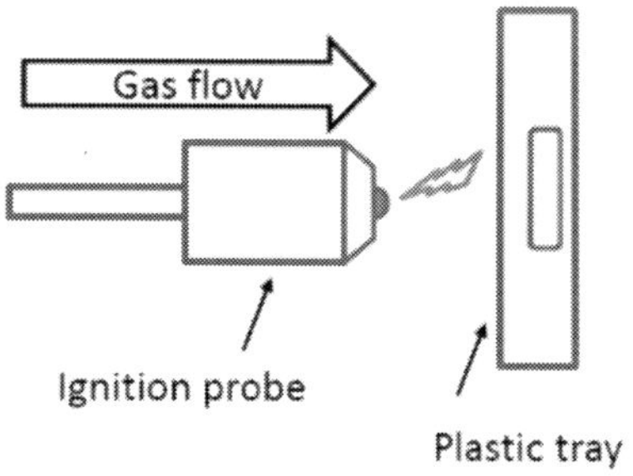

Figure 1. An overview of the experimental setup including the ignition probe and the charged object (plastic tray). The gas flow through the probe is indicated.

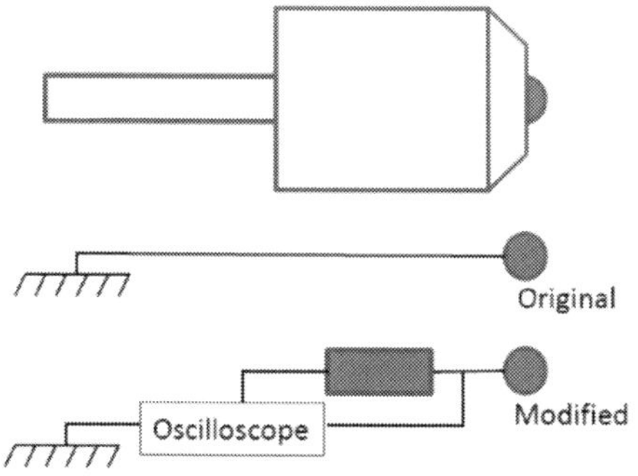

Figure 2. Overview of the original probe (Original) and the modified probe (Modified). The oscilloscope is connected to the probe with a coaxial cable.

The plastic tray used in these experiments had the following measures 40 cm width, 60 cm length and 12 cm depth and was made of polypropylene. The back of the tray was charged using a homemade woolen mitten. The potential of the charge surface was estimated with a field meter to be around 25 kV. The distance between the ball electrode and the charges surface was in between 2 cm and 7 cm. It was the plastic tray that was moved not the ignition probe, since it was easier to handle the tray than the probe.

In figure 2 we show how the ignition probe was modified. The ground connector had a series resistance connected in between the ball electrode and ground. The oscilloscope was connected to the ignition probe via a 50 Ω coaxial cable and the 50 Ω input of the oscilloscope was used. The series resistance is obtained by connecting 8 resistors in parallel to distribute the current evenly to the coaxial cables ground shell. The charge transfer was calculated by integration of the digitized waveform over the elapsed discharge time.

Additional experiments were made in a controlled atmosphere with a stripped probe without the gas mixture apparatus present. This was done by removing everything except the ball electrode and the resistance from the ignition probe. The measurements with the stripped probe were done in the same way as the ones with the modified ignition probe, however the plastic tray was charged with a polyester cloth instead of a woolen mitten. This may affect the state of charging of the ESD source. One should take care in interpreting the amounts of charge transferred to the probe, in these two sets of experiments.

3. Results

The peak discharge currents and the amount of charge transferred in the brush discharges related to the probe resistance are presented here in table 1 and figure 3 and 4.

Table 1. Average Peak Current and Transferred Charge (including Standard Deviation)

Experiment	Temperature / RH	Resistance	Average Peak Current	Average Transferred Charge
1	20 °C / 60 %	7 Ω	-(0.37±0.18) A	-(19.0±8.7) nC
2	20 °C / 30%	1 Ω	-(0.84±0.24) A	-(63±22) nC
3	23 °C / 12%	9 Ω	-(0.51±0.09) A	-(39±9) nC
4	23 °C / 12%	1 Ω	-(1.9±0.3) A	-(24±5) nC

In figure 3 and 4 we present, both current (full line) and charge (dashed line) as function of time, for a two different resistances values 1 Ω (figure 3) and 9 Ω (figure 4) at 12 % RH. The charge as function of time is monotonic in the case of the 9 Ω resistance value, but not in the case of the 1 Ω resistance. For this reason we defined a new parameter, the peak transferred charge, as is exemplified in figure 3. The peak transferred charge was -(33±6) nC in the fourth experiment where the average transferred charge was -(24±5) nC.

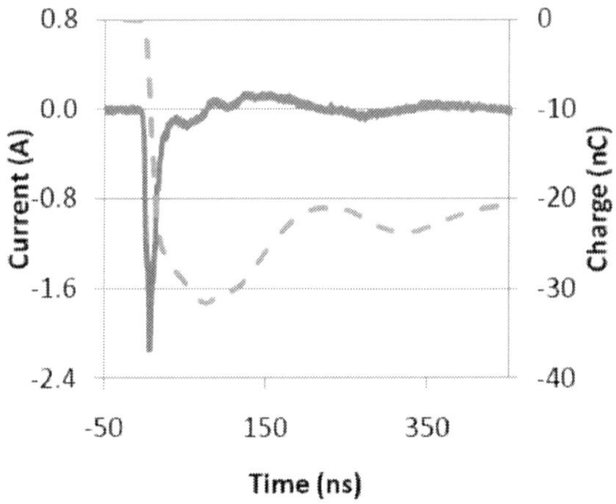

Figure 3. The discharge current (dark) and the transferred charge (light) as function of time for 1 Ω resistance and 12 % RH

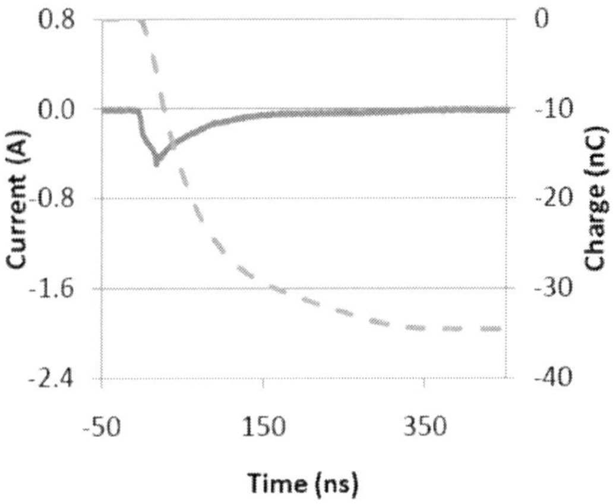

Figure 4. The discharge current (dark) and the transferred charge (light) as function of time for 9 Ω resistance and 12 % RH

In figure 5 and 6 we present the peak current as function of the transferred charge. In figure 5 we have discharges obtained with the modified ignition probe; triangles (1 Ω 30% RH) and dashes (7 Ω 60% RH). In figure 6 we have discharges taken with the stripped probe; diamonds (peak transferred charge) and crosses (transferred charge) (1 Ω and 12% RH) and circles (9 Ω and 12% RH). In figure 5 and 6 we also present the 1σ standard deviation from the average value of both the peak current and the peak transferred charge as ellipsoids, however any correlation between the variables is neglected.

In figure 5 we can clearly see that the charging level is different for the two data sets where the charging mechanism was the same (wool), but the humidity and probe resistance different (30%RH and 1 Ω triangles; 60% RH and 9 Ω dashes). It is possible that this difference depends on humidity [4] so we performed the last two sets of discharge experiments under constant humidity and temperature conditions, see figure 6. This implies that the peak current is strongly affected when the probe resistance goes from around 1 Ω to around 8 Ω.

If graphs presenting the discharge currents as function of time for many waveforms are viewed, it is very difficult to distinguish other characteristics than those already mentioned [5]. The discharge duration time is around 100 ns and the rise time around 10 ns.

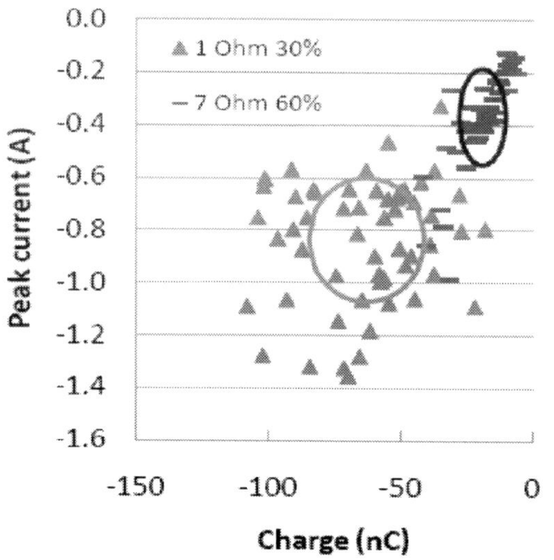

Figure 4. Peak current as function of transferred charge. The ellipsoids indicate the 1σ standard deviation from the mean value.

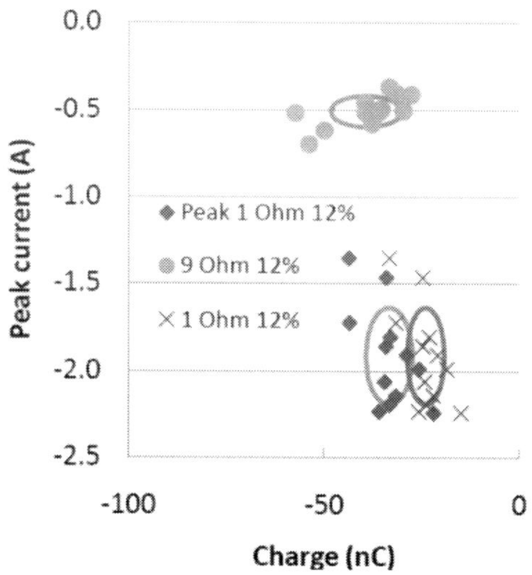

Figure 5. Peak current as function of transferred charge. The ellipsoids indicate the 1σ standard deviation from the mean value.

4. Summary

Peak current was reduced when a higher resistance probe was used. The charging level of the ESD source may have been affected by humidity and the material of the rubbing cloth used. We believe this may have affected the transferred charge and transferred peak charge. Where the peak current was determined at constant humidity with two different resistance values (1 Ω and 9 Ω) their ratio was around 3.7. However the amount of charge transferred in the discharge doesn't seem to differ that much when comparing the different resistance values of the discharge probe, when environmental and charging conditions are similar. Further investigations are needed to establish the relation between the peak current and the transferred charge for while fixing all known influential parameters.

These experiments demonstrate that the ESD source and measurement systems interact and it is not possible to separate them from the ignition process. In ESD ignition testing we must consider the whole system, not just the ESD and gas mixture. ESD ignition is a product of the gas mixture, ESD source, the test probe and the experimental situation.

References

[1] M.Glor, K.Schwenzfeuer, Direct ignition tests with brush discharges, J. Electrostat. 63 (2005) 463-468
[2] IEC 61340-4-4 Standard test methods for specific applications - Electrostatic classification of flexible intermediate bulk containers (FIBC)
[3] J.L. Davidson, T.J. Williams, A.G. Bailey, G.L. Hearn, Characterisation of electrostatic discharges from insulating surfaces. J. Electrostat. 51–52 (2001) 374–380
[4] C. Buhler, C. Calle, S. Clements, M. Ritz, J. Starnes, Test methodology to evaluate the safety of materials using spark incendivity. J. Electrostat. 64 (2006) 744–751
[5] H. Salmela, J. Paasi, T. Kalliohaka, L. Fast, Measurements of air discharges from insulating, electrostatic dissipative and conductive materials with different ESD probes. J. Electrostat. 63 (2005) 539–544

Complex system for risk assessment in ESD hazardous processes with unusually high risks

S V Szabo[1], I Kiss, B Németh, I Berta
Budapest University of Technology and Economics, Dept. of Power Engineering
18 Egry J. u., 1111 Budapest, Hungary

E-mail: szabo.sandor@vet.bme.hu

Abstract. Many types of granulated substances and liquids that are heavily used in industry may become easily inflammable when they are adequately mixed with air either in fine particle or vapour form. Many of these substances have very low conductivity which means that the particles can easily acquire electrostatic charge during transport or mixing operations, and the resulting electrostatic discharge may ignite the atmosphere leading to damage and even loss of life. The probabilities with which these events occur are hard to quantify. The authors present a case study and estimate the probability of ignition using FFTA (fuzzy fault tree analysis.) The result of the calculations is affected by several parameters, such as process temperature and humidity, which using suitable instruments can be easily observed. The authors thus propose a **new approach where these parameters can be continuously monitored, and the fault tree evaluated in real time** and early warning of the danger can be provided.

1. Introduction

In many industrial processes the management of electrostatic charge accumulation and discharge are very important. These include processes in which inflammable atmospheres are created when a low conductivity fluid vapour or fine particles of granular substance are created, as these can be easily ignited by the energy of the discharge, resulting in fire or explosion, possibly causing damage to apparatus and sometimes even resulting in loss of life. Such applications include the transportation and mixing of certain liquids and powders highly susceptible to ignition like flour, pharmaceutical substances or gasoline. [1] [2]

It is clear from the above that it is vitally important that the risk due to electrostatic discharges is properly managed, which in turn relies on the ability to accurately estimate risk. [3]

Charge accumulation can be modelled by the simple parallel circuit of a capacitor and a resistor being charged by a current source. Such a circuit is shown in figure 1. The capacitor in the circuit represents the capacity inherent in the system geometry, the resistance provides the relaxation if the system including both inherent and artificial grounding, while the current source provides the current caused by the separation of charges. The spark gap illustrates that once a sufficiently high voltage is accumulated, a discharge shall occur.

When designing solutions for suppressing discharges, usually lowering the grounding resistance is the method of choice.

---

[1] To whom any correspondence should be addressed.

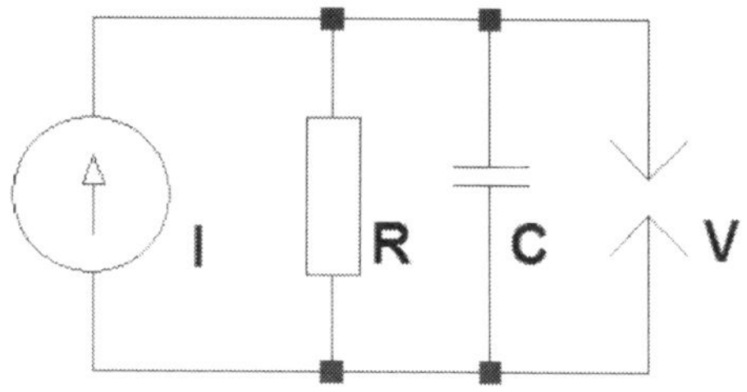

Figure 1. The simple circuit model for studying charge accumulation.

2. Fuzzy fault tree analysis

One of the methods widely employed in the industry for estimating the probability of damage due to electrostatic discharges is the fault tree, which relies on decomposing the top event, the probability of which we seek to determine, into certain simpler events, the probabilities of which are better known or are easier to estimate. [4]

In case of fuzzy fault tree analysis (FFTA), the probabilities of basic events are not simply numbers, but instead they are membership functions. This allows us to incorporate the uncertainty of estimates into the calculations. The resulting probability for the top event is in this case also a fuzzy membership function, which allows us to estimate the error of the result, or inversely, to determine which base probabilities need to be better estimated to decrease the uncertainty of the estimate.

In our case study we shall use triangular membership functions, of which an illustration is provided in figure 2. Throughout the calculation we shall assume that the product of two such numbers has the same form with the minimum, peak and maximum points being the respective products of the operands.

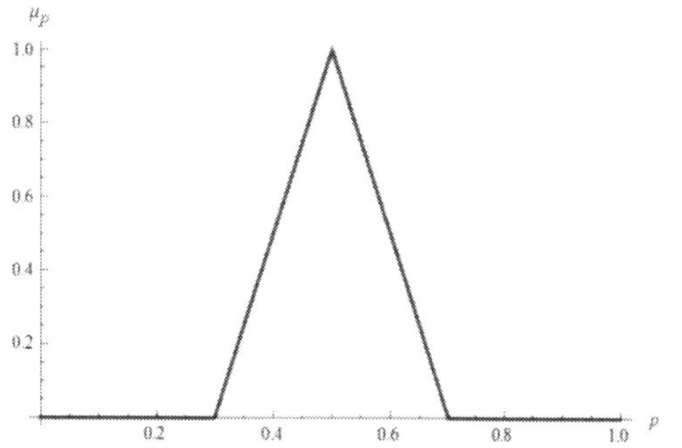

Figure 2. Triangular membership functions are used in the computation.

3. Case study

In the article we shall examine a prototypical process in detail. In this process a powdered substance is pneumatically transported into a container for further processing. A discharge may arise between the destination container and the end of the tube used for filling it. The container, the tube used for transportation and the destination container are all earthed. For the purposes of this study, it will be assumed that the destination container is permanently well grounded, however the source container, as it is mobile, shall be grounded by workers each time it is replaced and therefore we have to account for failures in this grounding path, which has four connections, each of which have to be intact in order to dissipate the accumulating charge. For additional safety, the worker handling the tube wears conductive clothing, socks and shoes and stands on a conductive floor.

It is clear, that while the grounding of the tube is intact, no risk of discharge is present, however should this fail, the accumulated charge has to be conducted to earth via the clothing of the worker and through the floor. This grounding path may fail due to a number of reasons, including low humidity in the plant, that causes an increased resistance on the interfaces between the tube and the clothing of the

Journal of Physics: Conference Series **301** (2011) 012037 doi:10.1088/1742-6596/301/1/012037

worker, or the conductive shoe and floor; low conductivity of the transported medium and in connection to this, the low temperature of the container located outside also increase the risk of discharge, and high process temperatures provide a powder that is more easily ignited. The fault tree for the process is presented in figure 3. For this study, the probability of the presence of an inflammable atmosphere is assumed 1, and the probability of ground failures is a triangular membership function with a minimum at 1%, the peak at 2% and the maximum at 3%.

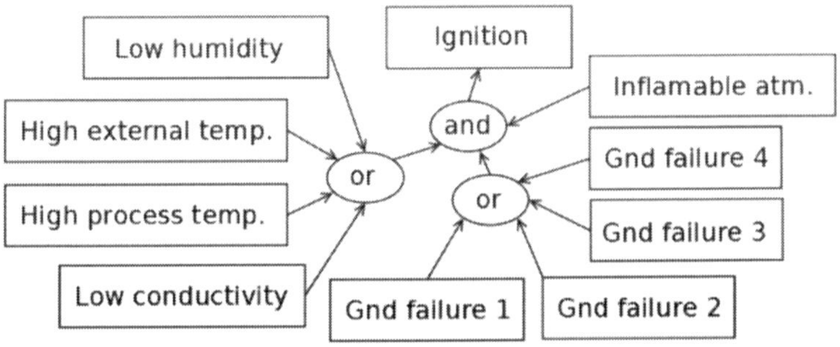

Figure 3. The fault tree for the process.

4. On-line monitoring

The parameters that the probability of the top event depends on can be easily measured and monitored, which allows us to evaluate the fault tree at each time instant. Let us assume that the parameters vary according to the diagrams presented in figure 4.

For evaluating the fault tree, we obtain parameter values for each instant by using an average of past values with exponentially decreasing weights, with 0.5 weight assigned to the values measured 15 minutes ago. Likewise, the standard deviation of the values is computed, and for the resulting probabilities a triangular membership function is used extending one standard deviation is each direction from the average. The computation results in the top event probabilities shown in figure 5.

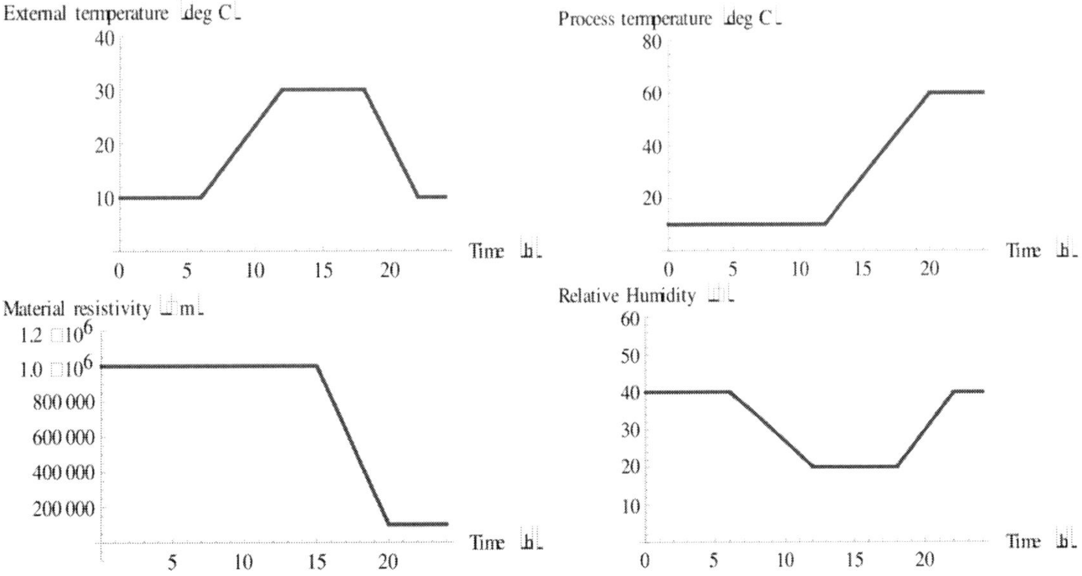

Figure 4. Variation of process parameters.

For comparison, a computation using the averages and standard deviations for the entire period was carried out, which results in an estimate of (0.01, 0.06, 0.112). It can be readily seen that the estimate using the global statistics is in good agreement with the results obtained by using time-dependent data, however the uncertainty of the former is much higher.

The time-dependent analysis can be used to observe at which stage of the process the significant risk arises, and this can aid in providing optimal measures for the reduction of the risk associated with the process.

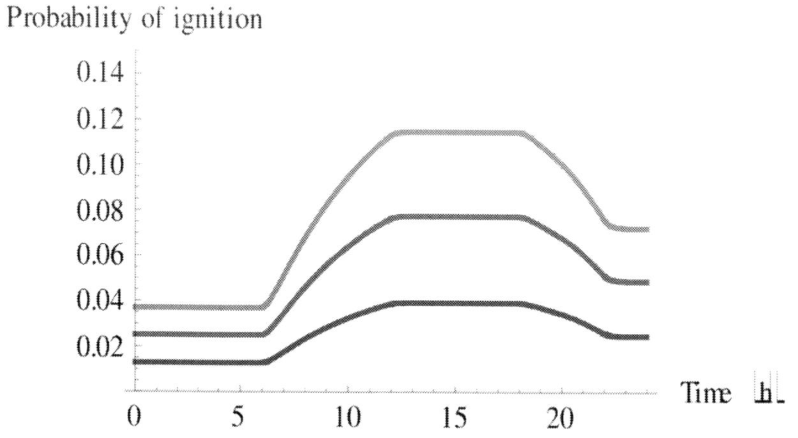

Figure 5. Probability of the top event. Minimum, peak and maximum of membership function shown.

5. Conclusion

The authors propose and demonstrate that when the input parameters of fuzzy fault trees are in close relationship with physical parameters whose time dependent variations are easily measured, additional information can be obtained by evaluating the fault tree in continuous time. This may be done based on logged data, in order to optimize the process in a targeted manner, focusing on operations where the probability of damage is highest; and possibly also in real time, to alert operating personnel to parameter constellations that carry unusually high risks.

6. References

[1] Berta I 2005 Static Control *Journal of Electrostatics*. 67:(2-3) pp. 297-300

[2] M. Glor 2007 Ignition risks from static electricity – problem solved? *Journal of Physics: Conference Series* 142 pp. 1-6

[3] Kiss I 2009 Increasing the reliability of risk assessment for systems endangered by electrostatic discharges. *Journal of Electrostatics* 67:(2-3) pp. 297-300

[4] Kiss I, Szedenik N, Berta I 2009 Electrostatic hazard and protection: expert system for fuel delivery modules *Journal of Electrostatics* 63:(6-10) pp. 495-499

Electrostatic hazards with underground plastic pipes at petrol stations

H. L. Walmsley

Harold Walmsley Electrostatics Ltd, 31 Fairways, Frodsham, Nr Warrington, UK

haroldw@nildram.co.uk

Abstract. The paper analyses some ignition incidents that have been reported with insulating, (non-conductive) underground plastic pipes in retail petrol stations. The occurrence of the incidents is compared with voltage measurements, observations of the typical spread of streaming currents recorded in gasoline handling and theoretical estimates of the voltages on the pipes. The comparisons suggest that neither incendive sparks from unbonded conductors nor incendive brush discharges from insulating pipe surfaces can be ruled out although both are expected to be rare. The hazards can be prevented by using pipes with earthed conductive or dissipative inner linings.

1. Introduction

Fuel arrives at petrol stations by truck and is transferred through buried pipes to underground storage tanks. From the tanks it is transferred through more buried pipes to the familiar "petrol pump" dispensers that fill customer vehicles. The buried pipes are increasingly being made of plastic (HDPE) to avoid corrosion. The plastic pipes may be either completely insulating ("non-conductive" type) or they may have a dissipative inner lining ("conductive" type). Connections to the pipes are made either above ground or in chambers ("fill boxes") or sumps let into the ground. With the non-conductive type of pipe there has been a small number of electrostatic ignitions (flash fires) either inside connection chambers or inside the pipes themselves.

This paper lists some reported electrostatic ignition incidents with non-conductive pipes and evaluates the possible ignition mechanisms by considering laboratory measurements of pipe voltage [1,2], observations of the typical spread of charging currents in hydrocarbon distribution systems [3] and new theoretical estimates of the voltages that result from liquid flow.

2. Incidents involving non-conductive plastic pipe systems at filling stations

Although a risk of escalation or injury is clearly present, the incidents have not usually been serious. Therefore, reporting may be somewhat haphazard and the list below is unlikely to be comprehensive. Nevertheless, there are relatively few incidents and, given the large number of vehicle fills that occurs globally at petrol stations, it is clear that the incident frequency is very low:

1) Hungary, 1998-2000: At least 5 fires in fill boxes at 4 sites. Unbonded conductors and exposed external pipe surfaces in the fill boxes.

2) UK, 2001: One minor ignition in a fill box when replacing the metal cap on an unbonded end-fitting after filling a tank. Unbonded end fittings and exposed external pipe surface in the fill box.

3) Sweden, 2002: 4 ignitions inside pipes immediately after disconnecting filling hoses. Pipes completely buried with no exposed external surfaces and no unbonded conductors. Discharges thought to have occurred from the inner surface of pipe.

4) Hungary 2005/2006: 6 fires in fill boxes. Unbonded welding sockets and exposed external pipe surfaces in the fill boxes. Flame arrestors on pipe ends. Discharges from isolated welding sockets suspected. Flame arrestors may have increased charging levels.

5) Australia, 2007: One fire when a newly fitted pipe was used before it was buried (backfilled).

6) Ireland, 2009: One fire, repeated audible and occasional visible discharges in a fill box. Unbonded and unplugged welding sockets and exposed external pipe surface in the fill box. Audible discharges reported to continue despite earthing all conductors although, subsequently, welding sockets neither capped nor earthed. Discharges eliminated by changing to conductive pipes.

3. Voltage measurements in non-conductive plastic pipe systems

Hearn [1] measured the voltages created by a 50:50 toluene/isooctane mixture in a laboratory mock-up of the piping that feeds the dispensers from the tanks. Some pipe sections were covered in conductive foil to represent burial whilst others were left exposed. The highest recorded voltage was -8.4 kV and Hearn concluded that this voltage was too low for hazards unless isolated conductors were present.

Other measurements [2] using E5 gasoline, alkylate gasoline and white spirit in long lengths of exposed plastic pipe have given voltages up to 90 kV. These would be hazardous even without isolated conductors. It is important to understand how these results differ from Hearn's.

4. Variations in streaming current and other factors

When performing laboratory voltage measurements to assess risk, the test conditions should either replicate the practical worst-case or allow it to be estimated. For fuel charging, the ideal procedure is to measure streaming current in the field, identify the maximum and then use laboratory tests with enhanced charging (e.g. charge injectors or filters [3]) to show that the current needed for incendive discharges is greater than the field-trial maximum. This approach is expensive and it is common to dispense with field trials and enhanced charging and simply do laboratory voltage measurements with representative liquids. This simplified procedure tends to underestimate the highest voltage that could occur but was adopted in [1] and [2]. We now examine the likely degree of underestimation.

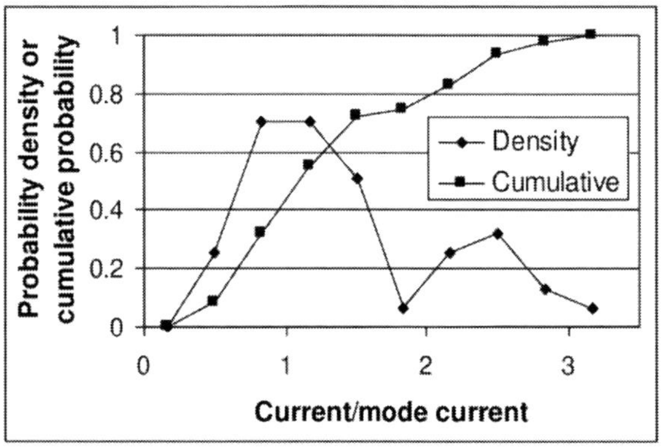

FIG 1: Example distribution of gasoline streaming current.

Figure 1 shows a typical gasoline streaming current distribution obtained by field measurement. The highest current is about three times the mode and the cumulative distribution shows that a randomly chosen sample has a 70% chance of giving less than half the maximum current. Gasoline is not unique in this respect: an even wider distribution of charging has previously been reported for diesel [3]. Typically the highest current is two to five times the mode and a randomly chosen sample is likely to give close to the mode current.

With a test liquid synthesised from pure hydrocarbon components, charging would generally be expected to be lower than in a real gasoline because the extra processing needed to isolate the pure components tends to remove the polar impurities that give rise to charging. Thus the streaming current of a synthesised test liquid is likely to be lower than that of a worst-case gasoline by more than the above factor of two to five.

A voltage measurement may also deviate from the worst case in other respects. The resistances of dissipation paths through both the liquid and the pipe walls increase at low temperature and humidity and no relevant measurements have been reported under these conditions. Also, voltages increase with time and in [1], the flow time was limited by the available fuel volume (recirculation was used in [2]).

5. Voltage calculations

We have done steady-state voltage calculations for exposed pipe lengths based on the Carruthers and Marsh expression [5] in order to identify the factors that promote high voltage. They show that the voltage on the exposed ends is proportional to the exposed length and to the square of the flow speed. Consequently, test rigs designed to assess operational risks must use the longest realistic lengths of exposed pipe (not always done) and the highest practical flow speeds (usually done).

We have also used an adapted form of equation (3) of reference [6] to calculate the time-dependent voltages that occur in a typical buried pipe (15 m long, 90 mm ID, 0.75 m exposed at each end, flow speed 2.8 m/s, streaming current 100 nA, flow duration 15 min). The streaming current approximately matches the maximum in the data referred to in [2] and is expected to be typical rather than worst-case. The worst-case current is probably higher by a factor in the range discussed above[1]. The flow time is typical of the time taken to fill an underground tank at a petrol station from a delivery truck.

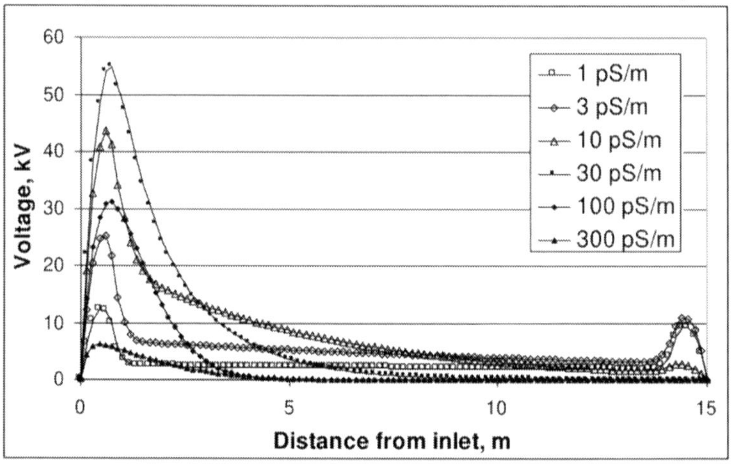

FIG 2: End-of-fill voltages in non-conductive pipes with different fuel conductivities using adapted form of equation (3) of [6].

Figure 2 shows the calculated voltage on the pipe at the end of flow versus distance along the pipe for different fuel conductivities. As the conductivity increases, potentials at the upstream end rise whilst those at the downstream end decline. A maximum voltage is reached at the upstream end when the conductivity is about 35 pS/m. Further increases in conductivity cause the voltage to fall at both ends because of increasing dissipation through the liquid column. The calculated maximum voltage on the exposed upstream end of the pipe considerably exceeds the lowest suggested threshold voltage for the occurrence of brush discharges that can ignite hydrocarbons (25 kV [7]) even though the assumed streaming current is not worst-case. Around the upstream end, a flammable atmosphere could occur either inside the pipe after disconnection of the delivery hose or, for pipes terminating in an enclosed chamber, outside the pipe due to spillage or vapour leaks.

Examination of the transient solutions shows that the maximum voltage at 15 min was approximately double that at 1 min (the duration of Hearn's tests [1]) and was still rising. Maximum voltages would have been higher if more fuel was delivered.

6. Discussion of hazards

Hearn [1] used a realistic pipe configuration but did not ensure that streaming currents, test-duration or dissipation conditions were worst-case. Consequently, his measured voltages are likely to be several times lower than the worst-case. When this is allowed for, his data suggest that brush discharges from

[1] The charging parameters suggested for hazard assessment in NFPA 77 [4] give a worst-case streaming current of 1.6 μA. This exceeds the currents estimated from [2] by rather more than our suggested uncertainty level. This is expected as the NFPA correlation represents a universal worst-case

insulating surfaces and sparks from unbonded conductors are both credible ignition sources. The KPS tests [2] confirm this but, as they used very long lengths of exposed pipe, the results are not directly relevant to the short exposed sections that exist in real site layouts. They do, however, demonstrate the need to ensure that pipes are properly buried to avoid incidents like Case 5).

Experimental measurements, incident investigations and calculations all confirm that voltages on exposed pipe ends in fill boxes can be high enough for incendive sparks from unbonded conductive objects. Regrettably, both incident investigations and site installation photographs (even from sites that have not had incidents) reveal multiple instances of welding sockets that are both uncapped and unbonded and of unbonded metal components such as flanges, end-couplings and hose clips.

The investigators of Case 4) linked that set of incidents to discharges from unbonded, uncapped welding sockets. Tests revealed a breakdown voltage of 20 kV for a similarly-located welding socket. The gaps between conductive items in a fill box are such that other unbonded items would typically have similar, or greater, breakdown voltages except for incidents, like that in Case 2), in which a manual activity such as replacing an end cap was involved. The occurrence of occasional incidents where breakdown voltages are high supports the view that worst-case voltages in the field can be considerably higher than those measured by Hearn [1]. In some of the incidents thought to involve welding sockets (e.g. Case 6), the gaps between the socket and the nearest earthed object were larger than in Case 4) and the breakdown voltage would be expected to have been higher. This supports the suggestion, arising from our voltage calculations, that voltages on exposed ends may also exceed the threshold of 25 kV [7] for incendive brush discharges from insulating surfaces. Therefore hazards from this mechanism cannot be ruled out. In addition, Case 3) shows that incendive discharges can be drawn from the inner surface of even a fully-buried pipe with no unbonded conductors.

The high voltages and associated ignition hazards that occasionally arise with non-conductive pipes can be prevented by using earthed conductive (dissipative-lined) plastic pipes. With this approach, it is important to ensure that all linings are earthed in order to avoid the creation of an isolated conductor.

7. Conclusions

1. Ignitions associated with non-conductive plastic pipes at petrol stations are rare. Most appear to be associated with sparks from isolated conductors but ignitions caused by incendive brush discharges from insulating pipe surfaces cannot be ruled out.

2. The voltage data in Reference [1] underestimate the worst-case, in-service voltage on non-conductive service-station pipes because testing did not establish the worst-case streaming current, dissipation conditions or flow time. However, the data probably does represent typical behaviour and suggests that ignitions will be rare provided conductive items are earthed.

3. The voltage test data in Reference [2], overestimates normal in-service voltages because unrealistically long unearthed sections of pipe were tested but they show the voltages that could arise on an unburied non-conductive pipe.

4. Voltages can be minimised by keeping exposed (unburied) pipe lengths as short as possible.

5. The electrostatic ignition risks arising with non-conductive plastic pipes can be avoided by using earthed conductive plastic pipes. If this approach is adopted, it is important to ensure that all linings are earthed in order to avoid the creation of an isolated conductor.

References

[1] Hearn G L 2002 *J Loss Prevent Proc* **15** 105-109
[2] 2010 Electrostatic tests on thermoplastic pipes for underground transportation of fuel *KPS report* Kungsors Plast AB Box 70 S-736 22 Kungsor Sweden
[3] Walmsley H L, Mills J S 1992 *J Electrostat* **28** 61-87
[4] NFPA 77 Recommended practice on static electricity 2007 National Fire Protection Agency
[5] Carruthers J A and Marsh K J 1962 *J I Petrol* **48** 169-179
[6] Walmsley H L 1996 *J Electrostat* **38** 249-266
[7] Britton L G and Williams T J 1982 *J Electrostat* **13** 185-207

Measurement of minimum ignition energy in hydrogen-oxygen-nitrogen premixed gas by spark discharge

Ayumi Kumamoto[1] , Hiroto Iseki[1] , Ryo Ono[1] , Tetsuji Oda[2]

[1]Department of Advanced Energy, The University of Tokyo, 5-1-5 Kashiwanoha, Kashiwa, Chiba 277-8561, Japan

[2]Department of Electrical Engineering, The University of Tokyo, 7-3-1 Hongo, Bunkyo-ku, Tokyo 113-8656, Japan

E-Mail: kumamoto@streamer.t.u-tokyo.ac.jp, ryo-ono@k.u-tokyo.ac.jp and oda@ee.t.u-tokyo.ac.jp

Abstract. We have measured the minimum ignition energy (MIE) of hydrogen-oxygen-nitrogen premixed gas by spark discharge. The gap distance of point-point electrode was changed between 0.3 and 1.0 mm. The effects of hydrogen and oxygen concentrations on the MIE were measured. It was 0.006 mJ for H_2 with O_2 / (O_2+N_2) = 0.35 mixture, while it was 0.017 mJ for H_2 with dry air mixture. For H_2–O_2 mixture, it was below 0.004 mJ, which was not precisely measured due to the limitation of our measurement.

1. Introduction

Hydrogen generates power by reaction with oxygen. It is clean and a promising energy source without the exhaust of carbon dioxide. We can reduce the dependence on fossil fuel by using hydrogen energy. Hydrogen is sensitive to electrostatic discharge (ESD) compared with other inflammable gases. The minimum ignition energy (MIE) of a hydrogen–air mixture is only 0.019 mJ, whereas that of other flammable gases such as petrol, methane, ethane, propane, butane, and benzene is usually on the order of 0.1 mJ according to Lewis and von Elbe [1]. Therefore, the assessment of the electrostatic hazard for hydrogen is urgent because hydrogen is used in fuel cells and fuel cell vehicles. However, the assessment has not yet been sufficiently achieved.

MIE is an important parameter in the assessment of electrostatic hazard. It has been measured for various flammable gases including hydrogen [1]. MIE is usually measured using capacitive spark discharge whose electrical circuit is shown in Fig. 1, because the capacitive spark is the most frequent electrostatic ignition source in practice. The capacitor C is charged up to voltage V, then the charge stored in the capacitor is discharged in a test chamber filled with an explosive mixture. MIE is defined as the lowest discharge energy required for ignition, where the discharge energy is defined as $CV^2/2$ [1]. In this study, we measure the MIE of hydrogen–oxygen–nitrogen premixed gas using the capacitive spark discharge.

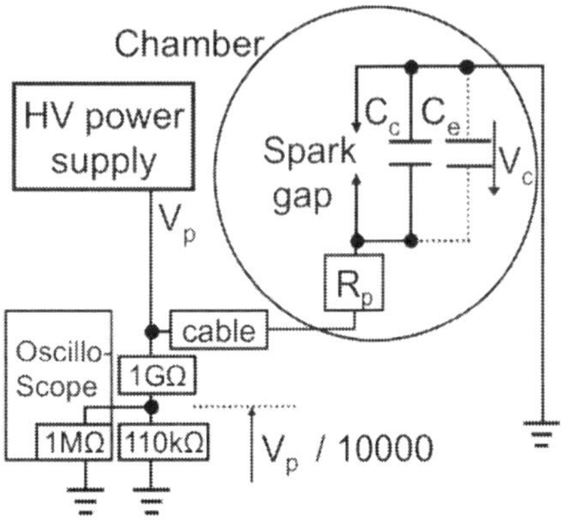

Figure 1. Electrical circuits for capacitive spark circuit.

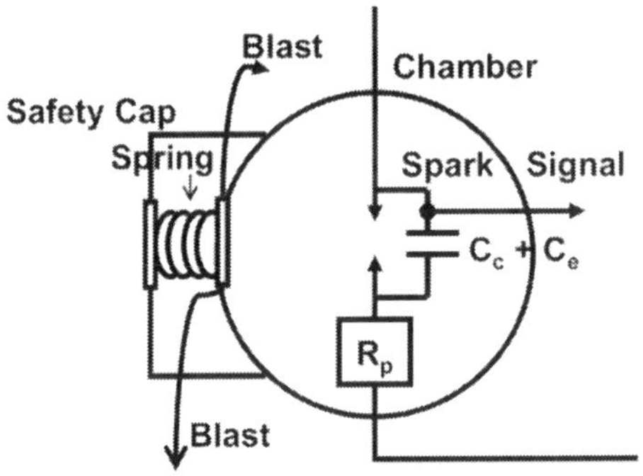

Figure 2. Safety cap on chamber.

2. Experiment

Hydrogen–oxygen–nitrogen premixed gas is ignited in a 110 mL stainless chamber. Hydrogen, oxygen, and nitrogen are introduced up to a total pressure of 1 atm (100 kPa) in the chamber. The concentration of each gas is determined from their partial pressures, which are measured with a Baratron vacuum gauge (MKS Instruments, Model 626A). The chamber has a safety cap with 4 cm diameter as shown in Fig. 2. When an explosion occurs, it releases the explosion energy.

The spark discharge occurs between needle-to-needle electrodes placed in the center of the chamber. The needles are made from tungsten, and have a 1 mm diameter and a 40° tip angle. The gap distance can be adjusted with a micrometer. MIE is measured using the capacitive spark discharge and compared with literature values [1-2]. Fig. 1 shows the electrical circuit for generating the capacitive spark. C_c is the capacitance of the ceramic capacitor connected in parallel to the spark gap and C_e is the capacitance of the electrode. The charge stored in the capacitor $C = C_c + C_e$ is discharged at the spark gap. C_c is measured with an LCR meter (Kokuyo, KC-536). It is provided to cover the range from 0.67 to 470 pF.

The HV power supply is connected to the spark gap through a cable and a high-resistance R_p = 1 GΩ. The voltage of the HV power supply, V_p, increases from 0 to 5.5 kV at a low rate of increase (i.e., 0.1 kV/s). As V_p increases, the capacitor voltage V_c increases with a time constant of CR_p. When V_c exceeds the breakdown voltage of the spark gap, a discharge occurs at the spark gap. The resistance R_p is chosen so that the charging time constant becomes $CR_p > 4$ ms, which is much longer than the spark duration (< 100 ns). Therefore, almost no charge stored in the stray capacitance of the power supply cable flows into the spark gap during discharge, thus the stray capacitance of the cable has no influence on the discharge. When a discharge occurs, the HV power supply detects the spark noise and sends trigger signals to the HV power supply and oscilloscope for synchronization. Upon receiving the trigger signal, the HV power supply reduces its output to 0 V within 0.5 ms to prevent the occurrence of subsequent sparks.

The spark voltage, V_s, is obtained from V_c just before the occurrence of the spark. However, the measurement of V_c with a high-voltage probe is undesirable because the input impedance of the probe affects the discharge. In the present experiment, $V_c = V_p$ is applicable because the increasing rate of V_p is sufficiently low. Therefore, V_s can be obtained from V_p (= V_c) just before the occurrence of the spark. V_p is monitored with an oscilloscope (Tektronix, TDS3034B, 300 MHz) after being reduced by a factor of 10,000 with a voltage divider. The discharge energy is defined as $CV_s^2 / 2$, which can be changed by varying C and V_s.

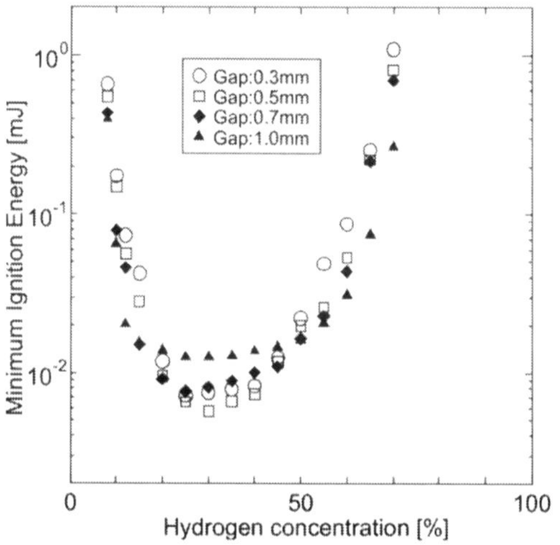

Figure 3. MIE of H_2-O_2-N_2 mixture with fixed ratio of O_2 / (O_2+N_2) = 0.35 for various H_2 and O_2 concentrations and four gap distances.

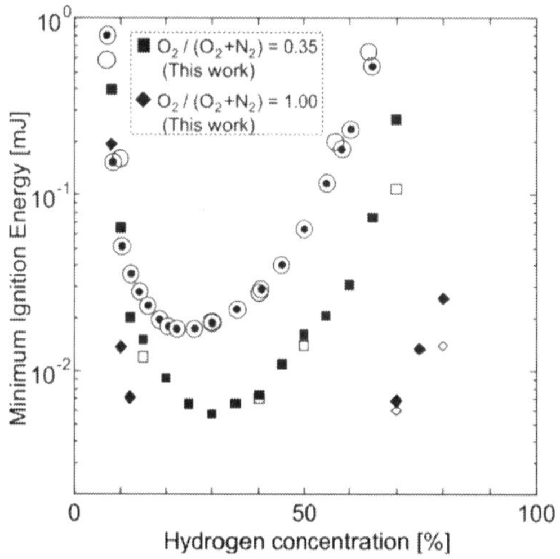

Figure 4. MIE of H_2-O_2-N_2 mixture for various H_2 and O_2 concentrations: Our previous measurement [2] under O_2 / (O_2+N_2) = 0.21 is symbol ◉ . Lewis and von Elbe measurements [1] under O_2 / (O_2+N_2) = 0.21 is ○ symbol, O_2 / (O_2+N_2) = 0.35 is □ symbol, and O_2 / (O_2+N_2) = 1.00 is ◇ symbol.

C_e is determined using the relation $(C_e + C_c)V_s = \int I(t)dt$, where *I(t)* is the discharge current, assuming that the electrical charge in C is almost completely discharged by the spark discharge. *I(t)* is measured using a 2.3 Ω carbon resistor connected in series between the capacitor C and the spark gap. The voltage drop of the 2.3 Ω resistor is measured with the oscilloscope. C_e is determined to be 1.7pF with an error of 10% when it is measured using various C_c between 0-3.3 pF. The above assumption that the electrical charge in C is almost completely discharged by the spark discharge is proved as follows.

We measured C_e of an almost identical electrode in another larger chamber (1 L) using the relation $(C_e + C_c)V_s = \int I(t)dt$. Under a fixed voltage of V_s = 3.6 kV, C_e was determined to be 1.8 pF with an error of 10% when various C_c between 5-100 pF were used. Next, C_e was measured using the LCR meter which also gave a value of 1.8 pF. This agreement verifies that almost all the electrical charge in C_c + C_e was discharged. In the present chamber (110 mL) used for the MIE measurement, C_e cannot be measured using the LCR meter because the volume of the chamber is too small to insert the probe of the LCR meter. However, we consider that the electrical charge in C_e is almost completely discharged by the spark discharge because (i) it is true at least in the larger chamber and (ii) the estimated C_e of the present chamber (1.7 pF) is in agreement with that measured using the larger chamber (1.8 pF) within an error of 10%. Therefore, we use C_e = 1.7 pF. The error of C_e is estimated to be 10%.

3. Results and discussion

The MIE of the hydrogen-oxygen-nitrogen premixed gas is measured using the capacitive spark. The procedure for determining the MIE is as follows. First, the ratio of O_2 / (O_2+N_2) was fixed at 0.35, which is the equivalent ratio to the experiment of MIE by Lewis and von Elbe [1]. MIE depends strongly on the gap distance between the electrodes [1, 2]. First, it was fixed to 0.5 mm. According to our previous work [2], the minimum of MIE of hydrogen–air mixture is obtained at 20-30% hydrogen concentration with 0.5 mm gap distance. Then the gap distance was changed between 0.3 and 1.0 mm. The results are shown in Fig. 3 for gap distances of 0.3 mm, 0.5 mm, 0.7 mm, and 1.0 mm for various hydrogen concentrations. Then, the minimum of these resultant curves are taken as the MIE for each hydrogen concentration. The MIE determined using this procedure is plotted in Fig. 4. It reaches a minimum of 0.0057 mJ at a hydrogen concentration of 30%. The MIE is also measured for hydrogen with O_2 / (O_2+N_2) = 1.00 mixture. It means hydrogen-oxygen premixed gas without nitrogen. This is plotted in Fig. 4, where the MIE obtained in this work is compared with those of previous studies [1, 2]. In this work, our experimental setup has a lower limitation of 0.004 mJ because our setup has a minimum capacitance C_e. Therefore, we could not measure the MIE for 20-65% hydrogen concentration with O_2 / (O_2+N_2) = 1.00.

The MIE obtained in this work is in good agreement with previous values. The differences are probably caused by differences in the configuration and material of the electrodes. These factors generally cause differences in MIE within a factor of 2-3 [3].

Fig. 3 shows that the minimum of MIE for O_2 / (O_2+N_2) = 0.35 is obtained at H_2 concentration of 30%. The effect of gap distance in this mixture is measured in detail. Fig. 5 shows the result. The minimum of MIE is obtained at around 0.4-0.5 mm gap distance.

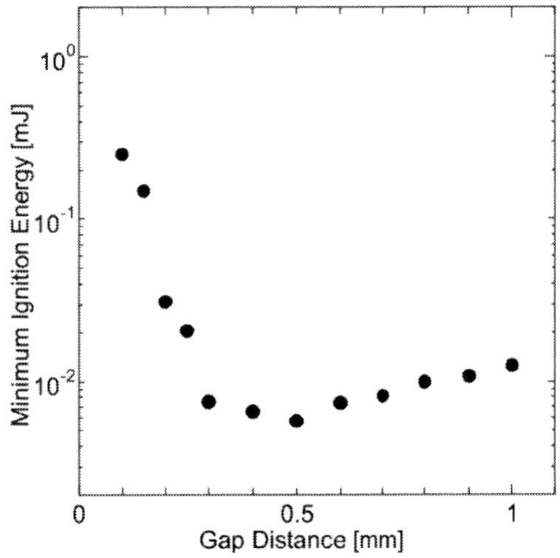

Figure 5. MIE of H_2-O_2-N_2 compounding ratio was fixed O_2 / (O_2+N_2) = 0.35; for each gas concentration was H_2 30%, O_2 24.5%, N_2 45.5%.

4. Conclusions

The MIE in hydrogen-oxygen-nitrogen premixed gases of O_2 / (O_2+N_2) = 0.35 and 1.00 was measured using a capacitive spark discharge. The minimum of MIE was 0.006 mJ for H_2 with O_2 / (O_2+N_2) = 0.35 mixture, and below 0.004 mJ for H_2 with O_2 / (O_2+N_2) = 1.00 (H_2 - O_2 mixture). The latter value could not be measured because it is much lower than the lowest possible spark energy of our experimental setup (0.004 mJ), which was determined by C_e = 1.7 pF and the minimum spark voltage V_s = 2.2 kV. We need lower C_e and V_s than the present experiment to measure the minimum of MIE H_2 - O_2 mixture.

References

[1] B. Lewis and G. von Elbe: Combustion, Flames and Explosions of Gases, New York, Academic Press, 3rd ed.,1987

[2] R. Ono, M. Nifuku, S. Fujiwarab, S. Horiguchib, and T. Oda, J. Electrostat., 2007, **65,** 87-93

[3] H.E. Rose, T. Priede: Seventh Symposium on Combustion, Butterworths, London, 1959, **63,** 436-445

Journal of Physics: Conference Series **301** (2011) 012040 doi:10.1088/1742-6596/301/1/012040

MIE experiments and simultaneous measurement of the transferred charge

T Langer, G Gramse, D Möckel, U von Pidoll, M Beyer

Physikalisch-Technische Bundesanstalt (PTB), Braunschweig, Germany

Email: tim.langer@ptb.de

Abstract. Using the apparatus commonly utilized for the determination of the minimum ignition energy a series of experiments have been carried out in ethene/air as a representative of explosion group IIB. Thereby, the transferred charge as a criterion to judge the ignition potential is determined to verify the threshold of 30 nC of transferred charge given in the standard IEC 60079-0. The stored charge in a capacitance before the discharge is compared to the transferred charge in the spark. Furthermore, the correlation of ignition energy and transferred charge is examined. Based on the results presented here, the threshold of the transferred charge for explosion group IIB is discussed. Moreover, the MIE value of an ethene/air mixture is reviewed taking into account the measurement uncertainty.

1. Introduction

The minimum ignition energy (MIE) is determined using a capacitive spark discharge [1]. Thereby, the energy of the spark discharge is assumed to have the amount of energy stored in the capacitance before the discharge. The MIE is determined by varying the capacitance, the charging voltage, the electrode diameter, the electrode gap width and the mixture composition in order to find the lowest energy necessary to ignite the given mixture.
Gibson and Lloyd [2] found in 1965 that there is a correlation between the incentivity of a brush discharge and the charge transferred by it. Eisfeld [3] proposed in 1984 a test method based on this finding and applied for a patent for a coulombmeter he had developed [4]. Later, in 1988, Gibson and Harper too stated that the charge transferred in a discharge provides guidance on its incentivity [5]. Based on MIE measurements, von Pidoll et al. [6] have calculated the charge stored in the capacitance in order to obtain an estimation of the transferred charge. By correlating the MIE with the lowest calculated charge, von Pidoll et al. [6] have suggested transferred charge thresholds in order to judge if electrostatic discharges are potentially incendive or not. These threshold values are now defined in the standard IEC 60079-0 [1]. Even if the criterion of transferred charge is more often used to assess the incentivity of brush discharges the criterion is also applicable and used for the assessment of spark discharges. However, in a recent study, Langer et al. found that these thresholds are sufficient concerning brush discharges regarding explosion groups IIA and IIC. But in case of explosion group IIB, the achieved results were close to the threshold of 30 nC. For the 8 % ethene/air mixture used, the lowest transferred charge leading to ignition of the mixture was 31.5 nC ± 1.0 nC [7]. Since spark discharges are known to be more incendive than brush discharges it is consequent to recheck the thresholds using spark discharges.

Since the validation of the transferred charge threshold can only be reliably performed near the MIE, another objective of this work is to recheck the MIE of ethene. Furthermore, using the "Guide to the expression of uncertainty in measurement" [8], the received values are examined concerning the measurement uncertainty. The considerations concerning the uncertainty should lead to an estimation as to how accurate the results given in the literature can be expected to be.

2. Experimental set-up

A schematic view of the test set-up is presented in figure 1. A defined charging voltage is given to the capacitance until spark discharges occur. The experiment runs until an ignition of the mixture is observed. The corresponding voltage is noted and the ignition energy is calculated. Additionally, the transferred charge is measured using the coulombmeter. Furthermore, the originally stored charge in the capacitor Q_c is calculated. Reducing the ignition energy was attempted by varying the capacitance, the charging voltage and the gap width, respectively. Thereby, for a given electrode gap and given capacitance the minimum voltage leading to a spark discharge was determined by varying the voltage. The diameter of both electrodes is 2.0 mm. For the tests the flammable gas/air mixture prepared was in the most ignitable range for ethene according to [9], since this is the most hazardous situation one can examine in matters of explosion protection. Hence, a concentration 8.0 % ethene in air was used for the experiments. A detailed description of the test set-up is given in [10].

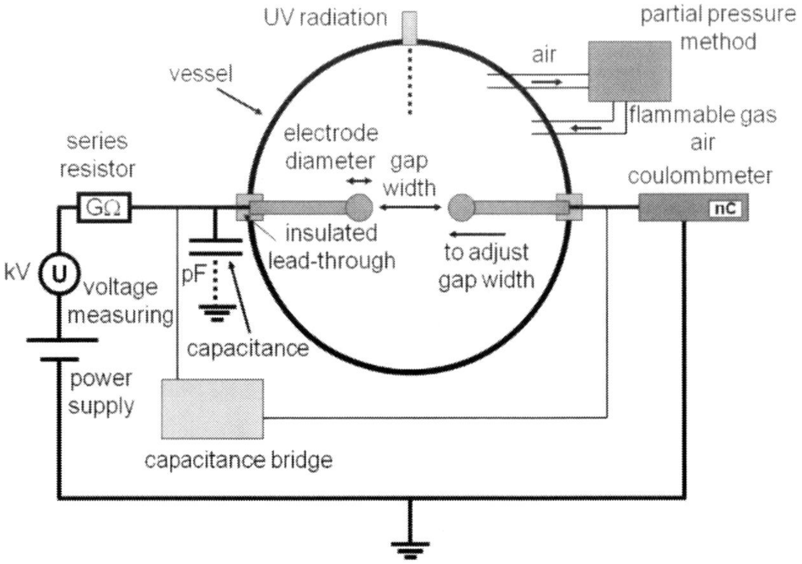

Figure 1. Schematic view of the experimental test set-up.

3. Results

In figure 2 the correlation between the transferred charge Q_t and the calculated charge Q_c is presented. Thereby, the transferred charge was measured in 8 % ethene/air mixtures during the ignition tests. The calculation of the measurement uncertainty is described in more detail by Langer et al. in [10]. The expanded measurement uncertainty (coverage factor k=2.0) is used here assuring that all results have a probability of 95 % of being within the resulting expanded measurement interval. The uncertainty in Q_t results from the coulombmeter, the uncertainty in Q_c is mainly influenced by the measurement of the capacitance. The straight line represents $Q_t = Q_c$. It can be seen in figure 2 that the measurement of Q_t gives reasonable values. The tendency of $Q_t \leq Q_c$ can be explained by the fact that there are residual charges in the capacitance [6]. This result shows, on the one hand, that the coulombmeter is suitable to measure the transferred charge during the discharge and, on the other hand, that Q_c is not an ideal value to determine the thresholds of transferred charge due to the fact that there are residual charges.

In the following, the transferred charge Q_t in dependence of the ignition energy for the observed ignitions of the 8 % ethene/air mixture is presented in figure 3. Thereby, the results are from different electrode gap widths. The criterion of the transferred charge can only be suitable for judging the safety of test items when - for all ignition energies - a transferred charge higher than the threshold value given in [1] is found. In figure 3 the threshold value according to the explosion group given in the standard [1] is drawn. Additionally, the hatched box represents the interval of ignition energy which is under the MIE given in [11].

For ethene/air mixtures which represent explosion group IIB, sparks with a transferred charge less than the threshold of 30 nC ignited the mixture. The lowest transferred charge Q_t igniting the mixture was 26.6 nC ± 1.1 nC. The corresponding calculated charge Q_c was 29.5 nC ± 3.5 nC. For this result, a gap width of 1.2 mm at a voltage of 6.00 kV and a capacitance of 4.90 pF was used.

In the last part of this paper, the MIE value of ethene given in [11] is reviewed. Thereby, the measurement uncertainty, in particular, is the focus of this study. The ignition energy is calculated using $W=0.5 \cdot C \cdot U^2$. For 8.0 % ethene/air mixtures (figure 4), an MIE of 80.8 μJ ± 9.7 μJ was found at a gap width of 1.2 mm at a voltage of 5.80 kV and a capacitance of 4.80 pF. The biggest influence with respect to the expanded measurement uncertainty originates from the determination of the capacitance. The transferred charge was 26.9 nC ± 1.1 nC, the calculated charge 27.9 nC ± 3.3 nC. Regarding the measurement uncertainty, the MIE is the same as the value of 82 μJ given in [11].

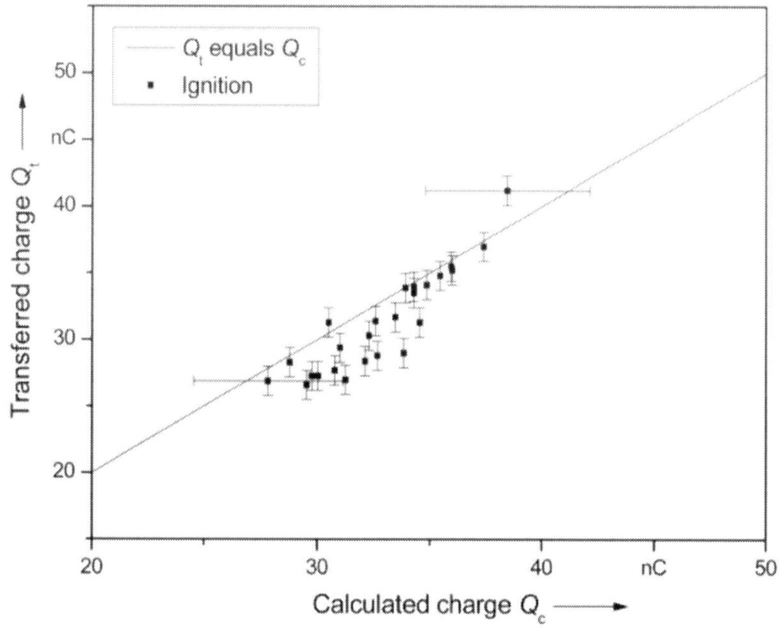

Figure 2. Correlation between transferred charge and calculated charge.

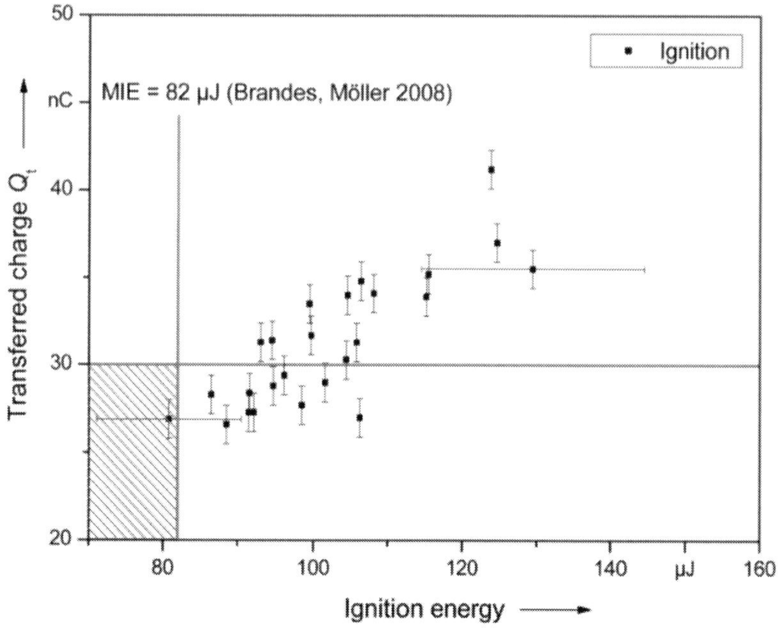

Figure 3. Transferred charge in dependence of the ignition energy for an 8.0 % ethene/air mixture.

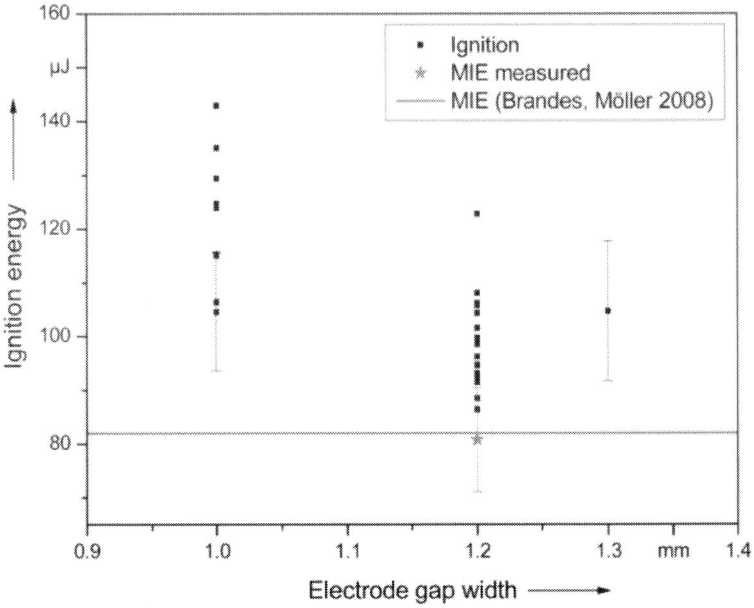

Figure 4. Ignition energy in dependence of the electrode gap width using an 8.0 % ethene/air mixture.

4. Conclusion

Using the apparatus commonly utilized for the determination of the minimum ignition energy (MIE), a series of experiments have been carried out in the most ignitable 8 % ethene/air mixture. Thereby, the transferred charge as a criterion to judge the ignition potential is determined to verify the thresholds of transferred charge given in the standards. The stored charge in the capacitance before the spark discharge is compared to the transferred charge in the spark. The measurement of the transferred charge performed with a coulombmeter gave reasonable values. It was found that the transferred charge is less than or equal to the stored charge within the measurement uncertainty budget. Furthermore, the correlation of the ignition energy and the transferred charge is examined. For ethene/air mixtures - as a representative of explosion group IIB - a minimum transferred charge of 26.6 nC ± 1.1 nC led to ignition which is lower than the current threshold of 30 nC. The MIE of ethene/air mixtures was determined to be 80.8 μJ ± 9.7 μJ which matches well with the given value of 82 nC in the literature. In summary, especially concerning the determination of MIEs and their corresponding transferred charge, it seems essential to give the corresponding measurement uncertainties in order to compare the achieved results. The measurement of the capacitance seems to be the main challenge in this context.

References

[1] International Standard IEC 60079-0 Ed 5.0: Explosive Atmospheres - Part 0: Equipment – General requirements 2007
[2] Gibson N and Lloyd F C *British Journal of Applied Physics* **16** 1619
[3] Eisfeld D *Glückauf-Forschungshefte* **45** 116
[4] Westfälische Bergwerkschaftskasse (applicant), Eisfeld D (inventor),patent application 3202702
[5] Gibson N and Harper D J *Journal of Electrostatics* **21** 27
[6] von Pidoll U, Brzostek E and Froechtenigt H-R 2004 *IEEE Transactions on Industry Applications* **40** 1467
[7] Langer T, Möckel D and Beyer M 2010 accepted for publication in *Journal of Electrostatics*
[8] ISO/IEC Guide 98-3:2008: Uncertainty of measurement - Part 3: Guide to the expression of uncertainty in measurement 2008
[9] Hattwig M and Steen H 2004 *Handbook of Explosion Prevention and Protection* (Weinheim: Wiley-VCH)
[10] Langer T, Gramse G, Möckel D, von Pidoll U and Beyer M 2011 submitted to the *Journal of Electrostatics*
[11] Brandes E and Möller W 2008 *Safety characterstic data - Volume 1: Flammable liquids and gases* (Bremerhaven: Wirtschaftsverlag N.W. Verlag für neue Wissenschaft)

Interpretation of charge transfer measurements of brush discharges

Kanchan Chowdhury*, Ulrich von Pidoll, Dieter Moeckel**, Tim Langer** and Michael Beyer****

* Cryogenic Engineering Centre India, Institute of Technology, Kharagpur 721302, West Bengal, India; ** Physikalisch-Technische Bundesanstalt (PTB), Bundesallee 100, 38116 Braunschweig, Germany, E-Mail: ulrich.v.pidoll@ptb.de

Abstract. In the present work, experimental results on the measurement of the total charge on a charged insulating sheet before and after a provoked brush discharge, their difference "C", the induced charge "A" when approaching an earthed microprocessor operated hand-Coulombmeter, and the transferred charge "B" at the instance of the discharge are presented. "B" is identical with the value measured by the hand-Coulombmeter within the expected measurement uncertainty. Due to observed corona losses and multiple brush discharges independent of each other, "B" correlates better with the incendivity than "C". The quotient B/C was closer to 1 than calculated in the literature but shows all predicted trends. The results obtained can be used for correct estimation of the incendivity of brush discharges between 10 nC and 90 nC. There is no need to change the existing threshold limits of 60 nC, 30 nC and 10 nC for the explosion groups IIA, IIB and IIC hitherto used in standards for zone 1.

1. Introduction

For the sake of safety to man and property, it is necessary to put the components or equipment for use in explosive atmospheres, which have the possibility of acquiring static electricity, through electrostatic tests under worst conditions in order to determine if the transferred charge of such discharges could ever exceed the safe threshold limits of 60 nC, 30 nC and 10 nC in order to prevent ignition of Gas Group IIA, IIB and IIC materials, respectively [1,2].

However, calculations in literature [3,4] had the result that due to charge induction only 20 % to 56 % of the actual transferred charge is expected to be recorded with a Coulombmeter. If these calculations would be true, such a significant underestimation of transferred charges may pose severe threat of fire and explosions. For example, if the actual discharge measured is 30 nC this may give the false impression to be well within the limits of 60 nC for gas group IIA.

Although brush discharges transferred from non-metallic surface such as PTFE have been found to need transferred charges measured by a Coulombmeter somewhat higher than the existing threshold limits for ignition derived from those of spark discharges from metal surfaces [5], this is clearly not sufficient if the calculated measuring efficiency would be as low as 20 %. For this reason, the transferred charges measured with a microprocessor operated Coulombmeter have been extensively investigated in a Faraday pail to know the realistic measurement errors by Coulombmeters either to revise the hitherto used threshold limits or to give sufficient experimental data to revise the calculations.

2. Experimental set-up

Charge measurements were made using a virtual earth charge measurement amplifier (JCI 178 Charge Measurement Unit from John Chubb Instrumentation, United Kingdom). The Faraday pail was JCI 247, having a 260 mm diameter 410 mm high galvanized bucket mounted on 3 insulators from the base of a converted galvanized iron dust bin, 480 mm diameter, 650 mm tall. A 260 mm diameter aperture was cut in the lid for entry and exit of samples. The pail was connected by a wire link to a BNC connector in the outer wall of the inner bucket of the bin and from this by a BNC coax cable to the charge measurement unit and a 2-channel color digital phosphor oscilloscope 300 MHz 2,5 GS/s (TDS 3032B from

Tektronix, US). The measuring deviation between oscilloscope and JCI 178 was found to be neglectable.

The transferred charges of the provoked brush discharges were additionally measured by a microprocessor operated hand-held Coulombmeter (HMG 11/02 from Schnier, Germany) with a ball probe 25 mm in diameter. It consists of a 100 nF capacitance coupled to a Voltmeter of 20 kΩ input resistance and a triggering microprocessor, measuring every 10 µsec. After a sudden increase in voltage on the capacitor exceeding a trigger level the instrument records the voltage and stops measuring if no further increase in voltage does occur in the following measurements. The display then freezes the differences of charge one measuring point before the trigger and its maximum value. The instrument was calibrated with low voltage yielding 8,5 % lower results than the factory calibration with high voltage. One measurement was made with a shielded probe 25 mm in diameter as described by Chubb [6] instead of the unshielded standard probe but showed no significant improvement.

PTFE samples were placed on a dissipative wood table outside the Faraday pail, charged by approaching and removing a corona charger, and then carefully placed on the distance pieces inside the Faraday pail. The charge measurement unit was then set to zero and the oscilloscope started in "run" mode with appropriate scales of amplification (10 – 20 nC/div) and time (2 s/div). The Coulombmeter or the earthed probe was approached towards the charged PTFE sheet. Once the crackle of a discharge was heard, the move of the probe or the Coulombmeter was stopped and, after a few seconds, taken out of the Faraday pail and the oscilloscope was put in "hold" mode and its record saved on a disk. The reading of the Coulombmeter was also noted.

This procedure was chosen because removing the sample out of the Faraday pail, discharging and replacing it yielded unpredictable charge losses of 3 nC to 90 nC whereas the disturbance of the electric field due to the walls of the Faraday pail seems neglectable when comparing their distance with the distance of the discharge electrode.

3. Interpretation of the oscilloscope recordings

Fig. 1 shows a typical oscilloscope record in case of an unearthed probe hanging on a plastic cord when approaching to a charged PTFE sheet within the Faraday pail. In this kind of experiment all charge transfers are occurring in a closed system without any contact to outside. "A" is the induced charge on the approaching probe at the instance of the provoked brush discharge. "B" is the transferred charge of the brush discharge as recorded at the instance of the discharge. "C" is the lost charge on the charged sheet caused by the discharge process after the probe has been removed out of the Faraday pail while the charged sheet is still within the pail.

Because of the charge maintenance law one would expect that within an ideal Faraday pail the induced charge "A" and the charge "B" lost during the charge transfer from sheet to probe electrode are both zero. However, as the used Faraday pail had an opening on its top a significant part of the charge may get undetected when shifted closer to the opening as a consequence of charge movement.

In the experiments performed with unearthed probes "B" was always 5 % of "C". As "A" is reversible this attributes the used Faraday pail a charge collecting efficiency of maximal 95 %. This value was supported by experiments with a normal Coulombmeter as a probe which usually displayed 95 % of the value of "C" after its removal out of the Faraday pail (e.g. 60 nC in Fig. 1).

Figure 2 reproduces a typical actual oscilloscope recording when approaching an earthed object to a charged sheet within the Faraday pail. As in the case of the unearthed probe "A" is the induced charge on the approaching probe, "B" is the value of the transferred charge as recorded at the instance of the discharge and "C" is the lost charge on the charged sheet caused by the discharge process after the probe has been removed out of the Faraday pail. Note that this type of system is no closed system as charges may move out of the system via the earth line between the Coulombmeter, which is within the Faraday pail, and earth.

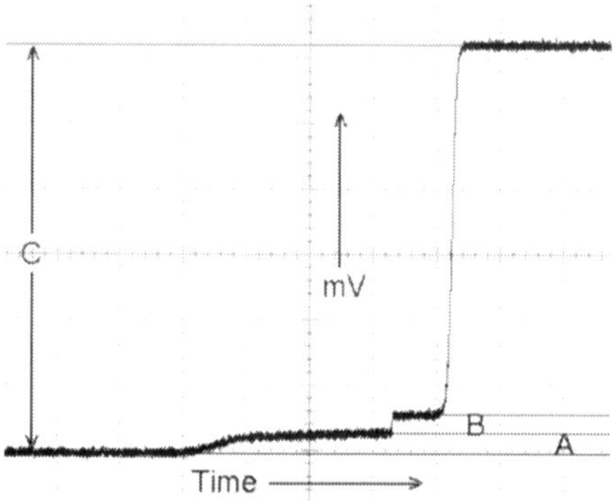

Figure 1: Oscilloscope record of the charge losses "C" of 62 nC (10 nC/div) when approaching an unearthed probe 25 mm in diameter to a charged sheet provoking a brush discharge (microprocessor operated hand-Coulombmeters don't work correctly under these conditions).

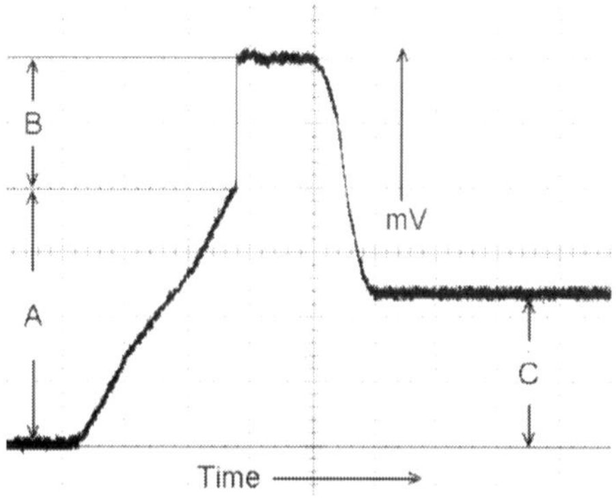

Figure 2: Oscilloscope record of the charge losses "C" of 44 nC (20 nC/div) when approaching an earthed probe 25 mm in diameter to a charged sheet provoking a brush discharge "B" of 40 nC (hand-Coulombmeter displays 38 nC).

All experimental results obtained show that "B" is in between 20 nC and 70 nC equal within +1 nC to -6 nC (± 3 nC when factory calibrated) with the reading of the microprocessor based Coulombmeter after the discharge has occurred to its input sphere. "C" should be zero if the probe is removed from the Faraday pail without any discharge having occurred. Nevertheless we often observed a lost charge "C" without any notable discharge occurring. We interpret this result that a significant part of the charge may get undetected due to corona from the charged sheet to the approaching earthed electrode independent from any brush discharge. In fact, when approaching an earthed sharp needle to the charged sheet very strong charge lost "C" can be observed without any brush discharge being provoked (Fig. 3).

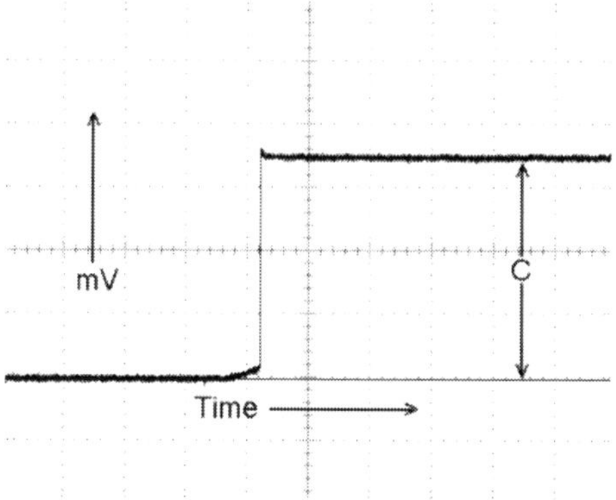

Figure 3. Oscilloscope record of the charge losses "C" of 170 nC (50 nC/div) when approaching and removing an earthed needle to a charged PTFE sheet without any brush discharge occurring (no hand-Coulombmeter present).

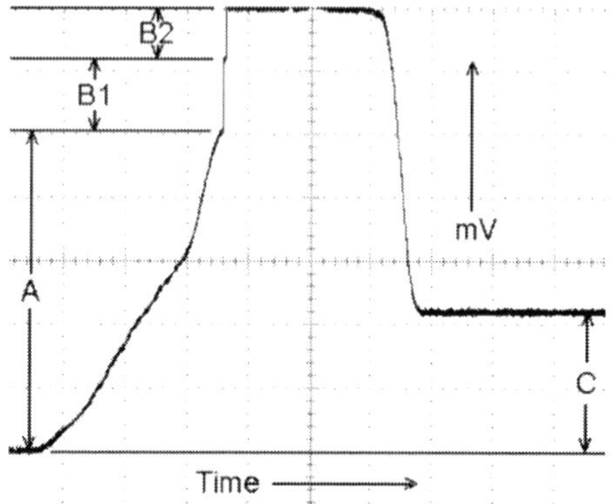

Figure 4. Oscilloscope record of the charge losses "C" of 42 nC (20 nC/div) when approaching an earthed probe 25 mm in diameter to a charged sheet provoking a double brush discharge "B1+B2" of 40 nC (hand-Coulombmeter displays 20 nC corresponding to "B1").

Additionally we sometimes observed the occurrence of double brush discharges B1 and B2 (Fig. 4). Due to observed time intervals of about 10 ms between both discharges when expanding the x-axis of the oscilloscope record only the stronger brush, in most cases the first of both, contributes to the incendivity. Our microprocessor based Coulombmeter was found to always record the first one B1 which seems appropriate in most cases.

4 Conclusions

As charge remains constant in a closed system it may be said that "B" represents the recorded transferred charge at the instance of the brush discharge, while "C" represents the lost charge caused by the whole discharge process. The value of "C" is always somewhat higher than that of "B". The difference of both values is often called the "absolute underestimation", while the ratio of "B" to "C" has been termed as "collection efficiency" in the literature although "C" may contain charge lost by non-incendive secondary discharges (Figs. 3,4).

Already a short glance on Figures 2 and 4 shows that the difference between measured transferred charge "B" and lost charge "C" caused by the discharge process is not as high as expected from literature calculations [3,4]. As additional experiments prove that a probe of 25 mm in diameter behaves like an average human thumb or knuckle, such a probe dimension is, therefore, suitable for discharge simulations [7]. For such a probe the measured transferred charge "B" was found to be in average about 70 % of the lost charge "C" after provoked brush discharges in the case of 60 nC transferred charge and more than 50 % in the case of 10 nC [7]. These values are significantly higher than those obtained from calculations in the literature [3,4].

One possible reason for this may be an underestimation of the induced charge after the brush discharge to simplify the calculations (typical assumption: homogeneous circular charge distribution after discharge). Experiments with Bürkers charge powder show, however, distinct discharge channels on the discharged surface leading to a centre [7]. This center and the mid of the discharge channels may contain opposite charge. The discharged surface may be circular symmetric or only a half circle and usually contains extended not discharged areas [7]. This is contrary to the assumptions made for the calculations. Against to the expectations shielded probes did not show a significant improvement concerning charge collection efficiency of microprocessor operated Coulombmeters (see [7] for more details).

As ignition experiments with brush discharges in 22 % hydrogen, 8 % ethylene and 5,2 % propane yielded minimum transferred charges "B" of 23 nC, 32 nC and 93 nC for ignition measured with the used microprocessor operated Coulombmeter [5] there is no need to change the existing threshold limits for transferred charges of 10 nC, 30 nC and 60 nC for zone 1. Due to the small safety gap for ethylene it is, however, reasonable to change the values to 0 nC, 10 nC and 30 nC for zone 0 as it is already proposed in the latest IEC DTS 60079-32 [8].

References

[1] International Standard *IEC 60079-0* 2007 Explosive Atmospheres - Part 0: Equipment – General requirements
[2] European Standard *EN 13463-1* 2007 Non-electrical equipment for use in potentially explosive atmospheres — Part 1: Basic method and requirements
[3] Walmsley H L 2009 Induced-charge Errors in Charge-transfer Measurement, *J. Electrostatics* **67** 320
[4] Walmsley H L 2010 Induced-charge Errors in Charge-transfer Measurement: Brush discharges between Charged, Insulating Discs and Earthed, Conductive Spheres, *J. Electrostatics* **68** 5
[5] Langer T, Moeckel D and Beyer M 2010 Transferred Charge of Brush Discharges in Explosive Gas Atmospheres - a Verification of the Ignition Threshold Limits, accepted for publication in *J. Electrostatics*
[6] Chubb J 2006 Measurement of Charge Transfer in Electrostatic Discharges, *J. Electrostatics* **64** 321
[7] Chowdhury K, von Pidoll U, Langer T, Moeckel D and Beyer M 2011, Experimental Investigation on the Measuring Uncertainty in Recording Brush Discharges from Insulating Surfaces by Coulombmeters, to be published
[8] International Draft Technical Specification *IEC DTS 60079-32* 2010 Explosive Atmospheres - Part 32: Electrostatics. In development and subject to change. Final version is expected in 2011.

Cold-electrode voltage fall for impulse arcs in argon between copper electrodes

O Diaz and V Cooray
Lightning Research Group, Division for Electricity, Uppsala University
Ångströmlaboratoriet Box 5234, 751 20, Uppsala, Sweden

E-mail: oscar.diaz@angstrom.uu.se

ABSTRACT: The full electric arc discharge in gases for short gaps in homogeneous electric field and pressure·distance (*pd*) below 150 Torr·cm, can be described as a transition between different discharge mechanisms such as: Townsend, glow, and arc. Once the arc is achieved the measured voltage drops to some volts and the current density increases several orders of magnitude. Depending upon the type of gas used, the electrode surface characteristics and type of electrical excitation, the cathode and anode voltage fall might change. The present work is directed to study the electrode fall (sum of anode and cathode falls) during a current impulse arc discharge between copper electrodes in ceramic tubes filled with argon between 0.01 and 6.5 Torr·cm. The copper electrodes were cleaned, degassed and hydrogen reduced. The arc voltages were measured with fast/slow rise times and short/long duration current impulses produced by a RLC circuit. An increasing variation of the electrode fall was found at the pressure·distance range analyzed.

1. Introduction

The discharge in gases for low $pd < 150$ Torr·cm can be explained as a transition between different mechanisms, from Townsend, to glow and finally arc. When the arc stage has been achieved, one can observe different phenomena at the cathode in comparison with the preliminary glow stage: the voltage fall changes from some hundreds of volts to some tens, the current density at the arc spot might increase from 10^2 to 10^6 A/cm^2 approx. and the discharge channel light emission begins to contain spectrum lines from the electrode material due to the presence of active electrode vapour in the cathode vicinity. The arc has a voltage drop at the cathode in the order of minimum ionizing or minimum exciting potential of the gas or electrode vapour, and a capability of withstand large currents by means of its own mechanism of electron emission from the cathode (field and/or thermoionic) [1, 2].

Several methods have been used in order to determine the electrode fall in glow and arc discharge stages. One of them consisted in measuring the voltage fall by changing the interelectrode distance in a constant current arc, similar to a circuit breaker opening process [3]. Other method consisted in establishing the arc by melting a thin wire between fixed separated electrodes; then the anode fall was estimated by measuring the current and amount of electrode mass eroded during one single full arc discharge [4]. An interesting method consisted in measuring arc voltages for different constant electrode distances and then to extrapolate the measured values to zero interelectrode distance, so the minimum possible voltage to be measured was approximately the sum of the anode and cathode fall

Published under licence by IOP Publishing Ltd

[5]. Depending upon the type and the characteristics of the gas used (electronegative, noble, impurity content, pressure), the electrode surface chemical composition (amount adsorbed gas, oxide layers, erosion), and the characteristics and type of electrical excitation, the cathode and anode voltage fall might change, being commonly reported to have a higher cathode fall [1, 3, 6].

During the first half of the 20th century, a lot of efforts were addressed to study in detail the different stages of the glow discharge and many authors have achieved good results while measuring the potential distribution inside the different glow regions [7]. Most of the studies related to electrical arcs were done after the 50's, with the improvement of experimental techniques and numerical/theoretical models. For the arc discharge the potential is distributed on the anode and the cathode falls, the transition regions between the electrode and the gas and along the positive column. The positive column is characterized by having an almost constant voltage drop along its length. The electrode fall (anode + cathode) have been also measured for different electrode materials and gas types, pressures close to 760 Torr (1 atm) and interelectrode distances of some millimetres [3-5].
For this work different current impulses were applied to ceramic gas filled tubes with copper electrodes in order to create a full arc discharge. Arc voltage and current were registered for the estimation of the electrode fall.

2. Experimental Setup

2.1. Gas tubes

The gas tubes were made of a pair of copper electrodes and cylindrical ceramic tubes of 3 mm internal radius, having interelectrode distances of 0.6 and 3 mm. Four tube samples for each pressure and interelectrode gap distance were tested. The electric field inside the tube assured enough insulation between the discharge area and the tube walls and a homogeneous distribution over the discharge area (better for the shorter gap); as it is shown in figure 1, where the light colours represent the high electric field intensity regions between the electrodes.

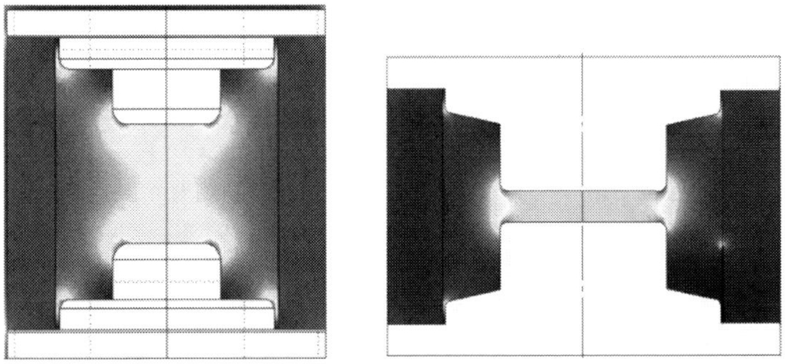

Figure 1. Electric field distribution for the 3 mm (left) and 0.6 mm gaps (right).

Argon with impurity concentrations lower than 2 ppb was used for filling the tubes at pressures between 0.21 and 21 Torr. Oxygen-free copper electrodes were first washed with acetic acid, rinsed in deionised water, dried, degassed in low vacuum (700 °C and $7.6 \cdot 10^{-7}$ Torr) and finally were reduced in a hydrogen atmosphere at 700 °C. These different tasks aimed to reduce the amount of impurities on the electrode discharge surface. However, the authors are aware that not all the copper oxide layers were completely removed, different situation to other works where more complex processes as ion bombardment or controlled/localized glow sputtering were used (see effects of clean and dirt surfaces in discharges in argon [6]). Once the tubes were sealed with eutectic welding, they were conditioned by applying of current impulses of ca. 100mJ in rounds of 1 s, five shots each tube side.

2.2. Current impulse generator

Two different RLC series circuits were used to produce the full arc discharge. The circuits were charged with a DC source at different voltages until a maximum of 5kV. The fast circuit had a main capacitance of 10 μF and an output current impulse of 1.5/40 μs and 130 A peak. The slow circuit main capacitance was 60 μF and its current impulse was 15/500 μs and 700 A peak (current impulse times: time to peak time / half amplitude time).

The measuring system consisted of a Lecroy PPE6kV voltage probe, a Pearson current transformer and a digital store oscilloscope (DSO) Tektronix TDS1002. The bandwidth of the measurement system was limited to 20MHz because the fast transients during the initiation of the discharge were not of interest of this work. Since the desired voltage value to be measured is in the order of some volts, the DSO 8-bit vertical resolution was set to obtain the best results while measuring. For circuit 1, the minimum recorded voltage resolution was 620 mV and for circuit 2 was 220 mV. Typical voltage and current waveforms recorded for both circuits used are presented in figure 2.

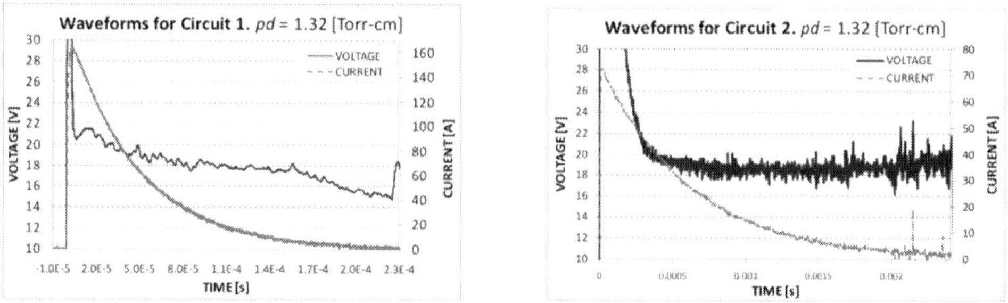

Figure 2. Typical current and voltage measured for the same *pd* product. Note for both cases the time scales and the current amplitude on the left vertical axis.

3. Results

Some general characteristics can be pointed out from the results obtained:

- *Arc voltage stability.* The measured arc voltage reached a noticeable stable value with the long time impulse from circuit 2, after the half amplitude time (~ 500 μs). For the fast impulse circuit 1, the arc voltage tended to stability as the current impulse was close to its minimum value (t ~ 200 μs), making it difficult to determine an arc voltage value.
- *Arcs at low pressure.* For both circuits and the lowest pressure analysed 0.21 Torr, the arc voltage was unstable for the current values below 10A. This behaviour is opposite to what was observed for higher pressures where the arc voltage did not deviate much from a mean value at lower currents values.

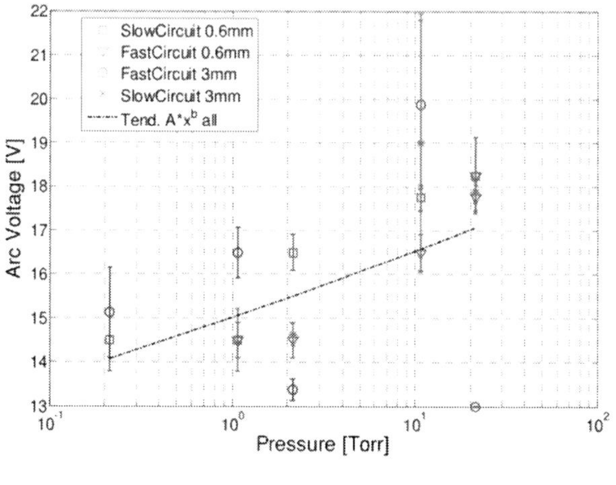

Figure 3. Arc voltage measurements. The bars represent the standard deviation obtained for each pressure value

The arc voltages reported correspond to the average and standard deviation of four different measurements for each circuit and *pd* value. It was observed that the arc voltages for *pd* below 0.2 Torr·cm were close to a mean value of 14.5 V, and tend to deviate to a higher voltage value as presented in figure 3.

4. Discussion

Some authors have done measurements of the electrode fall for configurations similar to the ones used in this work (copper anode/cathode electrodes, argon at a pressure ≤ 760 Torr, maximum current ~ 100 A for t < 1 ms): a) $V_{fall} = V_{anode} + V_{cathode} = 5 + 12$ V [3]; b) $V_{fall} = 16.5$ V [5]; where both the values were measured in Argon at 760 Torr (1 atm). Given that not all authors include information related with the electrode cleaning, setup dimensions or detailed characteristics of the electrical circuit used, a direct comparison in the same conditions from preliminary works is not possible.

Considering that: a) the values of pressure and current amplitude used in this work do not affect drastically V_{fall}, as concluded in [5]; b) the value of the positive column potential distribution is uniform and is approx. 10 V/cm for Argon, as it was measured in [8] and c) the transition regions from the electrodes to the gas are concentrated in the cathode and anode fall (with lengths in the orders of magnitude of 10^{-7} m and 10^{-4} m respectively [3]); it is possible to assume that the measured arc voltage corresponds to the electrode fall. One can observe in figure 3 that values for electrode fall below *pd* = 1 Torr·cm are in the vicinity of 14.5 V, showing a slight increment for the interval 1 < *pd* < 10 Torr·cm reaching a maximum of 19.5 V. The measured electrode fall values were more stable tendency for the 0.6 mm gap.

5. Conclusion

The measured electrode fall value showed an increasing tendency (from ~14 V at 0.2 Torr to ~17 V at 21 Torr) for argon discharge between copper electrodes under the conditions of this work.

6. References

[1] von Engel A 1965 *Ionized gases* (Oxford: Oxford University Press) chapter 9 pp 259-289
[2] Holmes R 1978 *Electrode Phenomena (Electrical Discharges in Gases)* ed J Meek and J Craggs (Norwich: John Wiley & Sons) chapter 11 pp 839-867
[3] Dickson B and von Engel A 1967 *Proc. Roy. Soc. Lon.* A **300** 1462
[4] Hemmi R Yokomizu Y and Matsumura T 2003 Anode-fall voltage of air arcs between electrodes of copper, silver and tungsten at currents up to 1500A *Technical meeting on electrical discharges* IEE Japan 2003
[5] Yokomizu Y Matsumura T Henmi R and Kito Y 1996 *J. Phys. D.: Appl. Phys.* **29** 1260
[6] Phelps A V and Petrović Z L 1999 *Plasma Sources Sci. Techn.* **8** (1999) R21-R44
[7] Francis G 1956 *The glow discharge at low temperature (Encyclopedia of Physics vol. XXII Gas Discharges II)* ed Flügge S (Berlin: Springer-Verlag) chapter 2 pp 53-203
[8] Busz-Peuckert and Finkelnburg W 1956 *Z. für Phys.* **144** 244

A Study of Electrostatic Charge on Insulating Film by Electrostatic Force Microscopy

K. Kikunaga, K. Toosaka, T. Kamohara, K. Sakai and K. Nonaka

National Institute of Advanced Industrial Science and Technology, Tosu, Saga 841-0052, Japan

E-mail: k-kikunaga@aist.go.jp

Abstract. Electrostatic charge properties on polypropylene film have been characterized by atomic force microscopy and electrostatic force microscopy. The measurements have been carried out after the polypropylene film was electrified by contact and separation process in an atmosphere of controlled humidity. The negative and positive charge in concave surface has been observed. The correlation between concave surface and charge position suggests that the electrostatic charges could be caused by localized contact. On the other hand, positive charge on a flat surface has been observed. The absence of a relationship between surface profile and charge position suggests that the electrostatic charge should be caused by discharge during the separation process. The spatial migration of other positive charges through surface roughness has been observed. The results suggest that there could be some electron traps on the surface roughness and some potentials on the polypropylene film.

1. Introduction

The study of electrostatic charge on an insulating surface is most important for clarifying the elementary process of electrification, controlling of electrostatic charge and finding effective charge relaxation techniques. An insulator is electrified by contact and friction process between two materials and the resulting discharge during the separation process. The amount of charge and its characteristic are determined by materials or their surface characteristics [1]. Some correlation between surface voltage and surface roughness on the insulator has been reported [2, 3]. They suggest that the charge should be transported at localized contact positions. Also, they anticipate that the surface roughness should form some electron traps, which should effect a change in charge amount and characteristic.

Microscopic evaluations of electrostatic charge created by contact process and discharge have been attempted using Atomic Force Microscopy (AFM) and Electrostatic Force Microscopy (EFM) [4-6]. The AFM and EFM have been used to investigate the surface charge properties of semiconductors and insulators [6-14]. In this study, the charge properties on an insulating surface have been characterized by AFM and EFM. The samples were used polypropylene films, which provide excellent electrical insulation. It is expected that the charge feature would be unaffected for a few minutes and could be used to study the detailed surface phenomenon. The obtained AFM and EFM images suggest that the charge transport should be caused by localized contact. On the other hand, other AFM and EFM images suggest that the electrostatic charge on polypropylene films should be caused by discharge during the separation process. Also, the time evolution of some EFM images suggests that the charges from a discharge would migrate to stable positions on the surface.

2. Experiments

The charge and surface properties of polypropylene films were characterized by AFM and EFM at room temperature in air using Innova Scanning Probe Microscope (Veeco Instruments) with a Pt/Tr coated tip (SCM-PIT). Samples were 150 μm thick used polypropylene films. After they were electrified by contact or separation, the AFM and EFM measurement were carried out. In the system, the measurements were performed in a two-pass mode. The first pass (tapping mode AFM) gives the topography of the surface. The second pass (EFM) follows the track obtained in the first pass with a fixed lift-up distance and measures the phase shift caused by the electrostatic force while the tip is biased with a voltage of a few volts. The EFM measurement conditions were the tip bias voltage in the range of 0.1 to 1 V and the lift height in the range of 50 to 500 nm. The humidity was maintained at 50 percent during the measurements.

3. Results and discussion

3.1. Electrostatic charge by contact electrification

Figure 1 (a) and (b) show typical AFM and EFM images of the electrified polypropylene film with positive charge, and figure 1 (c) shows line profiles of the AFM topography and EFM signal intensity. The light color shows areas of positive charge in figure 1(b). Two positive charges in the 7 μm size range were observed on the concave surface. Figure 2 (a) and (b) show typical AFM and EFM image of the electrified polypropylene film with negative charge, and figure 2 (c) shows line profiles of the AFM topography and EFM signal intensity. The dark color shows an area of negative charge in figure 2(b). A negative charge in the 15 μm size range was observed on the concave surface. The scan areas in figure 1 and 2 are 30 μm x 30 μm. The results reveal the presence of a correlation between the concave surface and charge position, although the electrical characteristic is different. They suggest that the electrostatic charges on the polypropylene film could be caused by localized contact. Figure 3 shows a change of the measured EFM images after a period of time at 0 to 37 hours and line profiles of the EFM signal intensity. The scan area is at the same position as shown in figure 2. The decrease

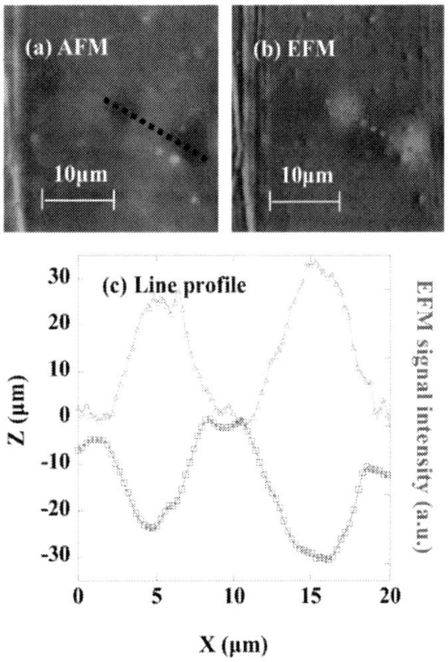

Figure 1. Typical (a) AFM and (b) EFM images of the electrified polypropylene film with positive charge, (c) line profiles of the AFM topography and EFM signal intensity.

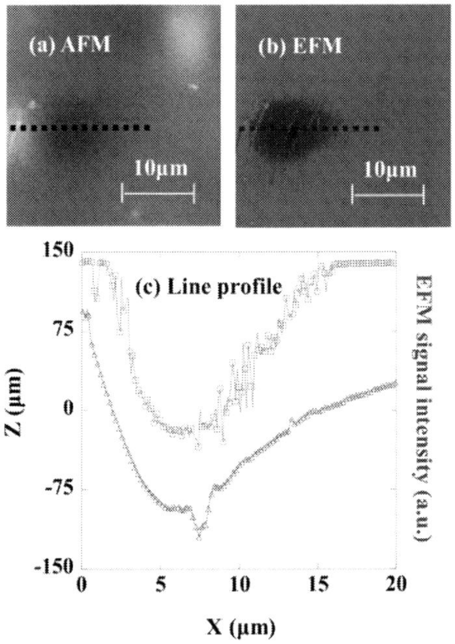

Figure 2. Typical (a) AFM and (b) EFM images of the electrified polypropylene film with negative charge, (c) line profiles of the AFM topography and EFM signal intensity.

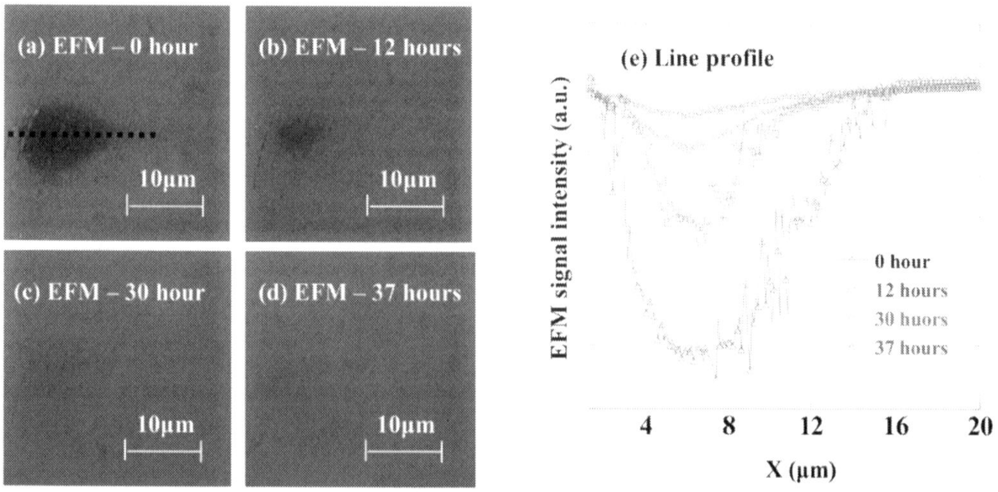

Figure 3. A change of the measured EFM images after a period of time at (a) 0, (b) 12, (c) 30 and (d) 37 hours and (e) line profiles of the EFM signal intensity.

of negative charge area and EFM signal intensity was observed. This result means the presence of trapped charge in the insulator migrating over time to cancel those in the surface. Though not shown here, AFM images on the same area remained unchanged during the measurements.

3.2. Electrostatic charge by discharge

Figure 4 shows a change in the measured AFM and EFM images of the electrified polypropylene film with positive charge after a period of time from 0 to 45 minutes. The scan area is 50 μm x 70 μm. In the same way as above, AFM images on the same area remained unchanged while measuring. A positive charge in the 40 μm size range was observed. This result reveals the absence of a relationship between surface profile and charge position, as shown in figure 4 (a) and (b). They suggest that the electrostatic charge on the polypropylene film should be caused by a discharge during the separation process. In the EFM image of the same area after 15 minutes, another positive charge was observed, as shown in figure 4 (c). Then the charges contacted through surface roughness, as shown in figure 4 (d). These results imply that the second positive charge had migrated. Although this should create a repulsive force between positive charges, the second charge could be pulled by a stronger attractive force. Also, the spatial migration of the second charge through surface roughness implies that there

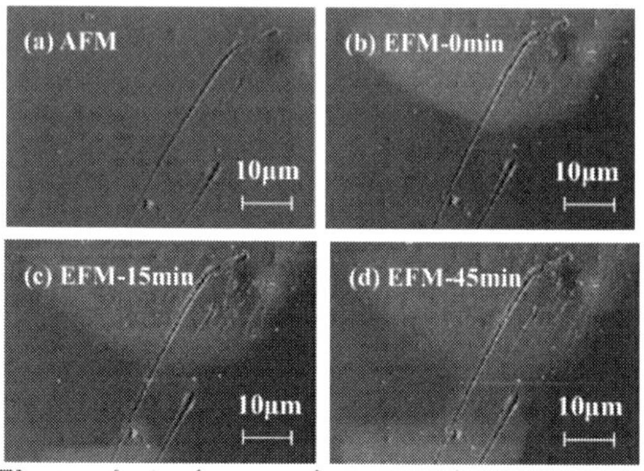

Figure 4. A change of measured (a) AFM and EFM images of the electrified polypropylene film with positive charge after a period of time at (b) 0, (c) 15 and (d) 45 minutes.

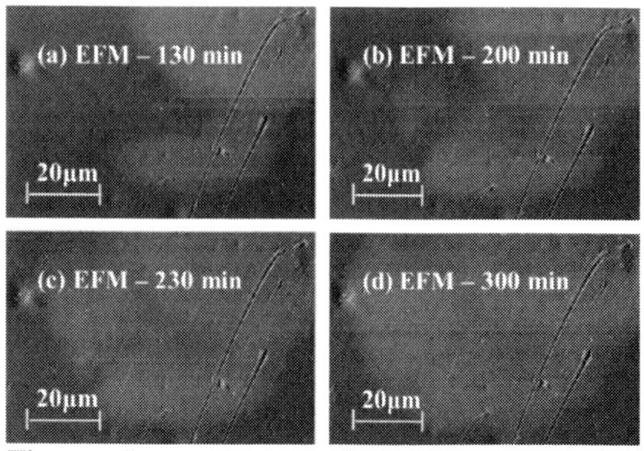

Figure 5. A change of the measured EFM images of the electrified polypropylene with positive charge after a period of time at (a) 130, (b) 200, (c) 230 and (d) 300 minutes.

could be some electron traps on surface structural defects [2, 3]. Figure 5 shows a change in the measured EFM images of the electrified polypropylene with positive charge after a period of time at 130 to 300 minutes. The scan area is wide range 70 μm x 100 μm at the same position as shown in figure 4. The view of second charge has migrated more, and the combined charge of the round shape was formed by the positive charges. The results suggest that there could be some potentials on the polypropylene film [15]. The shape of the combined charge remained unchanged after 300 minutes, and the charge disappeared with time.

3.3. Difference between charge by contact electrification and one by discharge

In the electrostatic charge properties by contact electrification, the charge on polypropylene film disappeared without migrating to other position. This indicates the possibility that surface topography is changed by localized contact. At the same time, electron traps could be formed at the same position. On the other hand, in the electrostatic charge properties by discharge, the charge on the polypropylene film changed position. This indicates the possibility that the charge was adsorbed on the surface after discharge. In that case, it is not always true that the charged position is a stable position for potential. Therefore, the charge could migrate slowly to a position with low potential for a few hours.

4. Conclusions

Electrostatic charge properties on polypropylene film have been characterized by electrostatic force microscopy and atomic force microscopy. The characterization of the charges by contact electrification and discharge indicate the possibility that there could be some electron traps on the insulating surface. Also, the characterization of the charge by discharge indicates the possibility that there could be some potentials on the polypropylene film. This present study suggests that there could be some charges of at least two types, and their characteristics could be difference. These results would support the understanding of the electrification phenomenon.

References

[1] Diaz A F and Felix-Navarro R M 2004 *J. Electrostat.* **62** 277
[2] Vladykina T N, Deryagin B V and Toporov Yu P 1985 *Phys. Chem. Mech. Surfaces* **3** 2817
[3] Murata Y 2005 *J. Surf. Fin. Soc. Jap.* **56** 436
[4] Terris B D, Stern J E, Rugar D and Hamin H J 1989 *Phys. Rev. Lett.* **63** 2669
[5] Schönenberger C and Alvarado S F 1990 *Phys. Rev. Lett.* **65** 3162
[6] Schönenberger C and Alvarado S F 1991 *Mod. Phys. Lett.* **5** 871
[7] Morita S, Fukano Y, Uchihashi T, Okusako T, Sugawara Y, Yamanishi Y and Oasa T 1993 *Jpn. J. Appl. Phys.* **32** L1701
[8] Morita S, Sugawara Y, Fukano Y, Uchihashi T, Okusako T, Chayahara A, Yamanishi Y and Oasa T 1993 *Jpn. J. Appl. Phys.* **32** L1852
[9] Ludeke R and Cartier E 2001 *Appl. Phys. Lett.* **78** 3998
[10] Tzeng D S and Gwo S 2006 *J. Appl. Phys.* **100** 023711
[11] Sugawara Y, Uchihashi T, Abe M and Morita S 1999 *Appl. Surf. Sci.* **140** 371
[12] Stark R W, Naujoks N and Stemmer A 2007 *Nanotechnology* **18** 065502
[13] Sun Z, Han B, Wang X and Lei Q 2009 *Proc. 9th Int. Conf. on Properties and Applications of Dielectric Materials* p942
[14] Kikunaga K, Kamohara T, Sakai K and Nonaka K 2010 *Proc. Annu. Meet. Institut. Electrostat. Jpn.* p55
[15] Kao K C, Hwang W 1981 *Electrical transport in solids* p235

Journal of Physics: Conference Series **301** (2011) 012044 doi:10.1088/1742-6596/301/1/012044

Implementation and interpretation of surface potential decay measurements on corona-charged non-woven fabrics

B Tabti[1, 3], A Antoniu[1], M. Plopeanu[1, 2], B Yahiaoui[3], B Bendahmane[3] and L Dascalescu[1]

[1]PPRIME Institute, UPR 3346 CNRS – University of Poitiers – ENSMA, France
[2]Electrical Engineering Department, University "Politehnica", Bucharest, Romania
[3]Laboratory of Electrical Engineering, University of Bejaia, Algeria

E-mail: btabti@ymail.com; lucian.dascalescu@univ-poitiers.fr

Abstract. The aim of this paper is to discuss the peculiarities of the surface potential decay (SPD) curves obtained for certain non-woven media. The experiments were performed on samples of non-woven poly-propylene (PP) sheets, which are typically employed in the construction of air filters for heat, ventilation and air conditioning. The samples were in contact with a grounded plane, in order to: (1) ensure better charging and measurement reproducibility; (2) simulate the worst situation of practical interest. They were charged using either a high-voltage wire-type dual electrode or a triode-type electrode arrangement. The aspect of the SPD curves depends on the electrode configuration. When the electric field is strong enough, it can activate charge injection at the insulator-metal interface and extrinsic conduction.

1. Introduction

Electret filters consist of non-woven fabrics made of dielectric materials that have a quasi-permanent electrical charge. Electric charging of the media improves their particle collection efficiency by the electrostatic particle capture mechanism that enhances the conventional mechanical filtration without causing an increase of the pressure drop [1]. Thus, the efficiency of an air filter is related to the initial level and the persistence of its electric charge [2, 3].

Surface potential decay (SPD) measurement techniques are widely used for the investigation of the electric charge on dielectric surfaces in a wide range of industry applications [4-9] Several interesting observations on the corona-charging characteristics of a non-woven poly-propylene (PP) sheet air filter are reported in [10]: the surface potential of the filter media is limited by the local discharges that occur inside the porous sheet and the relative humidity of the ambient air accelerates the charge decay. More recent studies [11, 12] confirmed the limitation of the surface potential attained by high-resistivity corona-charged fabrics when placed on conducting surfaces.

In two previous papers [13, 14], the authors employed the SPD technique to evaluate some factors that influence the corona charging of fibrous dielectrics. The critical issue concerning surface potential decay measurements is the interpretation of the curves, since these physical processes can lead to similar responses [4, 15, 16]. The rate of the potential decay dVs/dt is often considered as a better observable than the potential decay itself. The product of the derivative by the absolute time in seconds ($tdVs/dt$) versus ($log\ t$) characteristic can be used for data analysis in order to separate phenomena with different characteristic times and amplitudes [17, 18]. The aim of this paper is to analyse the peculiarities of this mathematical transformation in the case of non-homogeneous dielectrics, such as the nonwoven fabrics.

2. Materials and method

The experiments were performed on 100 mm × 85 mm samples of nonwoven sheets of PP (sheet thickness of 300 μm and average fiber diameter of 20 μm, as shown in Fig. 1) in ambient air (temperature of 18 °C–22 °C and relative humidity of 30%–50%). The PP fibres represent roughly 15% of the volume of the media.

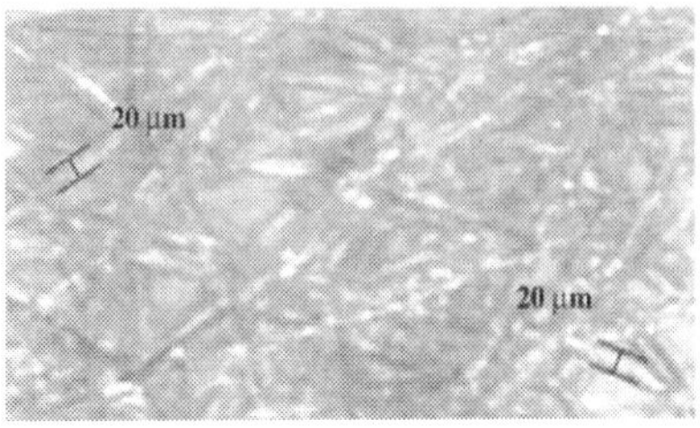

Figure 1. Photograph of the nonwoven PP media.

The samples were charged using the positive corona discharge generated by a high-voltage wire-type dual electrode [13], facing a grounded plate electrode (aluminium, 120 mm × 90 mm), as shown in Fig. 2(a). The high-voltage electrode consisted of a tungsten wire (diameter of 0.2 mm) supported by a metallic cylinder (diameter of 26 mm) and distanced at 34 mm from its axis. The wire and the cylinder were energized from the same adjustable high-voltage supply (model SL 300, SPELLMAN; rated voltage: 50 kV; rated current: 6 mA). Unless otherwise specified, the distance between the wire and the surface of the plate electrode was 30 mm. In some experiments, a grid electrode was interposed between the wire and the plate, to obtain a triode type electrode arrangement [14].

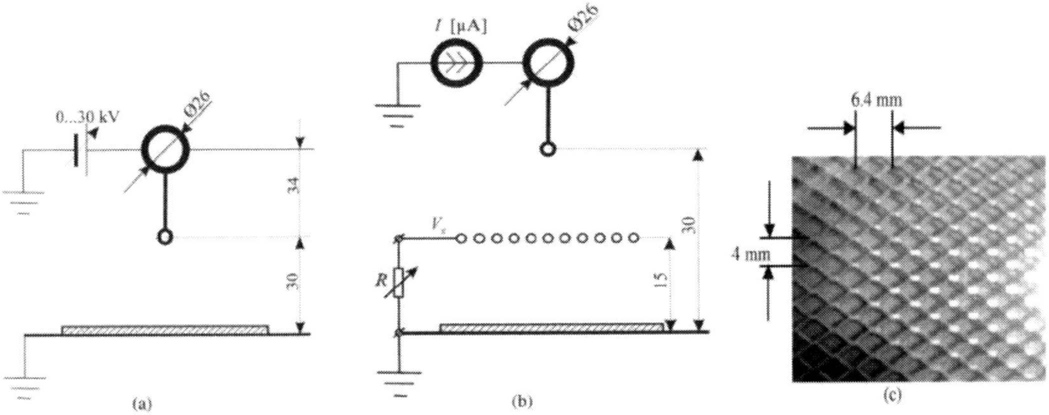

Figure 2. Electrode systems employed for the corona-charging of nonwoven media (all dimensions are in millimeters). (a) Wire-type dual electrode facing a grounded plate. (b) Triode-type arrangement. (c) Aspect of the grid electrode (grid-wire diameter: 1.18 mm).

In all the experiments, the samples were charged for 10 s (a duration beyond which no significant increase of the initial surface potential was noticed) by exposing them to the corona discharge, at various values of the high voltage applied to the dual electrode or at various grid potentials, in the case of the triode system. The samples were placed in contact to the grounded place so that to: (1) ensure better charging and measurement reproducibility; (2) evaluate the charge decay in the worst possible conditions. As soon as the high-voltage supply of the corona charger was turned-off, the conveyor belt transferred the samples from Position 1 to Position 2 (Fig. 3), where the surface potential was measured with an electrostatic voltmeter (TREK, model 341B, equipped with a probe model 3450, accuracy: ±(0.1% of full scale, drift with temperature 200 ppm/°C), calibrated before each set of measurements. The measured potential was monitored via an electrometer (Keithley, model 6514) connected to a personal computer (Figure 3). The processing of the data was performed using a virtual instrument in the LabView environment.

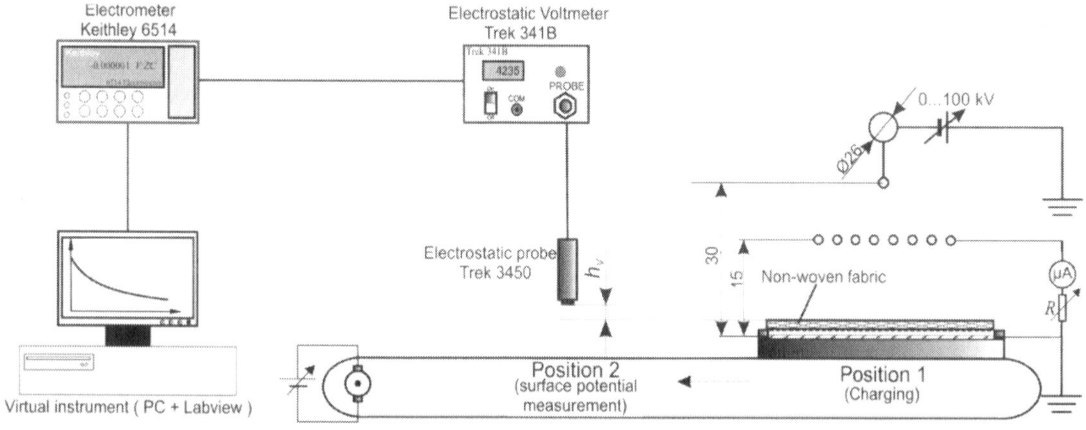

Figure 3. Experimental set-up for SPD measurement on non-woven fabrics.

3. Results and discussion

The results of the SPD experiments carried out with the wire-type dual electrode system for different values of applied high voltage and with the triode-type electrode arrangement for different values of grid voltage are shown in figure 4. The main feature of these curves is a slow decay for low surface potential values. At higher surface potentials, the slope of the curves becomes steeper, leading to the so called "cross-over phenomenon" [4], which is mainly due to charge injection. According to the models described in the literature [18], the crossover can be explained by the proportionality of the initial SPD rate to the square of the charge deposited on the surface the media.

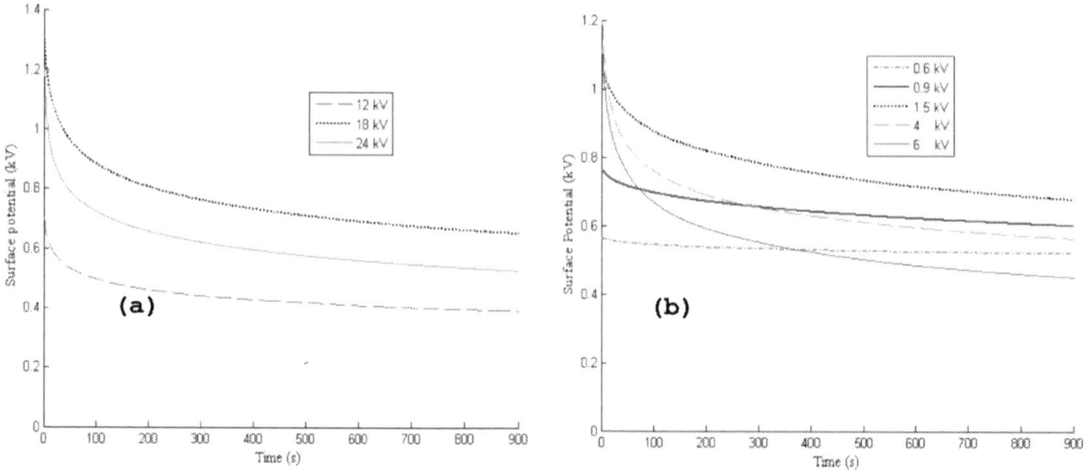

Figure 4. Surface potential decay curves for the "dual" corona electrode arrangement, at several applied voltages (a), and for the triode electrode system, at several grid voltage (b).

The representation $tdVs/dt = f(logt)$ of SPD measurements for corona-charged non-woven fabrics are shown in figure 5 for both electrode arrangements. In the case of the triode electrode system, the basic response curve is obtained for grid voltages of 0.6 kV and 0.9 kV. This intrinsic response of the media occurs in conditions of moderate temperature and electric field, which are factors of injection activation. The response amplitude increases with the grid voltage, and this corresponds to the occurrence of broad peaks on the 1.5 kV, 4 kV, and 6 kV curves (figure 5, b). At relatively high voltages, this broad peak – which is caused by charge injection from the grounded electrode [18] - is centred at about $10^{1.65}$ s, and is superimposed on the baseline response.

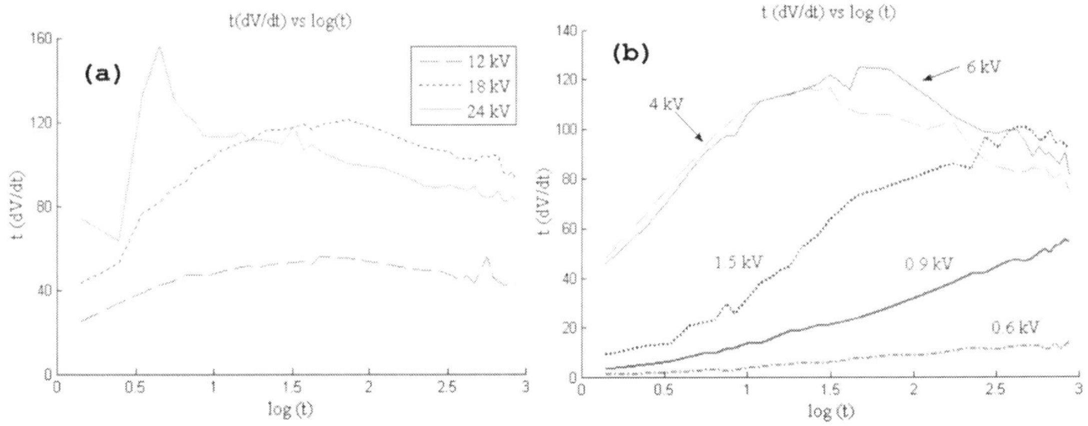

Figure 5: Plots of *tdVs/dt* curves for the "dual" corona electrode arrangement, at several applied voltages (a), and for the triode electrode system, at several grid voltage (b).

This type of injection, which occurs beyond a threshold voltage, is due to the increased electric field strength at the insulator-metal interface [4], so that the charge carriers can cross the potential barrier between the ground electrode and the surface states states of polypropylene fibers, to pass into the conduction band of the dielectric.

In the absence of the grid (figure 5, a), at applied voltages 12 kV and 18 kV, a broad peak replaces the basic response curve. The characteristic feature for this type of corona-charging is the steep peak with a long descent that occurres for an applied voltage equal to 24 kV, at around $10^{0.65}$ s. This can be explained by the intensity of the electric field and the non-uniformity of the charge deposit. When the electric field is strong enough, it can activate the injection of charges and extrinsic conduction [4]. At the same time, the fibrous and non-homogeneous structure of the media, associated with the non-homageneity of the charge deposite due to the absence of the grid, can cause paths to the ground electrode that will accelerate charge decay at the surface of the samples.

References

[1] Romay F J, Liu B Y H and Chae J S 1998 Aerosol Sci. & Technol. 28 224
[2] Tsai P P, Schreuder-Gibson H and Gibson P 2002 *J. Electrostat.* **54** 333
[3] Hutten I M 2007 *Handbook of Nonwoven Filter Media* (Oxford: Elsevier)
[4] Molinié P and Llovera P 2000 *Proc. Dielectr. Mater., Meas. Appl. (IEE Conf. Publ. No. 473)* pp. 253–258.
[5] Sahli S, Bellel A, Ziari Z, Kahlouche A and Segui Y 2003 *J. Electrostat.* **57** 169
[6] Xu Z, Zhang L and Chen Z 2007 *J. Phys. D: Appl. Phys.* **40** 7085
[7] Blacker RS and Birley A W 1991 *Polymer Testing* **10** 241
[8] Molinié P, Goldman M and Gatellet J 1995 *J. Phys D: Appl. Phys.* **28** 1601
[9] Chubb J 2002 *J. Electrostat.* **54** 233
[10] Oda T and Ochiai J 1988 *Proc. 6th Int. Symp. on Electrets* pp. 515–519
[11] Kacprzyk R and Mista W 2006 *Fibers and Textiles in Eastern Europe* **15** 35
[12] Hörenstein M 1995 *J. Electrostat.* **35** 31
[13] Tabti B, Dascalescu L, Plopeanu M, Antoniu A and Mekideche M R 2009 *J. Electrostat.* **69** 193
[14] Tabti B, Mekideche M R, Plopeanu M C, Dumitran L M, Herous L and Dascalescu L 2010 *IEEE Trans. Ind. Appl.* **46** 634
[15] Molinié P 2005 *IEEE Trans. Diel. & Elec. Insul.* **12** 939
[16] Lewis T J 1990 *J. Phys. D: Appl. Phys.* **23** 1469
[17] Mady F, Renoud R and Ganachaud J-P 2002 *Proc. 3rd SFE* (Toulouse: SFE)
[18] Llovera P and Molinie P 2004 *IEEE Trans. Diel. & Elec. Insul.* **11** 1049

'Stutter timing' for charge decay time measurement

John Chubb

Infostatic, 2 Monica Drive, Pittville, Cheltenham, GL50 4NQ, UK
(jchubb@infostatic.co.uk)

John Harbour

Hawthorne Technical Design, The Hawthornes, Startley, Chippenham, SN15 5HG,UK

Ian Pavey

Chilworth Technology Ltd, Beta House, Southampton Science Park, Southampton, SO16 7NS, UK

Abstract. The paper describes the approach of 'stutter timing' that has been developed to improve the accuracy of measuring charge decay times in the presence of noise in compact and portable charge decay test instrumentation. The approach involves starting and stopping the timing clock as the noisy signal rises above and falls below the target threshold voltage level.

1. Introduction

Assessment of the ability of materials to dissipate electrostatic charge requires measurement of the time for surface potentials to fall from an initial level to an end point level. The initial level is to be either the maximum initial peak voltage achieved by the charging action or, in the case of corona charging, may better be chosen to be a set time after the end of charging [1,2]. A delay, for example 0.1s, will mirror the time taken in tribocharging situations for the contacting surfaces to separate and for the surface voltages created to develop electric fields that will influence items nearby. The end point voltage level is set to be a defined fraction of the initial voltage – and the values of 1/e (37%) and/or 10% are recommended [1,2,3].

Surface voltages in charge decay studies are preferably measured, without contact, by a 'field mill' type of electrostatic fieldmeter [4]. The fieldmeter signal inevitably includes some noise. This may involve oscillatory components, related to the chopping frequency, and also quasi-random components. Noise will affect the accuracy with which decay times can be measured. The problem only becomes significant when dealing with materials having very long charge decay times, where the rate of decay may slow up significantly during the progress of charge decay, and where surface voltages are low so that signal to noise ratios are less than desirable – for instance where a high capacitance for surface charge depresses the surface voltage to a low level.

If, for example, the noise on the fieldmeter signal has a peak amplitude of N (volts) and the average rate of decay of surface voltage around the timing point is dV/dt (V s^{-1}) then if timing is judged simply by the signal's first crossing of the set voltage threshold, the end of timing will occur too early by a time T(s):

$$T = N/(dV/dt) \qquad (1)$$

Published under licence by IOP Publishing Ltd

In general, charge decay curves do not follow an exponential form but slow up progressively as decay proceeds. Where the rate of decay almost plateaus out towards the set threshold level, where dV/dt become very small, the measurement of decay time by first crossing of a target threshold level will be strongly affected by noise.

This paper is concerned with the technique of 'stutter timing' that was developed to overcome the problem of real time measurement of decay times in low power, self-contained, portable charge decay instrumentation suitable for use in industrial environments.

2. 'Stutter timing'

The basic technique of 'stutter timing' is for the timing clock, with a suitably short basic time step, to be started and stopped repetitively as the fieldmeter analogue source signal passes below and back above the selected target voltage level until it no longer traverses this level. The accumulated clock time shows the effective time at which the voltage level was crossed. The approach is a sort of simulation of manual drawing of a 'best fit' line through data points.

Figures 1 and 2 show spreadsheet modelling of noisy charge decay curves and the influence of noise on the measurement of decay time. The basic decay curve for both figures used a charge decay curve form that slows up progressively with time during the progress of decay [5]. The decay curve modelled had an initial peak voltage of 100V, an initial decay time constant of 0.03s, and a rate of increase in local decay time constant with time of 2s per second. The time to 10% of the initial peak voltage was then 1.5s. The random noise added was based on a maximum peak-peak voltage of 4V. Decay curves were calculated on a spreadsheet with time steps of 0.01s.

Figure 1 shows the basic decay curve and this same curve with added noise. It shows the accumulation of time until 10% of the initial peak voltage is reached for both the smooth basic curve and for the curve with added noise until the first crossing of the 10% level. With this level of random noise the end of timing, to the first transition across the target voltage threshold, occurs around 30% too early. Equation 1, above, would suggest the end of timing would be about 0.4s too early - from the value of dV/dt at the threshold level which may be compared the observed 0.45s.

Figure 2 shows the same noisy decay curve as Figure 1 but with 'stutter timing' used to measure the decay time. The top trace shows the timing clock starting and stopping as the signal varies above and below the 10% target threshold. Calculated decay times from various tests with different random number sequences for the noise, were within 3.5% too long.

3. Applications to practical charge decay measurement

Noise levels on fieldmeter signals have two main components: first, basic noise from operation of the fieldmeter sensing system, associated preamplifier circuits and any cross talk interference signals; second, noise arising from inadequate filtering of the signals after phase sensitive detection. Noise arising from digitisation of fieldmeter signals also needs to be minimised.

The output signal from phase sensitive detection includes noise mainly at twice the source signal chopping frequency. Filtering to reduce this level of noise to a suitably small fraction of the r.m.s. output signal level needs to be chosen to avoid prejudicing the response to fast changes in the source electric field signal.

For good identification of the initial peak surface voltage the response time of the fieldmeter needs to be much shorter than the time for opening the fieldmeter view of the test surface during removal of the plate carrying the corona charging electrodes. In practical charge decay instruments (for instance JCI 155 instruments) the opening time is 15-20ms so the fieldmeter response time needs to be 5ms or less [1,2,3]. This requires a fairly high chopping frequency in the fieldmeter, for instance around 2kHz. Careful design of the output signal filtering is needed to maintain good response to the initial peak voltage signal together with good rejection of phase sensitive detection noise. Care is also needed to minimise swashplating of the field mill rotor assembly.

Two facilities can be used to improve signal to noise ratios with digital signal processing:

a) the fieldmeter output can be simultaneously recorded by digitisation at several accurately related levels of sensitivity. The observations used in calculation of surface voltage at any time are then those from the highest sensitivity range that is not in saturation. This will give best use of the ADC resolution available.

b) the effective noise can be usefully reduced in the processing of digitised signals by averaging larger and larger numbers of data points during the progress of charge decay. This enables a fast response to be offered for short decay times with progressively reduced influence of noise as charge decay proceeds.

Stutter timing' has been used in the processing of fieldmeter signals as follows:

1) **Initial peak:** The initial peak is determined as the maximum median (or the average) of 4 successive fieldmeter readings. The peak signal is observed as a competition between the fieldmeter reaching maximum sensitivity, as the plate is fully retracted, and the decay of already deposited charge. The 'time zero' and the value of the 'initial peak voltage' are taken 2ms after occurrence of this peak. After choosing the initial peak voltage value the time zero can be calculated with better accuracy by stutter timing after the relevant range of readings has been collected
2) **Delayed start voltage:** The surface voltage a set time after end of charging (so corona charge decay observations match those after tribocharging [2,3]) is initially determined as the average of 2 readings before and 2 readings after that time. Time zero may then be more accurately calculated by stutter timing calculations as for the 'Initial peak'.
3) **End of decay timing:** The end of decay time, to a set % of the initial voltage, is determined by stutter timing by starting and stopping of the timing clock as the signal and noise fall through the end point voltage level.

Application of the above approaches for decay time measurement in practical charge decay measuring instrumentation (for instance JCI 155v5 and JCI 155v6 instruments) has shown that surprisingly good performance is achieved – even with quite noisy signals and slow rates of charge decay.

An important feature of this timing method when applied to instruments using processors with limited speed and program size is that it is easy to write compact code, needing no floating point arithmetic, and using a state machine concept to allow stutter timing to several different levels "simultaneously". There is no problem about a seamless transition between processing the initial fast changing part of a charge decay effectively offline, and the slower part of the decay when readings are taken less frequently and can be processed in available processor time as soon as the data is stored, i.e. effectively online. The method is self-optimising for different levels of noise.

It will be noted that if the noise peak level meets or exceeds the surface voltage level at which stutter timing is performed, the timing will in theory never stop. For this reason the method will not work on decay curves swamped with noise around the end of timing target level. It is not easy to determine what method could work in such a case – except perhaps off-line curve matching.

4. Conclusions

A method 'stutter timing' for measuring charge decay times has been described. This technique provides significantly better accuracy of decay time measurement in the presence of real signal noise than is achieved with measurement of decay time by first crossing of a target threshold level. The accuracy of decay time measurement may not be as high as can be achieved with curve fitting during post-processing of charge decay curve data, but it is a simple approach that provides real-time decay time measurement in low power, self-contained, portable charge decay instrumentation used to satisfy industrial test requirements for the electrostatic suitability of materials.

References:

[1] J N *Chubb* *"Test method to assess the electrostatic suitability of materials for retained electrostatic charge"* Document prepared in 2004 for discussion in GEL101 as prospective British Standard. Available in *"An Introduction to Electrostatic Measurements"* Annex 2 [2]

[2] J N Chubb *"An Introduction to Electrostatic Measurements"* 'Nova Publishers', 2010. ISBN 978-1-61668-251-4

[3] J N Chubb *"A Standard proposed for assessing the electrostatic suitability of materials"* J. Electrostatics 65 2007 p607-610

[4] J N Chubb *"Instrumentation and standards for testing static control materials"* IEEE Trans Ind Appl 26 (6) Nov/Dec 1990 p1182

[5] J N Chubb, H Walmsley *"Analysis of long corona charge decay times"* J. Electrostatics 68 2010 p469-472.

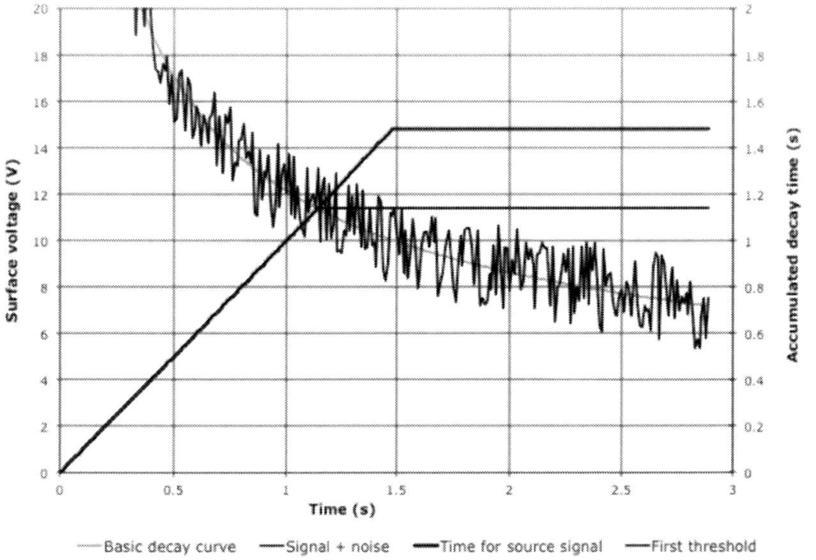

Figure 1: Model of decay timing to first crossing of target voltage threshold at 10V (10% of initial peak voltage of 100V) is 1.5s. First crossing of threshold by noise signal occurs 30% too early.

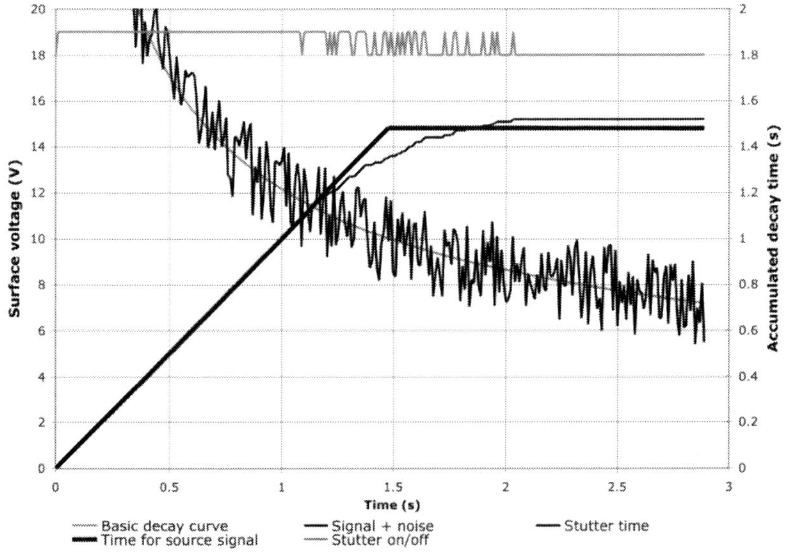

Figure 2: Model of decay timing with 'stutter timing'.

Electric field measurement using a non-perturbative method based on a calibrated electric potential sensor

A Aydin, P B Stiffell

Centre for Physical Electronics and Quantum Technology, School of Engineering and Design, University of Sussex, Brighton, BN1 9QT, UK

E-mail: a.aydin@sussex.ac.uk

Abstract. We present results of finite element analysis for simple test structures which demonstrate clearly that the measurement situation is complex. The test structure consists of an open geometry parallel plate capacitor within a screened enclosure. Indeed, the presence of earthed objects, even at considerable distances, is shown to have a significant effect on the field geometry close to the source. These simulations are compared with field measurements made using an ultra-high input impedance sensor, the Electric Potential Sensor. A single experimentally determined calibration factor is all that is required to achieve excellent agreement between experimental measurements and the results of the simulations. Given this, the sensor is capable of mapping accurately, and in a non-perturbative manner, the spatial potential both within and outside of the test structure.

1. Introduction

It is often assumed that the loading effect which an electrostatic field meter (ESFM) imposes on the electric field to be measured is well understood. In particular, that this may be accounted for using a simple capacitance calculation. In practice, the loading of the electric field by a meter is more complex than this, especially if the measuring electrode is intricate and/or within the unit. This can introduce measurement errors in excess of 100% [1], which can be a direct result of the meter or the method used to make the measurement. A non-invasive or non-perturbing way of measuring electric field is necessary in order to make reliable and accurate measurements.

There have been a limited number of specialized ESFMs developed over the years to measure electric field in a variety of environments but more so in industrial settings [2]. A recent development and addition is the Electric Potential Sensor (EPS or EP sensor). The EP sensor is a highly diverse ultra-high input impedance device with unique feedback circuitry, which achieves a high level of sensitivity. The EP sensor is a patented technology invented at the University of Sussex and has had a multiplicity of applications. These include imaging of active electrical circuits [3], electric field nuclear magnetic resonance signal detection [4], and imaging of charge spatial density on insulating materials [5]. Generic electrical specifications for an EP sensor are; input capacitance $<10^{-15}$ F; input resistance $>10^{13}$ Ω; bandwidth quasi d.c. to >100 MHz. It needs to be stated here that the EP sensor does not have a true DC response. However, static charge information can be recovered if the sensor is moved in a well-defined step motion [5]. The sensor specifications and response characteristics are described in our earlier publication [6]. This publication also outlines the high level of sensitivity of a calibrated sensor in a controlled electrical environment.

It is the intention of the authors to demonstrate that EP sensors may be used as non-perturbative field meters with extremely high sensitivity and the capability to work in unshielded environments. Progress has already been made using EP sensors in applications that bring them into open noisy environments [7]. A comparison of independent simulations and measurements are presented here.These findings present the possibility of making viable EP sensors as future ESFMs.

2. Experiment and simulation methodology

The experimental and simulation setup is similar to our treatment found in our prior publication [6]. An overview of this, together with a description of the measurement methods used is given here. The experiment is conducted within an aluminium Faraday enclosure, which sets the grounded boundary conditions. The physical enclosure has the dimensions 3.65 m (l) x 2.45 m (w) x 2.45 m (h). We adopt the Cartesian coordinate system to relate to the x-axis as the length, y-axis as the width, and z-axis as the height (see insert in figure 1). These dimensions and material parameters are modelled in the computational simulations of electric field distribution.

In the practical setup, a calculable electric field is generated within the enclosure using an aluminium parallel plate capacitor, measuring 1 m x 1 m with a non-metallic spacer producing 0.6 m of separation. The experiments carried out use an a.c. source to drive one of the plates with a voltage while grounding the other. For the practical experiment an a.c. voltage source with 10 V_{pp} at 158 Hz is used. This frequency is chosen to avoid multiples of the mains harmonics and to be within the bandwidth of the EP sensor. For the simulated excitation, a static 10 V is applied to the source plate. This produces the same instantaneous fall-off in potential across the plates as an a.c. signal and reduces the computation time required. The grounding of one plate and driving of the other is referred to as an *unbalanced excitation* mode.

The electric field measurements between the plates are made using the EP sensor. The EP sensor is mechanically supported externally to the parallel plate cavity, allowing a semi-rigid coaxial electrode to protrude into the desired measurement location. The unique electrical design of this EP sensor incorporates a guard driven outer coax of the electrode. This guard signal follows the measured potential to a high level of precision, with the effect of eliminating any perturbation on the generated electric field. This is only achievable by placing the electrode in a plane of near constant potential, namely the zx-plane. The electric field to be measured is coupled to a square copper electrode tip soldered to the inner coax, which has an area of 1 cm x 1 cm. This tip couples to the electric potential for a given measurement axis in either the y-axis through the centre of the plates, or along an x-axis that is at the mid-height of, and parallel to, the plates. For y-axis measurements the position is relative to the source plate. For the x-axis measurement, the coordinates are relative to the yz-plane.

For simulations, we make use of COMSOL Multiphysics to produce an electric field distribution to mirror the excitation and spatial conditions in the experiments. An identical wire-frame model of the enclosure and plates was constructed using relevant electrical parameters for aluminium and air.

Three different configurations have been devised for the electric field setup to demonstrate the effect on the field due to the boundary conditions. The first configuration places the source plate 0.2 m in front of the long wall (zx-plane) of the enclosure, with the remaining dimensions centralised. The second configuration places the plates precisely central to all axes in the enclosure. For these two positional configurations, a pseudo-colour spatial potential plot in the yx-plane is extracted from the simulations at the mid height of the plates. The final scenario is only achievable in simulation with the plates in infinite free space, ie, no grounded enclosure.

The x-axis measurements for all of these configurations are made 0.2 m, parallel to and in front of the source plate, i.e. between the plates. The practical measurement procedure was to position the EP sensor electrode, step out of the Faraday enclosure, seal the door, then to average the signal on a digital oscilloscope 64 times. The only active device in the enclosure was the EP sensor and its d.c. battery pack situated behind the grounded plate. The a.c. drive signal and EP sensor signal were carried through interface connectors in the enclosure wall.

3. Results

Figure 1 shows the simulated and experimental data for the y-axis measurements through the centre of the plates for the three configurations. These are a function of the output voltage for a given distance from the source plate. In the case of the EP sensor output, these are the averaged peak-to-peak output voltage measurements after a calibration factor is applied to compensate for the gain of the EP sensor in relation to the ideal results. An average calibration factor, of 0.92, was empirically calculated using a best fit to all the results. The EP sensor data is shown as points in figure 1 and the lines in the same figure are the matching simulation results. It can be seen that after the calibration factor is applied, the experimental results are in excellent agreement with the simulations. Figure 1 only shows results from 0.1 m, to the grounded plate at 0.6 m, due to the non-ideal behaviour of the EP sensor close to the active plate. This is due to direct capacitive coupling between the sensor electrode and the plate as was reported earlier [6]. It should be noted that a significant curvature is seen both in the simulations and the measurements. This result is mainly due to the aspect ratio of the structure. However, the proximity of earthed objects and finite boundary conditions also affects this behaviour as may be seen by comparing the three cases shown in figure 1.

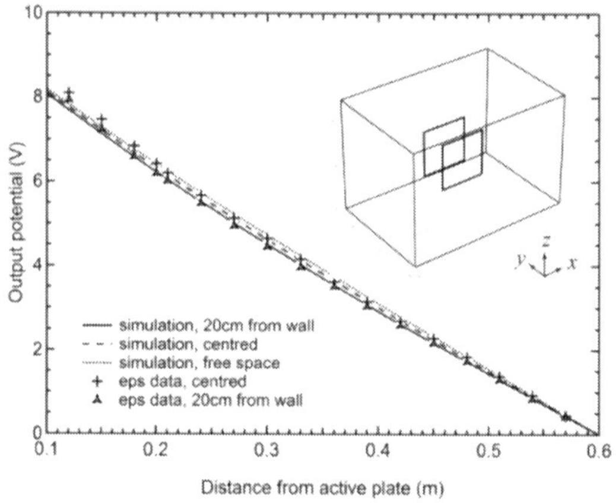

Figure 1. Simulation and experiment for y-axis. Configurations are; source plate 0.2 m from enclosure wall; plates centred; free space. The insert shows a wire frame model with axes.

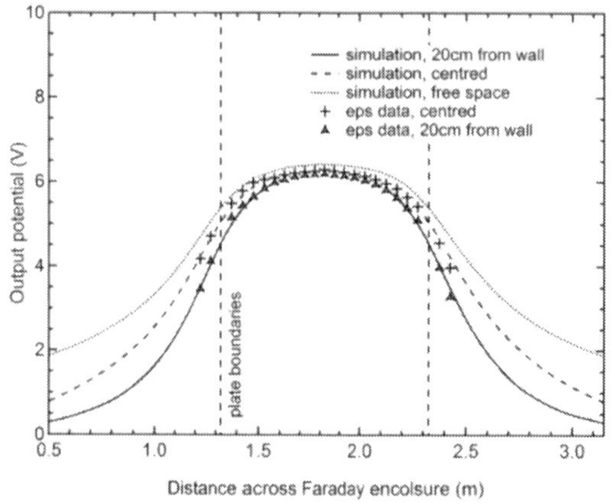

Figure 2. Simulation and experiment for x-axis at mid-height and 0.2 m in front of active plate. Configurations are; source plate 0.2 m from enclosure wall; plates centred; free space.

Figure 2 shows the data in the x-axis along the mid-height of and 0.2 m in front of the source plate. The output is a function of position with reference to the yz-plane edge wall of the enclosure. The vertical hatched lines in figure 2 show the edges of the plates. The lines represent the simulated data and the points the corresponding measurements. The curvature seen in figure 2 in the x direction is expected. The curvature persists within the plates due to the fringing created by the low aspect ratio of the parallel plate structure. Both the magnitude and field profile are in good agreement with the simulations.

Figure 3 shows a pseudo-colour plot of the simulation of the spatial potential between the plates, positioned 0.2 m from the long wall of the enclosure. The voltage data is extracted for the section through the mid-height of the plates, showing how the potential falls off to the grounded surroundings. The wire frame drawn depicts the enclosure boundary with a buffer margin outside this, used to simulate ground at infinity in the free space computations. The rate of voltage change over distance due to the nearby grounded wall suggests a confined field near the source plate. Figure 4 on the other hand, with the plates positioned central to all axes, shows a pseudo-colour spatial potential plot in the vicinity of the source plate, showing a less compressed field profile. As may be seen from the data presented in figures 1 and 2 there is a significant difference between the results for being centred within the enclosure and being in free space.

Figure 3. Simulated pseudo-colour voltage *yx*-plane plot through the mid-height of plates in unbalanced excitation with active plate 0.2 m from enclosure wall.

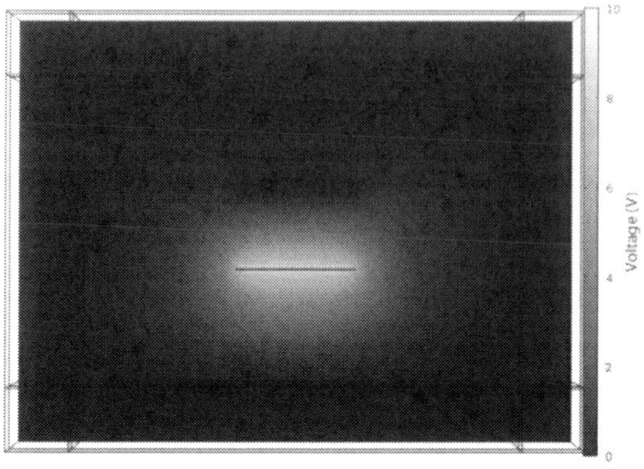

Figure 4. Simulated pseudo-colour voltage *yx*-plane plot through the mid-height of plates in unbalanced excitation.

4. Conclusions

The most interesting conclusion which may be drawn from these results is that the field is modified significantly, by the presence of the grounded enclosure, even when the test structure is carefully centralised on all axes. It is also evident that given a single sensitivity calibration the EP sensor is capable of making measurements both of the magnitude and of the field profile of the spatial potential without significantly perturbing the field. As such, this sensor is suitable for mapping of spatial potential, and hence the resulting electric field, with an ultimate sensitivity which has been demonstrated to be at the micro-Volt per metre level in an earlier publication [6]. Future work will concentrate on producing an open calculable electric field test structure which does not require the presence of a screened enclosure.

Acknowledgements

The authors would like to thank the Engineering and Physical Sciences Research Council for supporting this work under grant no: EP/E042864/1.

References

[1] Tishchenko V A, Tokatly V I and Luk'yanov V I 1998 Comparison and metrological investigation of electrostatic and low-frequency electric field meters *Meas Tech+* **41** 939-40
[2] Taylor D M 2001 Measuring techniques for electrostatics *J Electrostat* **51** 502-8
[3] Gebrial W, Prance R J, Clark T D, Harland C J, Prance H and Everitt M 2002 Noninvasive imaging of signals in digital circuits *Rev Sci Instrum* **73** 1293-8
[4] Prance R J and Aydin A 2007 Acquisition of a nuclear magnetic resonance signal using an electric field detection technique *Appl Phys Lett* **91**
[5] Beardsmore-Rust S T, Watson P, Prance R J, Harland C J and Prance H 2009 Imaging of charge spatial density on insulating materials *Meas Sci Technol* **20**
[6] Aydin A, Stiffell P B, Prance R J and Prance H 2010 A high sensitivity calibrated electric field meter based on the electric potential sensor *Meas Sci Technol* **21**
[7] Beardsmore-Rust S T, Prance R J, Aydin A, Prance H, Harland C J and Stiffell P B 2009 Signal specific electric potential sensors for operation in noisy environments - art. no. 012011 *J Phys Conf Ser* **178** 370

Journal of Physics: Conference Series 301 (2011) 012047 doi:10.1088/1742-6596/301/1/012047

Testing the electrostatic characteristics of polypropylene fabric with metallic yarns, intended for use in coal mines threatened by the explosion hazard. Part 1: Laboratory tests

M Talarek, L Orzech

KOMAG Institute of Mining Technology, Pszczynska 37, 44-101 Gliwice, Poland

E-mail: mtalarek@komag.eu

Abstract. The aim of this paper was to assess electrostatic safety of polypropylene fabric with metallic yarns intended for use in coal mines. Such fabrics have not been used in the Polish mining industry yet. The tests conducted have been divided into two subgroups: laboratory tests and tests in a coal mine. This paper presents the results of laboratory tests, which comprise charge transfer tests, impact of washing and mechanical stress on the resistance and resistance-to-ground at the manufacturer's site. Some problems with measuring the resistance are highlighted and discussed. The results obtained allow a reliable assessment to be made of the risk of using fabrics with metallic yarns in the explosive atmosphere, which often occurs in coal mines.

1. Introduction

Polypropylene fabrics with metallic yarns have not been used in the Polish mining industry yet and so far no criteria have been established to assess electrostatic safety of fabrics with metallic yarns. Consequently, a wide range of tests were arranged that have been divided into two subgroups: the laboratory tests and tests in a coal mine. The main objective of the work was to assess the risk of using fabrics with metallic yarns for mining applications. In this paper the results of laboratory tests are presented. Problems with measuring the resistance of fabrics with metallic yarns are highlighted and discussed. We also present the results of charge transfer tests, as well as the impact of washing and mechanical stress on the resistance and resistance-to-ground at the manufacturer's site.

The object tested was a composite, where conductive yarns made of stainless steel were present inside a matrix of highly resistive base fabric (polypropylene), as shown in Figure 1.

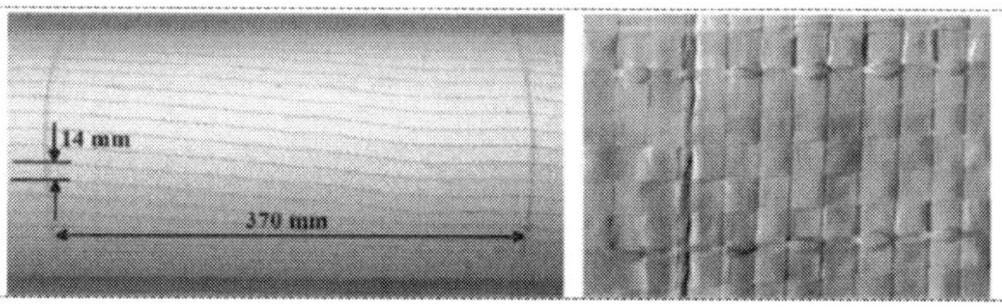

Figure 1. Structure of the fabric tested.

The EN 13463-1 Standard [1] describes the basic testing methods and requirements of non-electrical equipment to be used in a potentially explosive atmosphere and procedures for measuring surface resistace and charge transfer. To assess plastic fabrics intended for use in coal mines the EN 61340-2-3 Standard [2] describes procedures for measuring surface, volume and point-to-point resistance. Technical Report CLC/TR 50404 [3] recommends testing the impact of washing and the mechanical stress on the resistance of materials such as fabrics with metallic yarns. Requirements of flexible intermediate bulk containers (FIBC), which are usually made from materials such as the tested fabric, are given in the EN 61340-4-4 Standard [4], where the method of measuring resintance-to-ground is described.

2. Laboratory tests

2.1. Resistance tests

Surface resistance measured with the parallel stripe electrodes described in the EN 60079-0:2009 Standard [5], strongly depends on the direction of the electrodes in relation to the direction of the yarns [7]. The parallel electrodes are 100 mm long, 1 mm wide and separated by a distance of 10 mm. Tests were conducted with four configurations of electrodes in relation to the yarns. However, the results showed that this type of electrode arrangement is not applicable for materials such as fabrics with metallic yarns.

EN 61340-2-3 Standard [2] specifies a ring probe to measure surface resistance and point electrodes to measure point-to-point resistance as well as the resistance to a groundable point (see Figure 2).

Figure 2. Ring probe (left) and point electrode (right) (dimensions in mm).

As regards the surface resistance probe, it is also not applicable for fabrics with metallic yarns, because the outer ring of the probe which is 3 mm wide (Figure 3), does not make good contact with the yarns and the results can differ by eight orders of magnitude depending on the location of the probe, which is random. Only point electrodes make good contact with metallic yarns and can be used to measure point-to-point resistance as well as the resistance-to-ground. Tests of resistance-to-ground conducted using the point electrode are given in this paper later.

2.2. Charge transfer tests according to the EN 13463-1 Standard

The charge transferred in a single discharge was measured according to the EN 13463-1 Standard [1] by two independent accredited laboratories. Results are summarized in Table 1 [7].

Climatic conditions of the tests were as follows: temperature 23°C and relative humidity below 30%. The idea of the test is to measure charge Q transferred in a single discharge, as the ball electrode approaches the sample tested. Samples are charged by three methods: rubbing with polyamide cloth, method and the highest value measured was accepted as the final result. More about this subject can be found in [6]. Some differences between values obtained by the two laboratories have been noticed, although, all the results obtained met the criterion (60 nC [1]). Potential reasons for the differences have not yet been determined, because this was not the project objective; however possible factors include the following: human factor (personnel with individual features including speed and pressure of rubbing), relative humidity and different samples of material tested. Additional interesting information about errors in charge transfer measurement can be found in [8] and [9].

Table 1. Results of charge transfer tests according to EN 13463-1 Standard.

Laboratory	Sample	*Q* [nC]		
		Polyamide cloth rubbing	Cotton cloth rubbing	Corona charging
A	1	2.9	4.2	4.3
	2	5.9	3.8	5.7
	3	3.9	3.6	4.1
B	1	2.94	13.72	15.68
	2	34.30	30.38	16.66

2.3. Impact of washing and the mechanical stress on the resistance

Technical Report CLC/TR 50404 [3] recommends testing the impact of washing and mechanical stress on the resistance of materials such as fabrics with metallic yarns. Climatic conditions of the tests were the same as in section 2.1. The fabric was washed for 1 hour in water flowing at a rate of 1 dm^3/min. The resistance value after washing was the same as before.

To assess the impact of mechanical stress on the resistance of the fabric, the material was tested on the strength machine, where samples of fabric where elongated by 5%, 10% and 15% of their length for 60 seconds. The resistance value after the mechanical stress tests was the same as before. Besides, there was no rupture of the material, which could result in the appearance of isolated patches of the basic fabric.

2.4. Resistance-to-ground at the manufacturer's site

Resistance-to-ground was measured at the manufacturer's site for two cases: for empty bags and for bags filled with just prepared material, as shown in Figure 3.

Figure 3. Tests of resistance-to-ground at the manufacturer's site.

Climatic conditions of the tests were as follows: temperature 5°C and relative humidity 65%. Measurements were taken with a Metriso2000 resistance-meter with point electrode - resistance probe-model 850 (Figure 2) and with clip electrode. Both of the electrodes had good contact with the yarns and the results for both were consistent with most of order 10^4 Ω.

3. Results

Problems were encountered with using parallel stripe electrodes and the ring probe electrode to measure surface resistance of fabrics with metallic yarns. Neither is suitable for such kinds of fabric. Resistance-to-ground can be measured with point or clip electrodes. For this project we used point electrodes [2]. In the laboratory tests, charge transferred in a single discharge was tested in two independent laboratories using 5 different samples of materials. The worst result was 34.30 nC, which is well below the criterion of 60 nC [1]. Additional tests of washing and mechanical stress impact on the resistance were carried out in the laboratory, as recommended in Technical Report CLC/TR 50404 [3]. Post-test resistance values for both washing and mechanical stress were the same as before. Also, no rupture of the material was observed during the course of the tests. Resistance-to-ground was measured at the manufacturer site for empty bags and bags filled with just prepared material. Most of the results were about 10^4 Ω.

4. Conclusion

The tested object was the composition of conductive yarns made of stainless steel and highly resistive base polypropylene fabric. To assess the electrostatic safety of this kind of material charge transfer test, resistance to ground test and test of washing and mechanical stress impact on the resistance should be carried out in the laboratory. In our opinion it is also very important to verify the resistance-to-ground in real conditions of using fabric, which is a coal mine environment in this case.

5. Acknowledgment

Authors would like to express warm thanks to Mr. Stanislaw Makarski and Mr. Mariusz Rasek of Minova Ekochem S.A. for creative discussions and cooperation during the implementation of this project.

References

[1] EN 13463-1:2010 Non-electrical equipment for use in potentially explosive atmospheres Part 1: Basic method and requirements
[2] EN 61340-2-3:2002 Electrostatics - Part 2-3: Methods of Test for Determining the Resistance and Resistivity of Solid Planar Materials Used to Avoid Electrostatic Charge Accumulation
[3] CLC/TR 50404 Electrostatics. Code of practice for the avoidance of hazards due to static electricity
[4] EN 61340-4-4:2006 Electrostatics - Part 4-4: Standard test methods for specific applications – Electrostatic classification of flexible intermediate bulk containers (FIBC)
[5] EN 60079-0:2009 Explosive etmospheres - Part 0: Equipment - General requirements
[6] U. von Pidoll, E. Brzostek und H.-R. Fröchtenigt, Determining the incendivity of electrostatic discharges without explosive gas mixtures, IEEE Transactions on Industry Applications, 40 (2004), 1467-1475
[7] S. Makarski, M. Rasek, L. Orzech, M. Talarek, Badania tkaniny polipropylenowej z przeplotem metalowym stosowanej w wyrobiskach górniczych w aspekcie zagrożenia wybuchem związanym z elektrycznością statyczną, Wiadomości Górnicze, Katowice (Poland), 9/2009
[8] J. N. Chubb, Measurement of charge transfer in electrostatic discharges, J. Electrostatics, 64 (2006) 321-325
[9] H. L. Walmsley, Induced-charge errors in charge-transfer measurement, J. Electrostatics, 67 (2009) 320-325

Testing the electrostatic characteristics of polypropylene fabric with metallic yarns, intended for use in coal mines threatened by the explosion hazard. Part 2: Tests in coal mine

M Talarek, L Orzech

KOMAG Institute of Mining Technology, Pszczynska 37, 44-101 Gliwice, Poland

E-mail: mtalarek@komag.eu

Abstract. The aim of this paper was to assess the electrostatic safety of polypropylene fabric with metallic yarns intended for use in coal mines. Such fabrics have not been used in the Polish mining industry yet. The tests conducted have been divided into two subgroups: laboratory tests and tests in a coal mine. This paper presents the results of tests in a coal mine, where we have focused on the resistance-to-ground in some specific situations. Bags made of fabric at the roadway face were tested, as well as the roll of fabric during transport and carried by a miner. The results obtained allow the reliable assessment of the risk of using fabrics with metallic yarns in the explosive atmosphere which often occurs in coal mines.

1. Introduction

The main objective of the work was to assess the risk of using fabric with metallic yarns for mining applications. In the laboratory tests [1] the problems with measuring the resistance of fabric with metallic yarns were highlighted and discussed. We also presented the results of charge transfer tests and the impact of washing and mechanical stress on the resistance and resistance-to-ground at the manufacturer's site [1]. Resistance-to-ground was measured for two cases: empty bags and bags filled with just prepared material. Most of the results were of order 10^4 Ω. However, since the fabric is intended for use in coal mines, it is important to verify resistance-to-ground in the place where the fabric is to be used.

The material to be tested was a composite in which conductive yarns made of stainless steel were present inside a matrix of highly resistive base fabric (polypropylene), as shown in Figure 1.

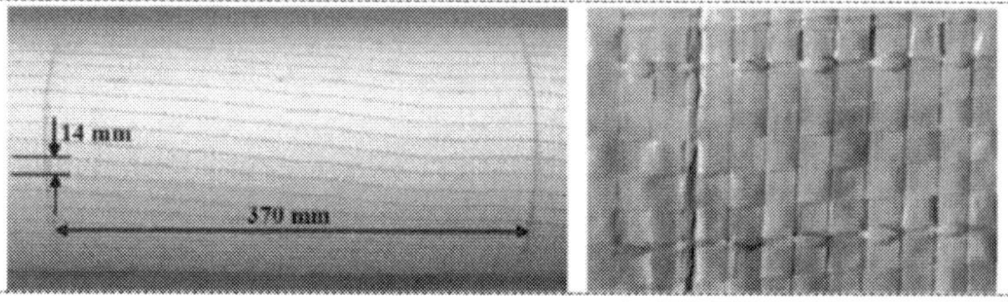

Figure 1. Structure of the fabric tested.

2. Tests in a coal mine

During the tests in a coal mine we focused on measurements of the resistance-to-ground in some specific situations.

Tests were carried out for two different cases in the roadway face: empty bags (1), bags filled with material just prepared (2), as well as bags filled with hardened material within 300 m of the roadway face (3). Environmental conditions where as follows: temperature 25°C and relative humidity 70% at the roadway face and temperature 29°C and relative humidity 70% 300 m from the roadway face. The above-mentioned conditions are shown in Figure 2. These conditions, as well as contamination of the fabric surface with carbon dust, had significant impact on the obtained results. Measurements were taken with Metriso2000 resistance-meter with point electrode - resistance probe-model 850. The resistance probe of weight 2.2 kg was located directly on the surface of the fabrics, which had not previously been cleaned. The results are summarized in Table 1. The worst results of resistance to ground were obtained for hardened material-filled bags, however all the results were below 10^6 Ω.

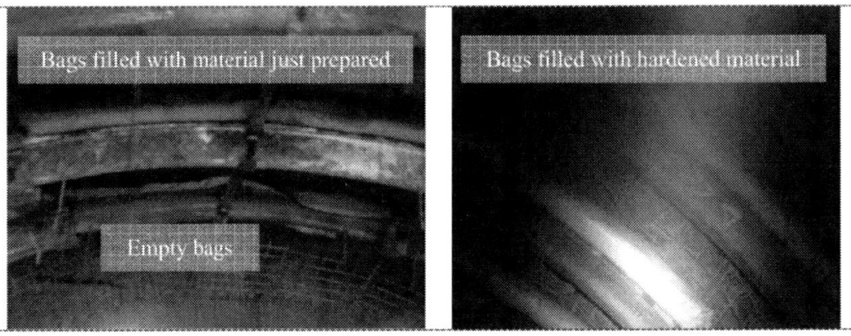

Figure 2. Conditions of resistance-to-ground measurements in the coal mine at the roadway face.

Table 1. Results of resistance-to-ground at the roadway face.

No of single result	R_g [Ω] Empty bags	Bags filled with material just prepared	Bags filled with hardened material
1	< $1.0\text{x}10^3$	$1.3\text{x}10^3$	$3.87\text{x}10^4$
2	< $1.0\text{x}10^3$	$1.1\text{x}10^3$	$2.10\text{x}10^3$
3	< $1.0\text{x}10^3$	< $1.0\text{x}10^3$	$1.46\text{x}10^5$

Both transportation of the roll of fabric on wheels and carrying it by a miner were found to give the worst cases for the resistance-to-ground, as shown in Figures 3 and 4. In the case of transportation, the fabric was placed on steel and on wood. When carried, the miner was standing just on the floor as well as on wood. Measurements were taken again with Metriso2000 resistance-meter with point electrode - resistance probe-model 850. The probe was placed on the roll of fabric at three points, spaced evenly along its length. Environmental conditions where as follows: temperature 9°C and relative humidity 64%. Results of measurements are summarized in Tables 2 and 3.

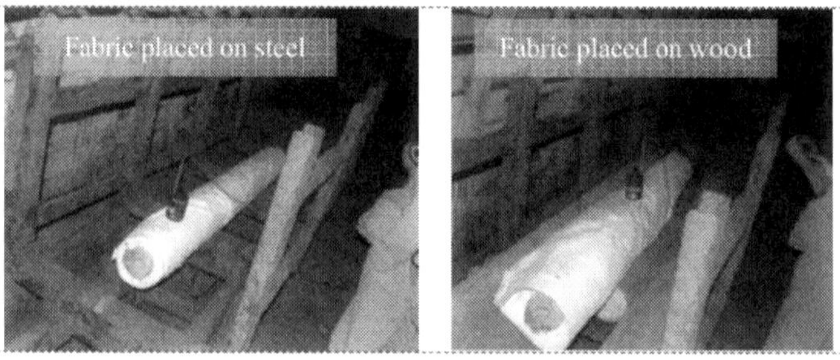

Figure 3. Roll of fabric in transportation placed on wood.

Table 2. Results of resistance to ground in transportation of the fabric roll.

Point number	R_g [Ω] Fabric placed on steel	R_g [Ω] Fabric placed on wood
1	3.74×10^6	4.04×10^7
2	1.73×10^7	3.36×10^6
3	2.12×10^7	7.18×10^7

Figure 4. Roll of fabric carried by a miner.

Table 3. Results of resistance-to-ground of fabric roll carried by a miner.

Point number	R_g [Ω] Standing on the floor	R_g [Ω] Standing on wood
1	3.13×10^6	4.33×10^6
2	3.08×10^6	4.49×10^6
3	2.75×10^6	5.64×10^6

When transported or carried by a miner all the measurements obtained from the fabric roll were above 10^6 Ω, but did not exceed 10^8 Ω. All the resistance-to-ground measurements obtained in the coal mine are presented in Figure 5.

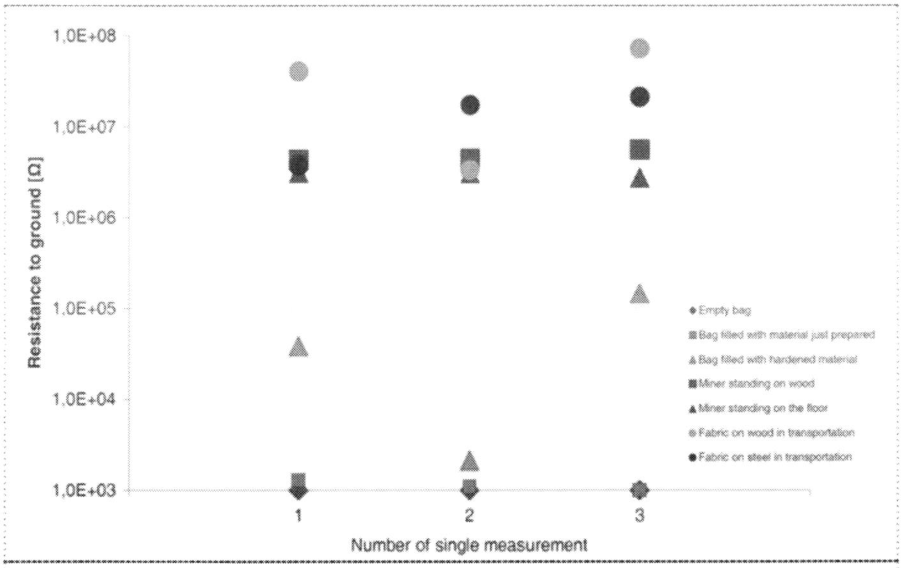

Figure 5. Results of resistance-to-ground in the coal mine.

3. Results

Tests were carried out in the roadway face for two different cases. Those tests gave results below 10^6 Ω. Both transportation of the roll of fabric on wheels and carrying by a miner were found to give the worst case for the resistance-to-ground. Results obtained in those situations were above 10^6 Ω, but did not exceed 10^8 Ω. It can be clearly seen that the results of resistance-to-ground measurements during material transport and carrying it by a miner are higher than the results obtained at the roadway face. This is caused by the higher relative humidity and higher coal dust contamination in the roadway face than in other places of the mine.

4. Conclusion

The tested material was a composition of conductive yarns made of stainless steel and highly resistive base polypropylene fabric. Verification of the resistance-to-ground for some specific situations in a coal mine, where the fabric is intended for use, was most important for us in assessing the fabric. The results obtained indicate that this fabric was reliably earthed and would not be expected to pose any threat associated with static electricity in the coal mine environment.

References

[1] M Talarek, L Orzech, Testing the electrostatic characteristics of polypropylene fabric with metallic yarns, intended for use in coal mines threatened by the explosion hazard. Part 1: Laboratory tests, JPCS (submitted)

[2] EN 13463-1:2010 Non-electrical equipment for use in potentially explosive atmospheres Part 1: Basic method and requirements

[3] CLC/TR 50404 Electrostatics. Code of practice for the avoidance of hazards due to static electricity

[4] EN 61340-4-4:2006 Electrostatics - Part 4-4: Standard test methods for specific applications – Electrostatic classification of flexible intermediate bulk containers (FIBC)

[5] S. Makarski, M. Rasek, L. Orzech, M. Talarek, Badania tkaniny polipropylenowej z przeplotem metalowym stosowanej w wyrobiskach górniczych w aspekcie zagrożenia wybuchem związanym z elektrycznością statyczną, Wiadomości Górnicze, Katowice (Poland), 9/2009

Journal of Physics: Conference Series **301** (2011) 012049 doi:10.1088/1742-6596/301/1/012049

Analysis of electroperforated materials using the quadrat counts method

E Miranda[1], C Garzón[1], C Martínez-Cisneros[2], J Alonso[2] and J García-García[1]

[1] Departament d'Enginyeria Electrònica, Universitat Autònoma de Barcelona, 08193 Bellaterra, Barcelona, Spain
[2] Departament de Química Analítica, Universitat Autònoma de Barcelona, 08193 Bellaterra, Barcelona, Spain

E-mail: enrique.miranda@uab.cat

Abstract. The electroperforation distribution in thin porous materials is investigated using the quadrat counts method (QCM), a classical statistical technique aimed to evaluate the deviation from complete spatial randomness (CSR). Perforations are created by means of electrical discharges generated by needle-like tungsten electrodes. The objective of perforating a thin porous material is to enhance its air permeability, a critical issue in many industrial applications involving paper, plastics, textiles, etc. Using image analysis techniques and specialized statistical software it is shown that the perforation locations follow, beyond a certain length scale, a homogeneous 2D Poisson distribution.

1. Introduction

Electroperforation is a well-known method used to generate a hole pattern in thin porous materials like paper webs, plastics, textiles, etc, with the objective of modifying their air permeabilities [1]. It basically consists in the formation of high-energy electrical discharges passing through the material, which often moves at a high speed in between arrays of opposite needle-like electrodes (see Fig. 1). The method is fast and reliable so that it is well-suited for industrial applications. However, a major limitation of this technique is that there is a limit to the number of holes that can be created in a certain area due to the fact that the already generated holes offer an easier discharge path between the electrodes. Because of this self-limited hole generation process the porosity cannot be further increased beyond a certain level even under longer exposure times (lower velocities) or higher discharge frequencies. Figure 1 illustrates a typical perforation system with needle-like electrodes.

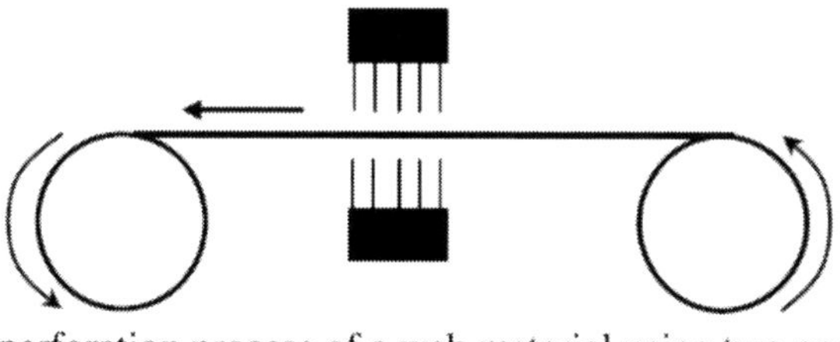

Figure 1. Electroperforation process of a web material using two arrays of electrodes.

Once the hole pattern has been created, it is necessary to characterize the hole distribution by means of exploratory statistical techniques. The hole pattern is regarded as a 2-D point process, which can be investigated using the Spatstat package for R language. R is a free software environment for statistical computing and can be downloaded from [2]. Spatstat can also be downloaded from the same website. Spatstat has been developed by Prof. A. Baddeley and Prof. R. Turner and is a specific package designed to be used for spatial data manipulation, exploratory data analysis, simulation, fitting and model diagnostics. In a few words, the aims of point process statistics are to describe the locations of randomly distributed objects in 1, 2, or 3-D space and, if possible, understand the mutual interactions that led to that distribution [3]. Numerous techniques have been developed to detect clustering (aggregation) or repulsion effects, as well as to investigate the distance scales involved with the purpose of inferring the physical underlying mechanism responsible for the pattern formation.

In this work, in order to characterize the hole distribution generated in a thin porous material, the quadrat counting method (QCM) is considered [3]. QCM is one of the most widely used tests in point pattern analysis and is included, together with a χ^2 (chi-square) goodness-of-fit test for homogeneous or inhomogeneous Poisson processes in the Spatstat package. In this method, the window containing the point pattern is divided into a grid of rectangular tiles or 'quadrats', and the number of points falling in each quadrat is counted. The quadrats are square by default but may have arbitrary shape. However, as it has been pointed out in literature, one major drawback of QCM is that the choice of the quadrat number is strongly linked to the spatial scale of the problem, which establishes a minimum quadrat size. This limitation has been investigated using the Morishita index (MI) [4]. Further analysis based on the Pearson residuals provided by Spatstat allows to evaluate the accuracy of having assumed a complete spatial randomness (CSR) model for the generation of holes.

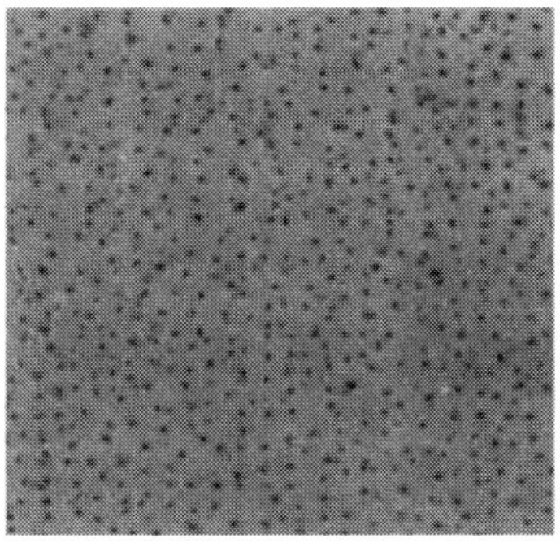

Figure 2. Photograph of an electroperforated material. The dots are holes pierced through the thin porous material.

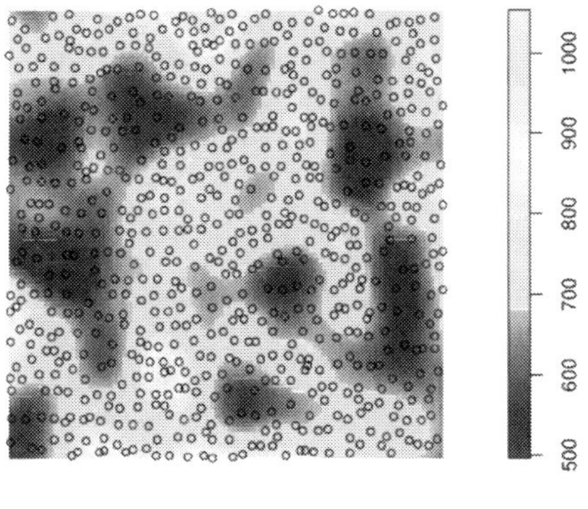

Figure 3. Density plot for a section of the electroperforated material similar to the one shown in Figure 2. N=695 circles (holes) were counted.

2. Exploratory analysis of the perforation pattern

Figure 2 shows a typical example of an electroperforated material. For simplicity, the size of the square has been normalised to unity. A first characterization of the hole distribution is given by the so-called intensity plot. The color map in Figure 3 corresponds to the magnitude of the local intensity estimator (number of points weighted by an isotropic Gaussian smoothing kernel), which is the average density of points, *i.e.* the expected number of points per unit area. Notice that for this particular realization the intensity is not homogeneous as it is expected for a Poisson process.

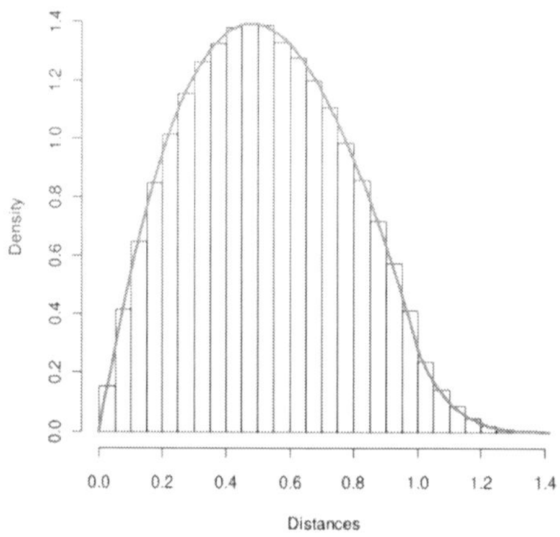

Figure 4. Histogram for the normalised distances between all holes. The size of the square has been normalized to 1x1. The histogram comprises $(N^2\text{-}N)/2=(695^2\text{-}695)/2=241165$ values. The solid line corresponds to the theoretical probability density function given by expression (1).

Quantitative information about the hole distribution can be achieved by analyzing the interpoint distances. The histogram in Figure 4 shows that the distribution of the normalised distances d between all circles in Figure 2 follows the theoretical probability density (solid line) given by the expression [5]:

$$f(d)=\begin{cases} 4d\left[\dfrac{\pi}{2}-2d+\dfrac{d^2}{2}\right] & 0\le d\le 1 \\ 4d\left[\arcsin\left(\dfrac{1}{d}\right)-\arccos\left(\dfrac{1}{d}\right)-1-\dfrac{d^2}{2}+2\sqrt{d^2-1}\right] & 1<d\le\sqrt{2} \end{cases} \quad (1)$$

This function is obtained under the hypothesis that the x- and y-coordinates of each data point are uniformly distributed in the range [0,1]. It must be noted that the agreement is very good, what at first sight seems to indicate that the distribution is consistent with a CSR process. However, the main problem with this approach is that the histogram representation is not sensitive to the details of the distribution, in particular with respect to what happens at the smaller length scales (lower tail).

In order to get further insight into the hole pattern distribution it becomes necessary to use an alternative exploratory technique such as the already introduced QCM. In this method, the window containing the electroperforation pattern (*N* points) is divided into a grid of Q quadrats. The number of points falling in each quadrat (i) is $n[i]$. One major drawback of the QCM is that the appropriate choice of the quadrat size is strongly linked to the spatial scale of the problem, which is often unknown. This apparent limitation can be overcome with the help of the Morishita index (MI):

$$MI=Q\sum_{i=1}^{Q}\frac{n[i]\left(n[i]-1\right)}{N\left(N-1\right)} \quad (2)$$

an index used to discern different scales of dependence in the point pattern data. The basic idea behind the QCM test is to find evidence against the null hypothesis of a CSR process. In what follows an illustrative example of the aforementioned analysis is presented. Figures 5 shows the perforation distribution (N=695) and a particular quadrat plot (Q=5x5) for the data shown in Figure 2. The choice of Q can severely affect the results of the analysis and therefore it must be taken into account to avoid misinterpretations. Figure 6 shows MI as a function of the size of the quadrats. For a completely random pattern MI≈1. In our example, CSR is clearly detected for Q≤25 (linear dimension approximately≥0.2). Values of MI<1 suggest repulsion or hard core potential effects [3].

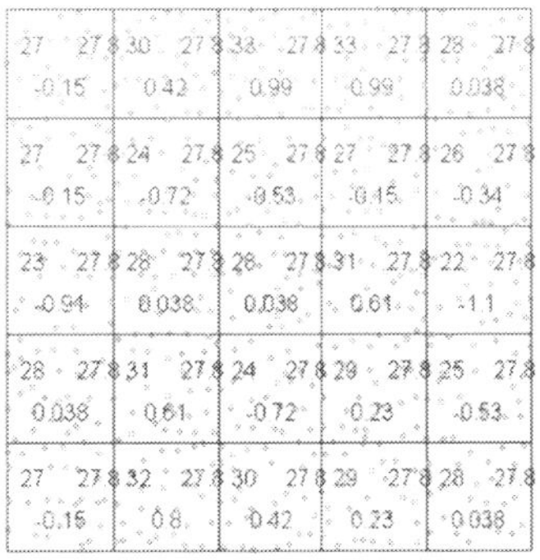

Figure 5. Electroperforation distribution and an example of a quadrat plot (5x5).

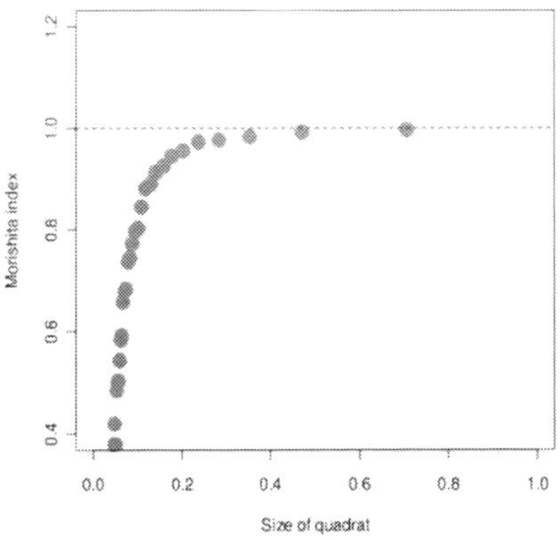

Figure 6. Morishita index as a function of the quadrat size.

Assuming that the $n[i]$ values are independent and identically-distributed Poisson random variables with the same expected value, the Pearson χ^2 test can be evaluated, which in turn can be used to reject or accept the CSR model. Figure 5 includes the observed number of points in each quadrat, the predicted number of points according to a fitted Poisson model (obtained by simulation with Spatstat) and the Pearson residuals (=[(observed)-(expected)]/√expected). A small Pearson residual indicates good agreement with the CSR model.

3. Conclusions

Electrical discharges are often used to perforate thin porous materials in industrial environments. In order to characterize the resulting hole distribution, statistical techniques are imperatively required. In this work, we have shown how the quadrat counts method in combination with the Morishita index can be applied to determine the minimum length scale from which the electroperforation pattern can be strictly regarded as a 2-D Poisson process.

The authors acknowledge the TRACE project from Ministerio de Ciencia e Innovación, Spain.

References

[1] García-García J, Miranda E, Martínez-Cisneros C, *et al*, 2010 *J. Elec.* 68 196-199
[2] http://www.r-project.org/
[3] Illian J, Penttinen A, Stoyan H and Stoyanet D, in *Statistical analysis and modelling of spatial point patterns*, 2008 Wiley
[4] Morishita M, Measuring of the dispersion of individuals and analysis of the distributional patterns. Memoir of the Faculty of Science, Series E2, Kyushu University, 215-235, 1959.
[5] Mathai A, *et al*, 1999 *Rendiconti del Circolo Matematica di Palermo* XLVIII 163

Correlation between electrical, mechanical and chemical properties of fresh and used aircraft engine oils

Juliusz B. Gajewski[1*], **Marek J. Głogowski**[1], **Maciej Paszkowski**[2], **Bogusława Czarnik-Matusewicz**[3]

[1] Wrocław University of Technology, Institute of Heat Engineering and Fluid Mechanics, Wybrzeże Wyspiańskiego 27, 50-370 Wrocław, Poland

[2] Wrocław University of Technology, Institute of Machine Design and Operation, ul. Łukasiewicza 5, 50-371 Wrocław, Poland

[3] University of Wrocław, Faculty of Chemistry, Joliot-Curie 14, 50-383 Wrocław, Poland

Email: juliusz.b.gajewski@pwr.wroc.pl

Abstract. In this paper the results are presented of measurements of electrical, mechanical and chemical properties of fresh and used aircraft engine oils. Oils were used in a four-stroke aircraft engine and their samples were taken after the 50-hour work of the engine. The resistivity, permittivity and viscosity of oils were measured as a function of temperature. Additionally, some measurements of the absorbance spectra and size of particles contained in the oils were carried out. The significant reduction in the resistivity of the used Total oil was observed. The relative permittivity of both used oils was slightly increased. The oil's relative viscosity depends on temperature of oil and given time that elapsed from the very first moment when the shear force was applied in a rheometer. The results obtained allowed one to identify more precisely the chemical and physico-chemical interactions occurring in the tested samples, as compared with a typical infrared spectroscopy.

1. Introduction

Lubricating oil protects engines from the effects of pressure, corrosion, heat, and oxidation. The oil creates a separating film between moving parts, reduces friction and wear, and cleans the interior of an engine by removing particle as a product of wear, carbon deposit and combustion contaminations. The rate of deterioration depends on operating conditions and loads of an engine. Aircraft engines are made of lightweight alloys, and operate over a wider temperature range than normal automotive engines which causes the rapid ageing of oil. In order to avoid any engine failure, the oil must be changed before it looses its protective properties. Changes in these properties can be seen as variations in the resistivity, the relative permittivity and the viscosity [1, 2, 3].

2. Experimental system

The measurements of the oil's resistivity ρ and relative permittivity ε were made for the steady and stable temperature. The temperatures T were changed from 20 to 100°C every 10°C. The resistivity and permittivity were measured using a capacitor KP-4 which was made according to the Polish technical standard PN-EN 60247 [4]. The measurements of the permittivity were performed with the use of a precision LCR meter 7600 Plus made by QuadTech, while a meter MIC-2500 made by Sonel was used to measure the resistivity. The measurements were repeated five times during each experimental session to get a certain statistics.

The rheological properties of the engine oil were measured with an Anton Paar Physica MCR-101 rotational rheometer which worked in the Couette system. The measuring system was equipped with a computer software RheoPlus for the data acquisition. The temperatures of the oil's samples were kept constant in a thermostat by the Peltier module at a given level. The temperatures ranged from 0 to 100°C every 20°C. The dynamic viscosity measurements were performed as a function of time.

The size of different particles contained in both the fresh oils and the used oils was measured with ZetaSizer Nano ZS made by Malvern Instrument. The aim of these experiments was an assessment of percentage of particles of different sizes in the volume of the oil being tested.

The FTIR-ATR spectra were recorded using a FT-IR Nexus-Nicolet spectrophotometer with a resolution of 1 cm^{-1} with 128 scans per spectrum in a frequency range of 4000–400 cm^{-1} at 25°C. The spectrometer was equipped with the DTGS detector. The transmission measurements were carried out in a 208 µm–KBr cell, while the ATR spectra were collected on the ATR system supplied with a diamond crystal (face angle: 45°, 1 reflection) (Graseby, Specac, UK).

3. Materials tested

The experiments were conducted for two aircraft piston engine oils: mineral Total Aero D 100 and synthetic AeroShell 15W50 whose kinematic viscosities were 182.0 and 19.0 $mm^2 \cdot s^{-1}$ as well as 146 and 19.9 $mm^2 \cdot s^{-1}$ at 40 and 100°C, respectively. Their viscosity indexes were 118 and 175, respectively. The oils were used in an air-cooled, carburetted, four-stroke gasoline aircraft engines, and the samples of the used oils were taken after the 50-hour work of an engine. AeroShell oil was used in a Lycoming O-360 engine while Total Aero oil was used in an AI-14RA engine whose powers were 135 and 191 kW, respectively. For the sake of simplicity both oils will henceforth be termed Shell and Total.

4. Experimental results and discussion

4.1. *Resistivity and relative permittivity*

The resistivity ρ and the relative permittivity ε as a function of the temperature T are shown in figure 1 for two aircraft oils.

The decrease in the resistivity with the increasing temperature is due to the higher mobility of all charged particles and molecules in the oil. A higher value of the resistivity of Total reflects the lower content of additives in the oil. The larger resistivity decrease of used Total may be caused by an increase in the contamination and acidity of the oil. The minimal change in the resistivity of used Shell may result from the ZnDTP content in oil, which protects moving metal parts of engines.

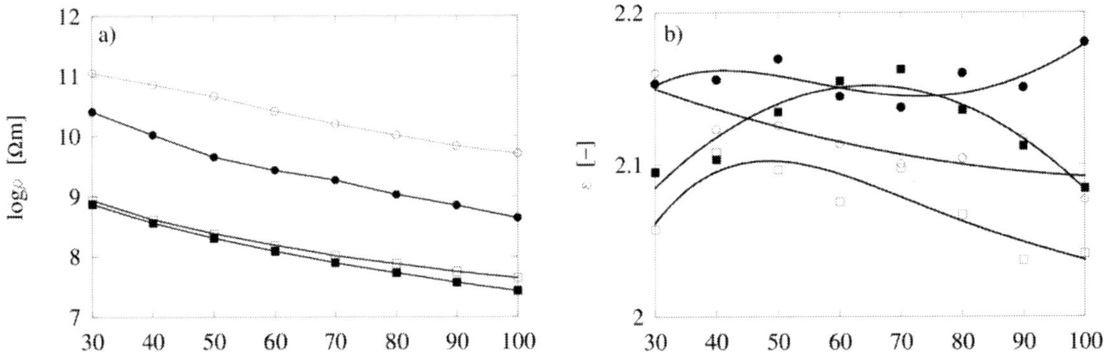

Figure 1. Resistivity (a) and relative permittivity (b) vs. temperature for oils: Total — ○ – fresh oil and ● – used oil, and Shell — □ – fresh oil and ■ – used oil.

The relationships between the relative permittivities and the temperatures of both oils tested are shown in figure 1b. It may be noticed that for the used oils values of the permittivity to increase. This may indicate the increase of water content in the oil tested; the water can be produced by, for example, the condensation of water vapour.

4.2. *Dynamic viscosity*

In figure 2 the sample graphs show the trends of the relative dimensionless value of the viscosity $\Delta\eta$ as a function of the temperature of both oils for the shear rate $D = 500\ s^{-1}$. The relative viscosity $\Delta\eta$ means the ratio of the viscosity of used oil to the viscosity of fresh oil.

The relative viscosity was determined for the measurement results obtained after a given time elapsed from the very first moment when the shear force was exerted in a rheometer. These times were: 1st, 60th, 120th, and 180th second from the start of a shaft's rotation and of the measurements. In figure 2 these times are marked with the following symbols: ○, ◇, △, and □, respectively.

The highest values of the viscosity for the used oils were observed at temperatures of 0°C and above 40°C. The values of the viscosity measured depend on the duration of measurements. An increase in the duration of measurement causes the relative viscosity to tend to the constant value.

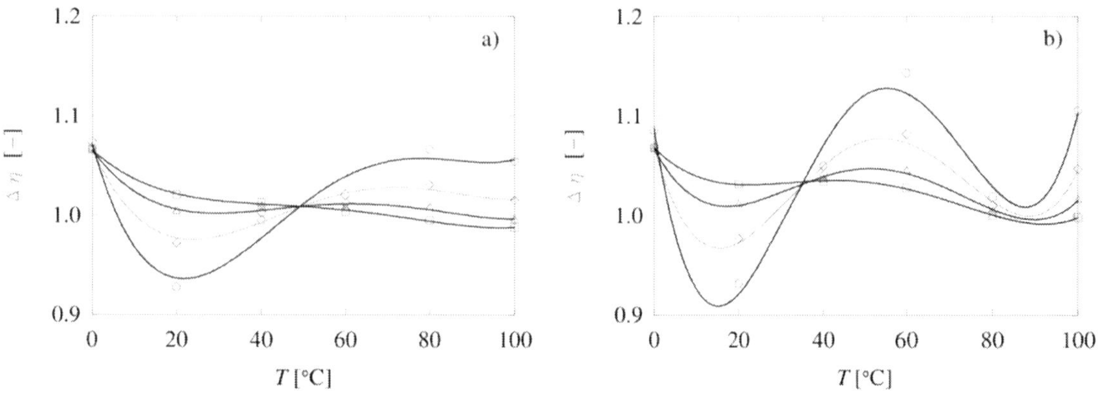

Figure 2. The sample plots of the relative dynamic viscosity vs. temperature of Shell (a) and Total (b).

Used oils seem to be more complex rheological liquids because they contain much more contaminations than fresh oils. All the oils tested also showed pseudo-plastic behaviour.

4.3. *Absorbance spectra measurements*

According to the figure 3a for the Shell oil in the spectrum of its used sample the concentration of aromatic components decreases. It can be caused by presence of such an additive as Irgalube 211 that is not present in the Total samples. In the Shell samples some component is present that contains an acid group and the concentration of the component is not changed in the sample of the used Shell oil. The Total samples are free of acid components. For both kinds of oils the concentration of components that contain the ester group increases.

For the spectra shown in figure 3b there are distinct changes in the two samples of the Shell oils in the range of two peaks which can be attributed to the ZnDTP content in the oils – ZnDTP is an antioxidant and antiwear extreme pressure additive. For the Total samples, free from this additive, the differences between different peaks are negligible. Only one effect can be observed that is the elevated absorbance for the sample of used Total. It was probably caused by a higher content of soot in the sample.

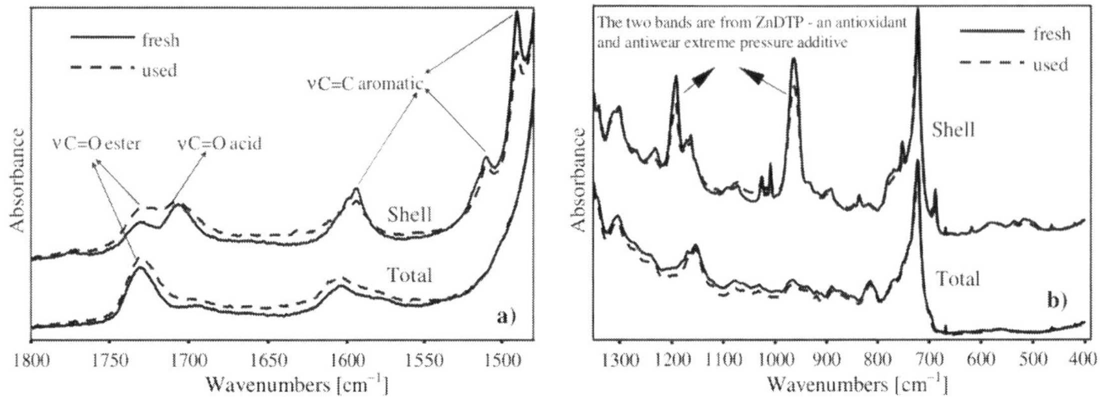

Figure 3. FTIR spectra of the oil samples tested.

4.4. *Granulometry measurement*

Additionally, some granulometric experiments and analysis have been carried out. Their results permitted one to know the particle size distribution in both samples of the used oils.

In the Shell sample the largest fraction of particles (7.7%) were particles of 412.5 nm in diameter. The smallest recorded size of the particles was 21.9 nm. This sample did not contain particles larger than 1152 nm within an optical range of 0.4–4000 nm for ZetaSizer Nano ZS.

In the Total sample the largest fraction of particles (10.6%) were 2780 nm diameter particles. The size of the Total particles ranged from 39.4 to 3728 nm. It should be noted that in both cases there were probably the particles larger than 4000 nm as well. These particles were outside the measuring range of the instrument used.

5. Concluding remarks

The research results presented here show the relationships between the electrical, mechanical and chemical properties of investigated oils. A decrease in the resistivity for the used oils depends on an increase in the metal particles produced by wear (wear debris) and in the amount of soot from a combustion processes – these particles and soot were getting into the oils during their normal, regular use in the aircraft engines. The used Total oil had more soot than a Shell one. The lowest value of the resistivity of Shell's oils was a result of the higher content of different additives.

An increase in the relative permittivity of the used oils was the result of increasing ester and acid groups in the volume of all the oils tested.

The increase observed in the viscosity of the used oils tested results mainly from soot particles that were introduced to the oil as the products of oil and fuel combustion as contaminants, as supposed.

References

[1] Gajewski JB, Głogowski MJ and Paszkowski M 2010 Electrical and tribological properties of engine oil during its operating time in the aspect of tribocharging *SFE 2010: Proc. of 7th Conference of the French Society of Electrostatics (Montpellier, France, 30 August–1 September 2010)* pp 341–344

[2] Olszewski W 2001 Możliwości oceny stanu jakości oleju silnikowego podczas eksploatacji za pomocą pomiaru własności dielektrycznych (The possibility of estimate quality state of using motor oil by measuring dielectrical properties) *J. of Kones. Combustion Engines* 8 (3–4) pp 84–90

[3] George S, Balla S, Gautam V and Gautam M 2007 Effect of diesel soot on lubricant oil viscosity *Tribol. Int.* 40 p 809

[4] Polish standard PN-EN 60247:2004

Understanding core conductor fabrics

D E Swenson

Affinity Static Control Consulting, LLC

2609 Quanah Drive, Round Rock, Texas, USA 78681

deswenson@affinity-esd.com

Abstract. ESD Association standard test method ANSI/ESD STM2.1 – *Garments* (STM2.1), provides electrical resistance test procedures that are applicable for materials and garments that have surface conductive or surface dissipative properties. As has been reported in other papers over the past several years[1], fabrics are now used in many industries for electrostatic control purposes that do not have surface conductive properties and therefore cannot be evaluated using the procedures in STM2.1[2]. A study was conducted to compare surface conductive fabrics with samples of core conductor fibre based fabrics in order to determine differences and similarities with regards to various electrostatic properties. This work will be used to establish a new work item proposal within WG-2, *Garments*, in the ESD Association Standards Committee in the USA.

1. Introduction

Core conductor fibres are being used in a wide variety of clothing applications, particularly where there is a need to control static electricity but surface conductive properties are viewed as being an electrostatic discharge (ESD) risk. The present industry test procedures that rely on surface or volume resistance or resistivity measurements do not provide appropriate or sufficient information when applied to fabrics with core conductors or other buried layer conductors. Work is underway within the ESD Association Standards Committee working group on garments in the USA to attempt to develop appropriate test methods for non-surface conductive materials used in garments.

Numerous fabrics were involved in this study including: plain cotton, polyester, conductive grid pattern on polyester base (conventional cleanroom garment), Chinese silk, and three (3) levels of core conductor fabrics (varying weights).

Several aspects of electrostatics were evaluated including: surface resistance, shielding/suppression of samples in free space with and without grounding; triboelectric charging with an inclined plane with quartz, PTFE and hard brass rollers; and a rotating drum triboelectric charging test using the various fabrics mentioned above.

2. Surface Resistance

The surface resistance of the samples was measured according to ANSI/ESD STM11.11. The core conductor fabrics were in excess of 3×10^{10} ohms and the conductive grid pattern materials measured 3.6×10^{6} ohms. The conventional cloth samples were greater than 1×10^{11} ohms.

Journal of Physics: Conference Series **301** (2011) 012051 doi:10.1088/1742-6596/301/1/012051

3. Shielding/Suppression Test

Fabric samples were supported in wood rings that were suspended by monofilament fishing line within another wooden ring. A small Van de Graaf generator placed 30 cm (12 inches) away from the samples under test was used as the electrical field source. A field mill (3M Model 711 Voltage Analyzer) was connected to a computer through a data acquisition system (Prostat Model PGA710 Auto Analysis system). Figure 1 shows a photograph of the test set-up. Figure 2 compares the grounded and ungrounded voltage allowed through the conventional cloth samples. Figure 3 shows the results for the core conductor and grid pattern cleanroom garment materials. Note the change in voltage level on the y-axis of the charts. Also note that "GRD" means attached to ground and "no GRD" means the ground lead is removed which electrically isolates the sample under test.

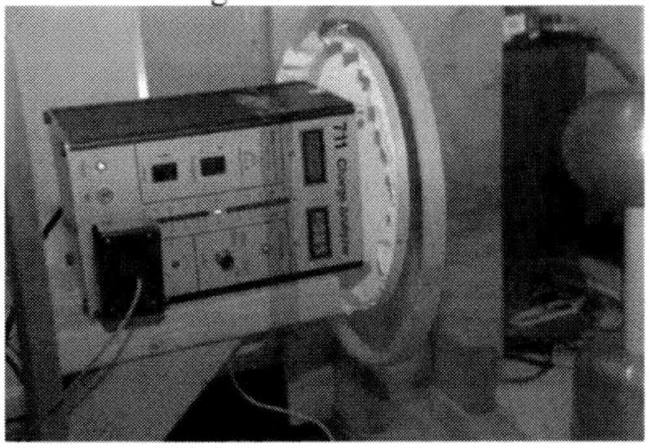

Figure 1. Shielding/Suppression Test Set-up

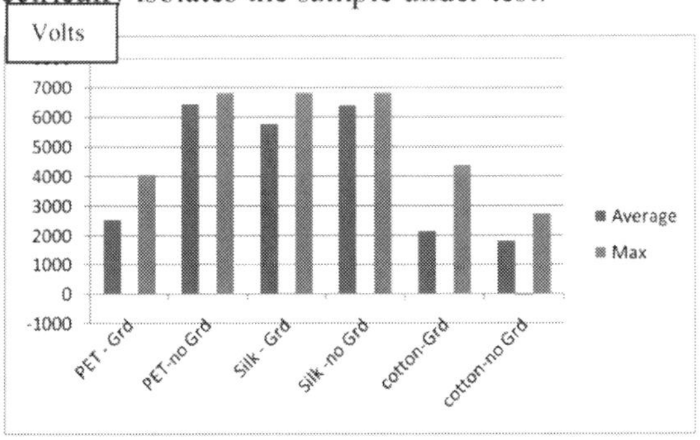

Figure 2. Voltage allowed –standard cloth

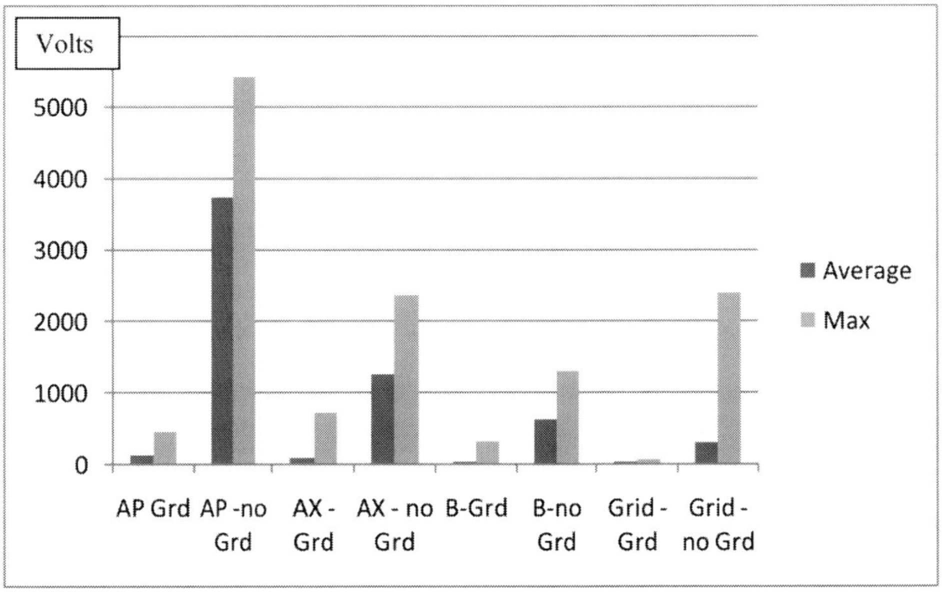

Figure 3. Voltage allowed – Core Conductors (AP, AX and B) and Grid cloth samples

None of the fabrics in Figure 2 provided much in the way of electrical field suppression although cotton seemed to perform better when the ground lead was not attached. This observation probably is a result of the electrical field actually interacting with the ground lead even though it was connected on the field mill side of the fabric. When a grounding clamp was attached to the fabric, all of the samples shown in Figure 3 were able to suppress the electrical field substantially. The core conductor suppression level varied according to the fabric weight – AP being the lightest weight and Sample B the heaviest. The grid pattern material had very good suppression ability when bonded to ground but when in a non-grounded state, sample B performed at a similar if not superior level.

4. Inclined Plane Triboelectric Test

Evaluation of the triboelectric properties of materials involves selecting the two (or more) interacting materials. The inclined plane test is highlighted in ESD ADV 11.2 – *Triboelectric Charge Accumulation Testing*[3]. A steel plate incline of 15° is used to support a sample under test. Cylinders made from quartz glass and PTFE are used as they represent the generally accepted extremes on a Triboelectric Series. Hard brass cylinders also provide significant information and insight into the charge generation characteristics of materials[4]. The cylinders roll down the inclined plane and are captured in a Faraday Cup and the charge in nanocoulombs recorded. The test is repeated with six (6) cylinders of each type and the charge averaged. Two of the core conductor samples, cleanroom grid material, cotton and silk were evaluated using the inclined plane method. Tests were conducted with the metal incline grounded and also with the metal incline insulated (base still grounded) to look for differences in the charge generation and accumulation properties. Figure 4 shows the summary results of the inclined plane tests.

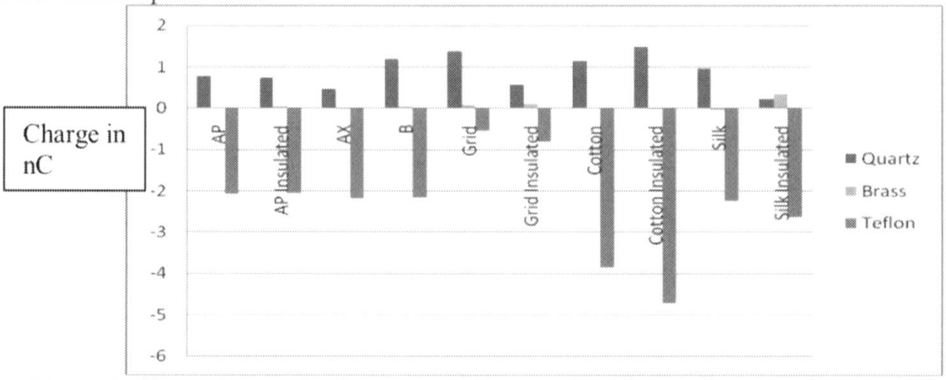

Figure 4. Charge in nanoCoulombs for rollers on grounded and insulated inclined plane supporting the fabric samples. Note: samples AX and B show only grounded metal plate values.

The charging level of the hard brass is negligible except against the ungrounded silk sample. All three of the core conductor materials had similar charge generation properties and were lower than the cleanroom grid sample for quartz and brass.

5. Rotating Drum Triboelectric Test

The conventional inclined plane test did not provide the differentiation needed to verify the performance properties of fabrics. Another procedure was developed whereby fabrics could be contacted in a controlled manner to a fabric sample under evaluation. The same wooden hoop arrangement used in the suppression test shown in Figure 1 was used to support the sample under test. A rotating drum from a rock tumbler ball-mill was prepared to hold a contacting fabric sample. Figure 5 shows the test set-up. Figure 6 shows the results of the triboelectric testing using the rotating drum. An additional grid pattern fabric was added to the test (Red Grid). Voltage recorded is for the peak value after approximately 60 seconds of contact time between the roller fabric and the test sample.

Figure 5. Rotating drum set-up

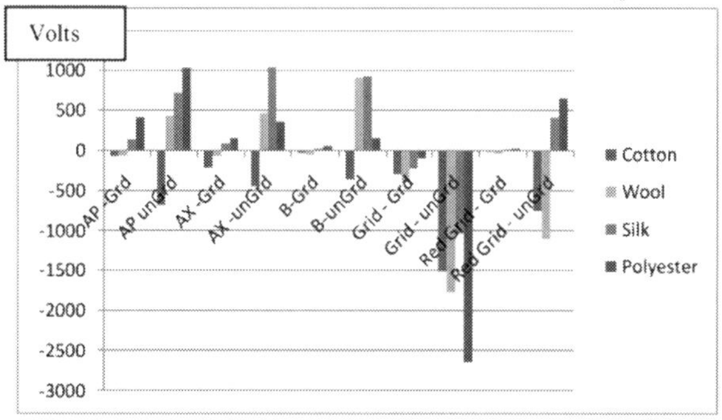

Figure 6. Core conductor and grid pattern peak voltage

When the fabrics were grounded, the charging levels were quite low and controlled and there is only minor difference between core conductor materials and the grid pattern fabrics. In an ungrounded condition, the core conductor fabrics actually charge less than the grid pattern fabrics. This property would lend itself for use in garments in environments where specific and intentional grounding was not possible or only incidental. Figure 7 and 8 show the actual data acquisition graphs for sample B in the grounded and ungrounded conditions with natural wool as the contacting fabric. Figures 9 and 10 show the grounded and ungrounded grid pattern fabric against natural wool.

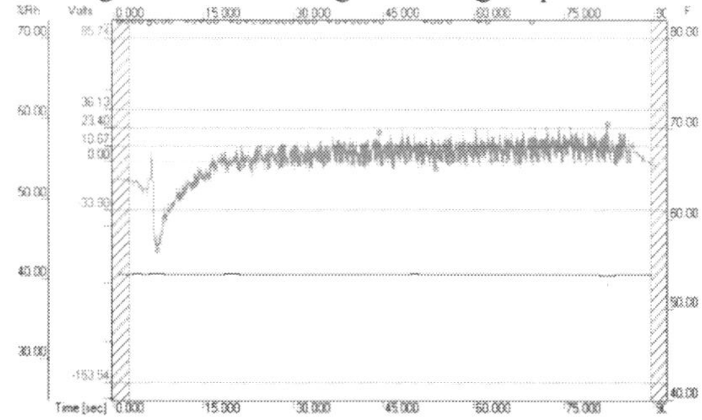

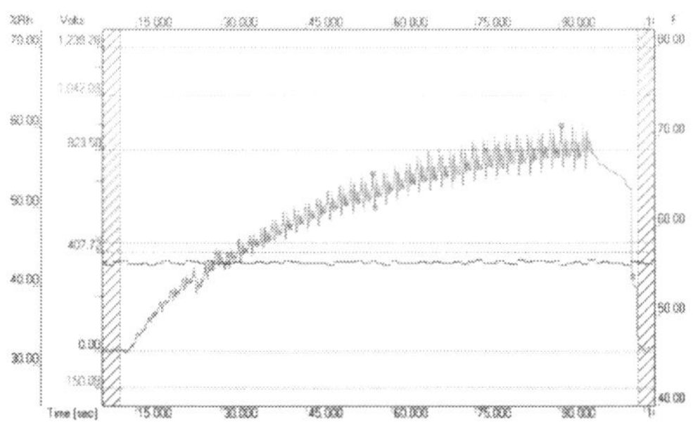

Figure 7. Sample B Grounded – Wool Roller **Figure 8**. Sample B ungrounded – Wool Roller

As can be observed, grounded sample B maintains a very low steady voltage while the ungrounded sample increases over time before eventually levelling out due to charge saturation.

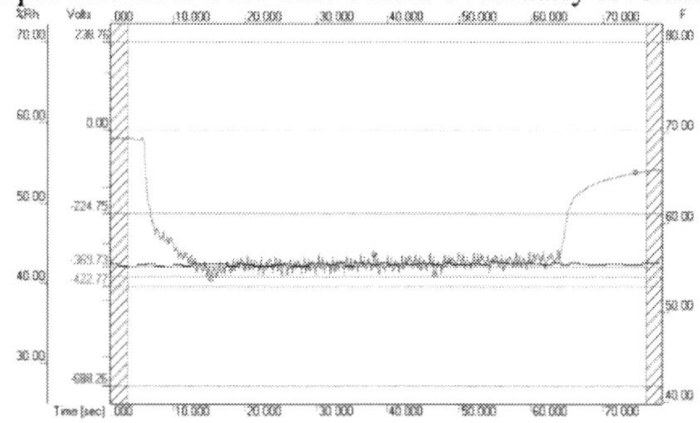

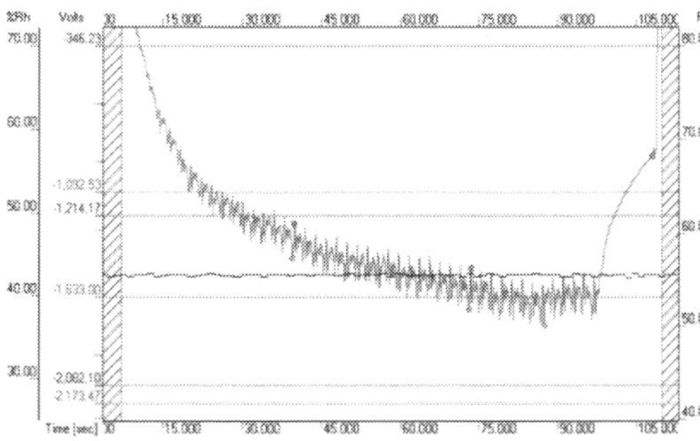

Figure 9. Grid pattern grounded – Wool Roller **Figure 10**. Grid pattern-ungrounded –Wool roller

The grid pattern cleanroom fabric shows opposite polarity and significantly higher overall charging levels against the wool roller than the core conductor sample B fabric shown above. At the end of each graph, the natural decay pattern is shown after the contacting drum rotation was stopped.

6. Conclusions

If taken at face value, the surface resistance of the core conductor fabrics would limit their application in some areas due to not meeting the arbitrary specifications established in some industry accepted garment standards. This study shows that the core conductor fabrics have lower charge generation propensity overall than conventional conductive grid fabrics. Especially important is the observation that when the core conductor fabrics are not bonded to ground, the charging characteristics are considerably lower than the conductive grid pattern materials. This feature would lend itself well to garments made for special applications, especially for flammable or other ignition hazard environments where exposed conductors may be considered a risk for ESD.

References

[1] Holdstock, P: September 7, 2009, Proceedings, EOS/ESD Symposium, *"Limitations of using Resistance Measurements to Qualify Garments for use in an EPA"*, 2B.4-1 to 2B4-6

[2] ANSI/ESD STM2.1, *Garments*, ESD Association, www.esda.org USA

[3] ESD ADV11.2 – *Triboelectric Charge Accumulation Testing*, ESD Association, 7900 Turin Road, Building 3, Rome, New York, USA 13440, www.esda.org

[4] Huntsman, J R, Proceedings, 1984 EOS/ESD Symposium, *"Triboelectric Charge: Its ESD Ability and a Measurement Method for its Propensity on Packaging Materials"*

13th International Conference on Electrostatics IOP Publishing
Journal of Physics: Conference Series **301** (2011) 012052 doi:10.1088/1742-6596/301/1/012052

Modeling of contact discharge between floating devices

Y Soda and T Oda
The University of Tokyo, 7-3-1 Hongo, Bunkyo-ku Tokyo, 113-8656 Japan

E-mail: y_soda@wd5.so-net.ne.jp

Abstract. The Giant-magnetoresistive (GMR) head suffers magnetic damage from ESD current on the order of 10 mA and thermal damage from ESD energy on the order of 0.5nJ. Contact discharge from an electrified object into a floating device was investigated as the peak current caused magnetic damage and as the energy-loss caused thermal damage. Experiments with discharges between various capacitors clarified that the discharge between 2pF-capacitors at 10V gave a peak current of 34mA and the discharge between 100pF-capacitors at 100V reached 1.5A. A 2D-plot of the peak current simulated by PSPICE indicated that the peak value arose at the order of pF and was approximately maximized at the same value of both capacitors. Theoretical energy-loss was estimated by the difference between the potential energies prior to and following the discharges. The energy-loss was 0.05nJ for 2pF-capacitors at 10V and increased in excess of 0.5nJ when both capacitors were grater than 20pF. The energy-loss reached 250nJ for 100pF-capacitors at 100V. 2D-plot of the energy-loss indicated a gradual increase as both capacitance-values increased. In order to decrease the risk of ESD failure, reducing the capacitance of the device was effective, because the transferred-charge and energy-loss were restricted.

1. Introduction

Magnetic heads such as MR/GMR have been found to be sensitive to electrostatic discharge (ESD). It has been reported that GMR heads have suffered magnetic damage from ESD current on the order of 10 mA having pulse widths on the order of nanoseconds [1]-[3]. Various protection methods and analyses of ESD hazards have been presented [4]-[9]. An example of the contact discharge is shown in figure 1. When an electrified tool such as a screwdriver comes into contact with a magnetic head, the discharge causes the element ESD failure. The discharge between two objects has been analyzed and reported [10]-[13]. A model of the contact discharge between floating devices is shown in figure 2.

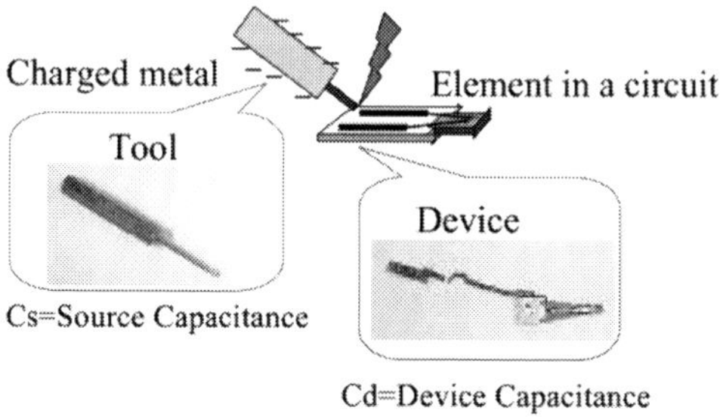

Figure 1. An example of the contact discharge from a tool to a magnetic head.

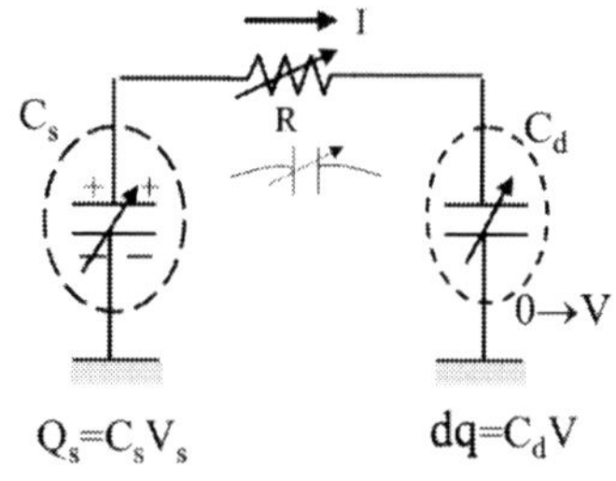

Figure 2. Conceptual diagram of the contact discharge between floating devices.

Published under licence by IOP Publishing Ltd

Generally the capacitance of the device depends on the distances to surrounding objects such as human bodies, machines and ground planes. At the ESD event the current-waveform and energy-transfer are determined at the moment of the contact by the total equivalent-circuit. In this paper the electrostatic discharge was investigated as the peak current caused magnetic damage and as the energy-loss caused thermal damage. Normally the capacitance of magnetic heads, depending on the area of lead pattern, is between 2pF and 20pF. The capacitance of tools depending on the size is the order of pF. A modified ESD tester was used to measure the current-waveform. A PSPICE circuit simulator was used to simulate the current-waveform. The energy-loss was estimated theoretically by the difference between the potential energies prior to and following the discharges.

2. Experiment

Figure 3 shows the photo and illustration of the experimental setup of the modified ESD tester. Figure 4 shows the equivalent circuit. The source capacitor (C_s), 2pF to 100pF, was charged by the voltage source (DCV). The charge was discharged to the device capacitor (C_d), 2pF to 100pF, through two 10Ω-resistors simulated as the contact resistance. The current waveform through the needle was measured using a ct-6 probe and a Tektronix TDS6604 oscilloscope (6 GHz BW, 20 GS/s).

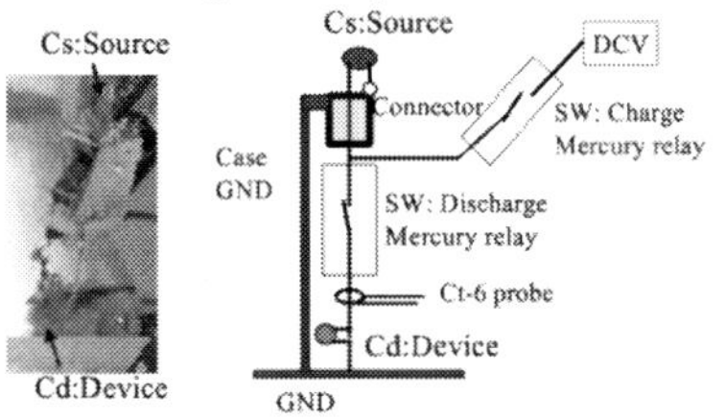

Figure 3. Experimental setup is a modified ESD tester.

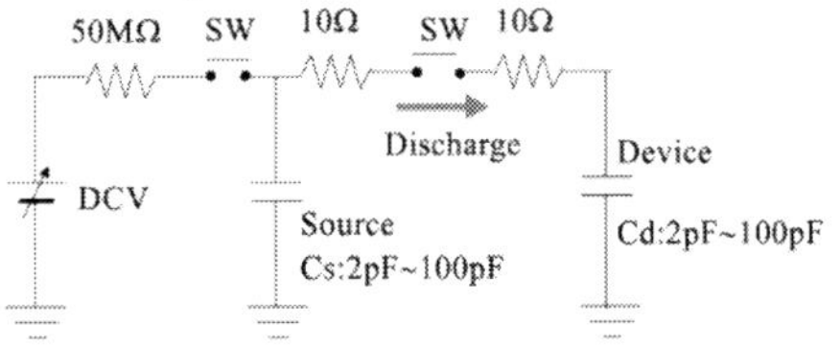

Figure 4. The equivalent circuit of the experiment.

3. Results and discussion

3.1. Peak current

Figure 5 shows typical current waveforms discharging (a) from 2pF to 2pF, (b) from 22pF to 22pF and (c) from 100pF to 2pF, where 10V was applied to the source capacitors. In these cases, although the voltage was the same, the charge increased fifty times from 2pF to 100pF. As can be seen, the peak current from 2pF to 2pF was 34mA and the duration of the 1st pulse was 0.7ns. The current from 22pF to 22pF reached 100mA and 2.7ns. Similarly that from 100pF to 100pF reached 160mA and 6ns. Meanwhile the current from 100pF to 2pF was as low as 36mA and 1.3ns. Thus, when discharging to the floating device, the current waveform was highly related to both source- and device-capacitances and their ratio. Figure 6 shows the peak-current Ip versus applied voltage, where C_s capacitances were (a) 2pF, (b) 10pF and (c) 100pF. In all curves the peak current increased linearly as the applied voltage increased. As C_s increased the capacitance 2pF, 10pF and 100pF, the gradients of 2pF- to 100pF-C_d increased. In this experiment, the peak current increased from 34mA at 10V to 1.57A at 100V.

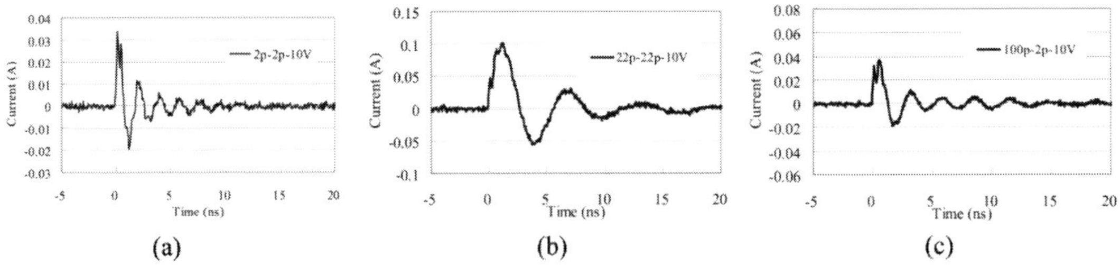

Figure 5. Current waveforms discharging (a) from 2pF to 2pF, (b) from 22pF to 22pF and (c) from 100pF to 2pF at 10V applied.

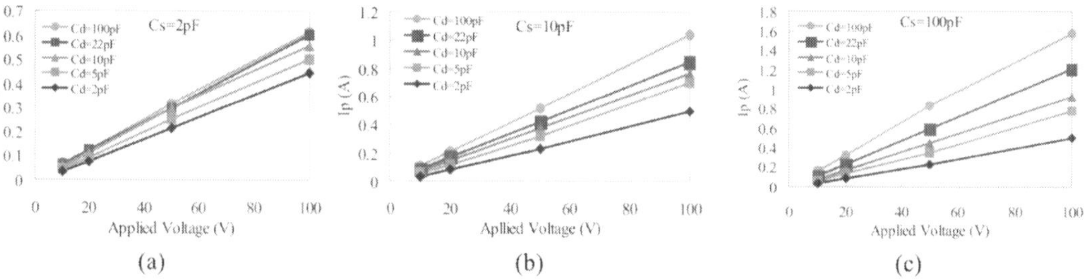

Figure 6. Peak current vs. applied voltage discharging from (a) 2pF-, (b) 10pF- and (c) 10pF-C_s source capacitors into 2pF- to 100pF-C_d capacitors.

A PSPICE circuit simulator was used to simulate the current waveforms. Figure 7 (a) shows the equivalent circuit. Figure 7 (b) shows the experimental and fitted waveforms of 10pF-to-10pF discharge at 10V. Then each current waveform was simulated by this circuit with various combinations of C_s and C_d. 2D-plot picked up the peak values was obtained as shown in figure 8. It was visible that the peak current tended to arise from the order of pF and to increase as both C_d and C_s increase together.

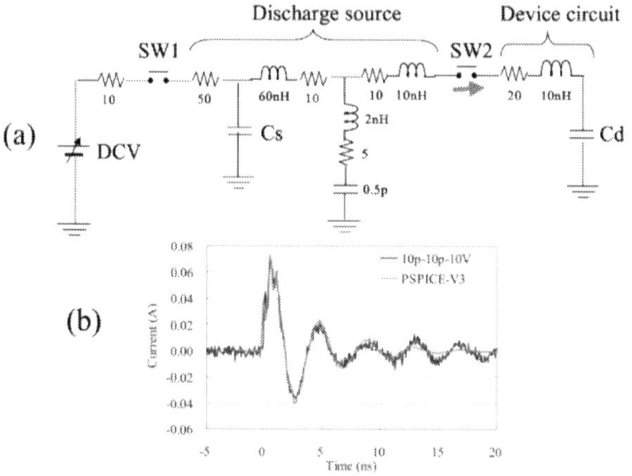

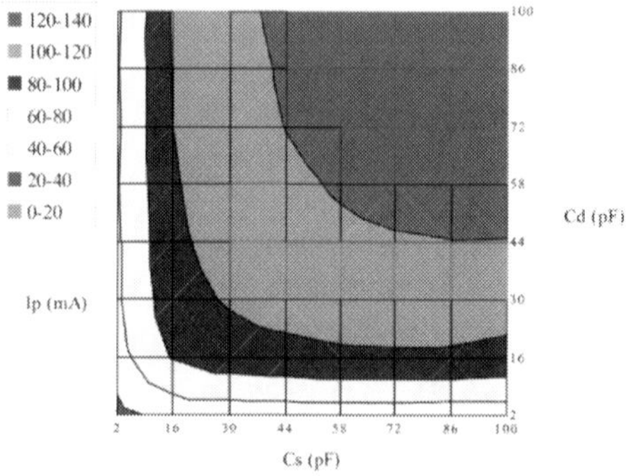

Figure 7. (a) Equivalent circuit simulated by PSPICE. (b) Experimental and fitted waveforms of 10pF-to-10pF discharge at 10V.

Figure 8. 2D-plot of the peak current as functions of C_s and C_d capacitances simulated by PSPICE.

3.2.Energy loss

Theoretical energy-loss due to the discharge is given by the difference between the potential energies of U_b before the discharge and U_a after the discharge.

$U_b-U_a= 1/2\, C_sV_s^2 -1/2\, (C_s+C_d)\, V^2$. Total charge is conserved, therefore $Q=C_sV_s= (C_s+C_d)V$. Finally, the energy-loss ratio is given by $dU/U_b=(U_b-U_a)/ U_b =1-U_a/U_b=n/(n+1)$, where $n=C_d/C_s$.

The energy-loss ratio increases as n of capacitance ratio increases. At the case of n=1,where $C_d=C_s$, half the initial potential energy would be lost as thermal energy and the remainer is shared equally as potential energy in both capacitors. Figure 9 shows dU/U_b as a function of C_d for 2pF- to 100pF-C_s capacitances. The energy-loss ratio increased as C_d increased. In addition, as C_s increased from 2pF to 100pF, the energy-loss ratio decreased gradually. Figure 10 shows an example of dU (nJ) calculated as a function of C_d at 10V. As can be seen, the energy-loss increased as C_d increased. However it saturated at a lower value as C_s decreased. Thus, the energy-loss remained less than 0.5nJ where C_s was less than 10pF. Figure 11 shows a 2D-plot of dU as functions of C_d and C_s at 10V. It was visible that the energy-loss tended to increase gradually compared to the peak current.

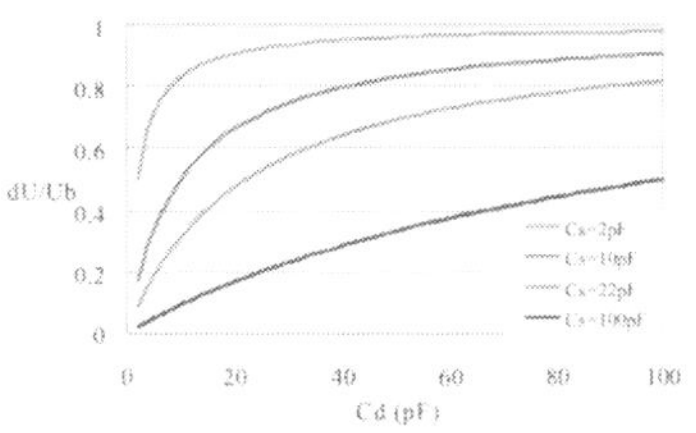

Figure 9. dU/U_b ratio as a function of C_d for 2pF- to 100pF-C_s applied 10V.

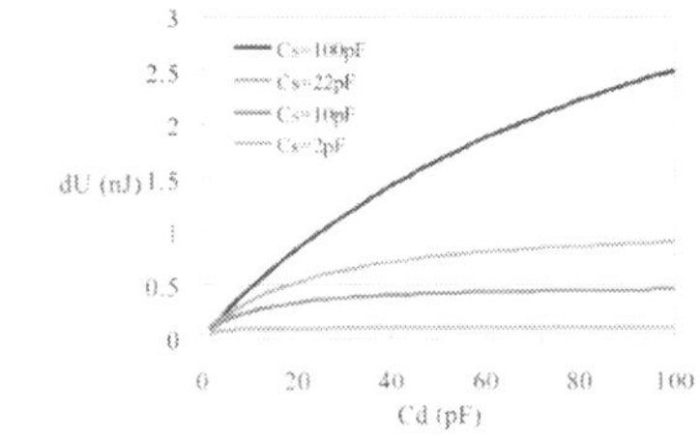

Figure 10. dU (nJ) as a function of C_d for 2pF- to 100pF-C_s applied 10V.

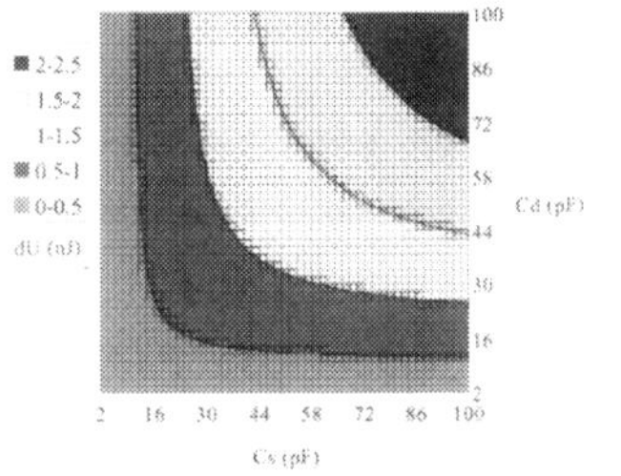

Figure 11. 2D-plot of dU as functions of C_d and C_s capacitances at 10V-applied C_s.

3.3. ESD failure model

Ip-plot indicated that the peak current tended to arise when C_s and C_d were of the order of pF. The peak current discharged from 10V-applied 2pF to 2pF was 34mA, which can cause magnetic damage to the GMR head. Whereas dU-plot indicated that the energy-loss increased gradually as C_s and C_d increased together. The energy-loss discharged from 10V-applied 2pF to 2pF was as low as 0.05nJ, which could not cause thermal damage. As both capacitances increased more than 20pF, the energy-loss increased over 0.5nJ and could bring thermal damage into the GMR head. Therefore, it can be concluded that for voltage as low as 10V, the discharge between capacitances of order pF tended to cause magnetic damage and as the capacitances increased more than 10pF the discharge increased the risk of thermal damage.

4. Conclusion

The contact discharge between floating devices was investigated experimentally and theoretically. The peak current and the energy-loss depend on the capacitance values and their ratio. Theoretical energy-loss due to the discharge was estimated by the difference between the potential energies prior to and following the discharges. From the 2D-plots of the peak current and the energy-loss, the increase of the capacitance tended to vary the ESD damage from magnetic to thermal. In order to avoid the risk of ESD failure, to reduce the capacitance of the device was effective, because the transferred-charge and energy-loss were restricted.

References

[1] Wallash A and Kim Y K 1998 *IEEE Trans. Magn.* **34-4** 1519
[2] Lam C F, Chang C and Karimi R 1998 *EOS/ESD Symposium Proc.* 360
[3] Ramaswamy S, Carter J, Stubbart J, Singh A, Krasnick R and Gocemen F, 2000 *EOS/ESD Symposium Proc.* 4B8 505
[4] Soda Y, Nagai N, Kasuga K, Shirai T, Kondo M and Ozue T 2001 *IEEE Trans. Magn.* **37-4** 1716
[5] Soda Y, Ono H and Ozue T 2003 *IEEE Trans. Magn.* **39-5** 2347
[6] Iben I E T 2004 *EOS/ESD Symposium Proc.* 370
[7] Soda Y, Ito Y, Kobayashi I, Takekuma T and Takahashi Y 2006 *EOS/ESD Symposium Proc.*2B5
[8] Soda Y 2007 *IEEE Trans. IAS* **43-5** 1144
[9] Soda Y and Sekine M 2010 *IEEE Industry Applications Magazine* **16-4** p48
[10] Maxwell J.C., "A Treatise on Electricity and Magnetism, Vol. 1, 3rd edition". Reprinted by *Dover Publications*, NY, 1974.
[11] Smythe W.R., "Static and Dynamic Electricity, 3rd edition". *McGraw-Hill Book Co.*, 1968.
[12] Greason W.D., *Research Studies Press Ltd.*, Taunton, Somerset, England, 1992.
[13] Huntsman J.R., *Proceedings EOS/ESD Symposium*, EOS-6, October 2-4, 1984, pp. 64-77.

On-chip detection circuit for protection design in display panel against electrical fast transient (EFT) disturbance

Cheng-Cheng Yen[1], Ming-Dou Ker [1,2], Wan-Yen Lin[1], Che-Ming Yang[3], Shih-Fan Chen[3], and Tung-Yang Chen[3]

[1] Institute of Electronics, National Chiao-Tung University, 1001 Ta-Hsueh Road, Hsinchu, 30010, Taiwan

[2] Department of Electronic Engineering, I-Shou University, No.1, Sec. 1, Syuecheng Road, Dashu Township, Kaohsiung, 84041, Taiwan

[3] Research and Development Center, Himax Technologies, Inc., 26 Zih-Lian Road, Tree Valley Park, Tainan, 74445, Taiwan

E-mail: p9011846@alab.ee.nctu.edu.tw

Abstract. A new on-chip detection circuit is proposed for electrical fast transient (EFT) protection design in a display system. For microelectronic products, electrical transient disturbances often cause upset or frozen states under the IEC test standard. The output signal of the proposed detection circuit can be used as a firmware index to execute system automatic recovery operations and to release the EFT-induced locked states in display panels. The circuit function to detect positive or negative electrical transients has been investigated in HSPICE simulation and verified in silicon chip. The experimental results have confirmed successful circuit performance under EFT tests. With hardware/firmware co-design, the immunity of a display system against electrical transient disturbance has been significantly improved.

1. Introduction

Electrical fast transient (EFT) event has become an important reliability issue to display systems equipped with CMOS driver integrated circuits (ICs) [1], [2]. This tendency results from not only the progress of more functions integrated into a single silicon chip but also from the strict requirements of reliability test standards, such as the EFT test. The microelectronic product must sustain the EFT voltage level of ±2 kV under EFT tests to achieve the immunity requirement of "level 4" in the IEC 61000-4-4 test standard [3]. When some real-world EFT events occurred, the power lines of the CMOS ICs in the microelectronic products no longer maintained normal voltage levels, but an exponential voltage pulse with the amplitude of several tens volts occurred. The waveform of a single EFT pulse has a rise time of about ~5 ns and the pulse duration of ~50 ns. The CMOS ICs inside the microelectronic products are very susceptible to electrical transient disturbance [4]-[8], even though they have passed the component-level ESD specifications.

In this work, an on-chip detection circuit is designed to detect electrical transient disturbance under EFT tests and verified in a 0.13-μm CMOS process. The proposed circuit can provide a hardware/firmware co-design solution to improve the susceptibility of microelectronic display systems against EFT-induced electrical transient disturbance.

2. On-Chip Detection Circuit

2.1. Circuit Implementation

In this work, an on-chip detection circuit has been proposed, as shown in Fig. 1. The CR-based circuit structure is designed to detect EFT-induced transient disturbance. The NMOS (M_{nr}) is used to set the initial output voltage levels at nodes V_{OUT} and V_A as 1.8 V with the V_{DD} of 1.8 V. During the normal operating condition, the node V_X is biased at 0 V by connecting to the V_{SS} line through a poly resistor. The two-inverter latch is designed to memorize the logic state before and after EFT events. The MOSFET capacitor is a PMOS device and connected between the V_{DD} line and the node V_X to sense transient disturbance coupled to the 1.8-V power line. For the buffer stage, there are three inverters designed in series to drive the 8-pF loading capacitance of the voltage probe at node V_{OUT}. Under the EFT tests with an overshooting transient voltage, the node V_X will be coupled with positive voltage by the MOSFET capacitor coupling. Then, the NMOS device (M_{nt}) will be turned on by the overshooting EFT voltage to pull down the voltage level at the node V_A from 1.8 V to 0 V. Therefore, the logic level stored at the node V_B can be further pulled up from logic "0" to logic "1" to memorize the EFT events. With the buffer inverters, the final output voltage of the proposed detection circuit is changed from logic "1" to logic "0" to detect the occurrence of EFT-induced transient disturbance.

The simulation conditions and measurement setup for the EFT test combined with the attenuation network is shown in Fig. 2. The EFT generator is connected to the V_{DD} line of the device under test (DUT) through -40dB attenuation network. The V_{DD} and V_{OUT} transient responses of the detection circuit are monitored by the digital oscilloscope.

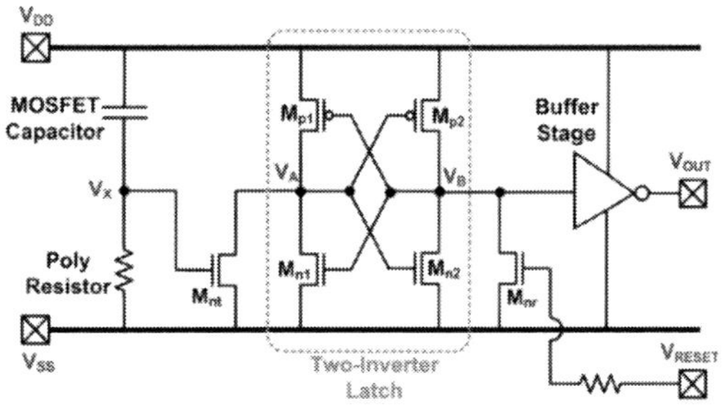

Figure 1. The proposed on-chip detection circuit.

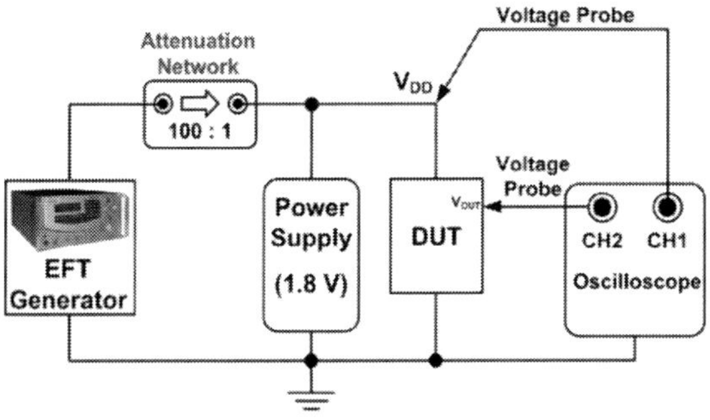

Figure 2. Measurement setup and simulation conditions for EFT test combined with -40dB attenuation network.

2.2. HSPICE Simulation

For HSPICE simulations, with the proper parameters (including the rise time constant τ_1 (=3 ns), fall time constant τ_2 (=25 ns), rise time delay t_{d1} (=200 ns), fall time delay t_{d2} (=210 ns), initial dc voltage value V_1 (=1.8 V), and exponential pulse voltage value V_2), the exponential voltage pulse can be constructed to simulate the EFT-induced disturbance under EFT tests. The rising edge of this exponential time-dependent voltage pulse is expressed as

$$V_{p(rise)}(t) = V_1 + (V_2 - V_1) \times \left[1 - \exp\left(-\frac{t - t_{d1}}{\tau_1}\right)\right], \text{ when } t_{d1} \le t \le t_{d2}. \tag{1}$$

The falling edge of this exponential time-dependent voltage pulse is expressed as

$$V_{p(fall)}(t) = V_1 + (V_2 - V_1) \times \left[1 - \exp\left(-\frac{t - t_{d1}}{\tau_1}\right)\right] + (V_1 - V_2)\left[1 - \exp\left(-\frac{t - t_{d2}}{\tau_2}\right)\right], \text{ when t} \ge t_{d2}. \tag{2}$$

Journal of Physics: Conference Series **301** (2011) 012053 doi:10.1088/1742-6596/301/1/012053

The simulated V_{DD} and V_{OUT} waveforms of the proposed on-chip detection circuit with positive and negative exponential pulse transient disturbance on V_{DD} line are shown in Figs. 3(a) and 3(b), respectively. From the simulated waveforms, V_{DD} begins to change rapidly from 1.8 V and acts with positive or negative exponential voltage pulse waveform during the simulated transient disturbance on power line. During EFT zapping conditions, V_{OUT} is disturbed simultaneously during the EFT-induced disturbance in the simulation. After V_{DD} finally returns to its normal voltage level of 1.8 V, V_{OUT} will be changed from 1.8 V to 0 V, as shown in Figs. 3(a) and 3(b). In future works, the variation of transistor parameters should be taken into design considerations.

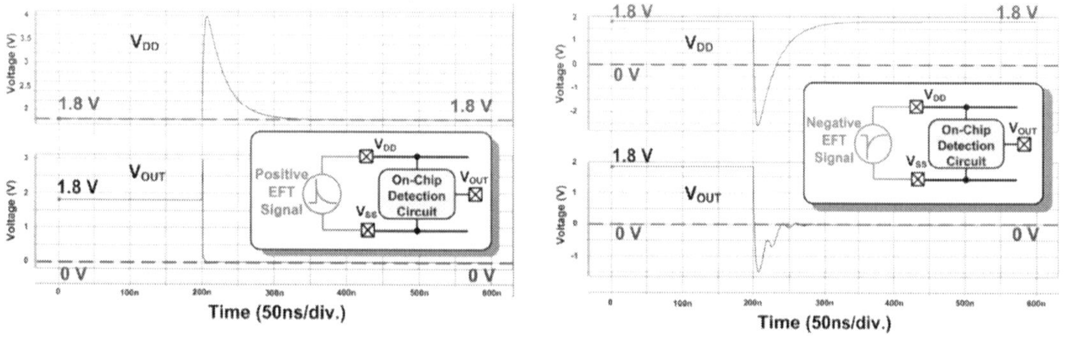

(a) (b)

Figure 3. Simulated V_{DD} and V_{OUT} waveforms of the proposed on-chip detection circuit under EFT tests with (a) positive, and (b) negative, exponential voltage pulse waveforms coupled to V_{DD}.

3. Experimental Results

The measured V_{DD} and V_{OUT} waveforms of the on-chip detection circuit under EFT test with EFT voltage of +200 V and -350 V are shown in Fig. 4(a) and 4(b), respectively.. As shown in Fig. 4(a) (Fig. 4(b)), under the EFT test with positive (negative) EFT voltage, V_{DD} begins to increase (decrease) rapidly from 1.8 V. Meanwhile, V_{OUT} begins to greatly increase (decrease) with positive (negative) exponential voltage pulse coupled on V_{DD} line. Finally, after the EFT-induced transient disturbance, the final output voltage (V_{OUT}) of the detection circuit is changed the logic "1" to logic "0."

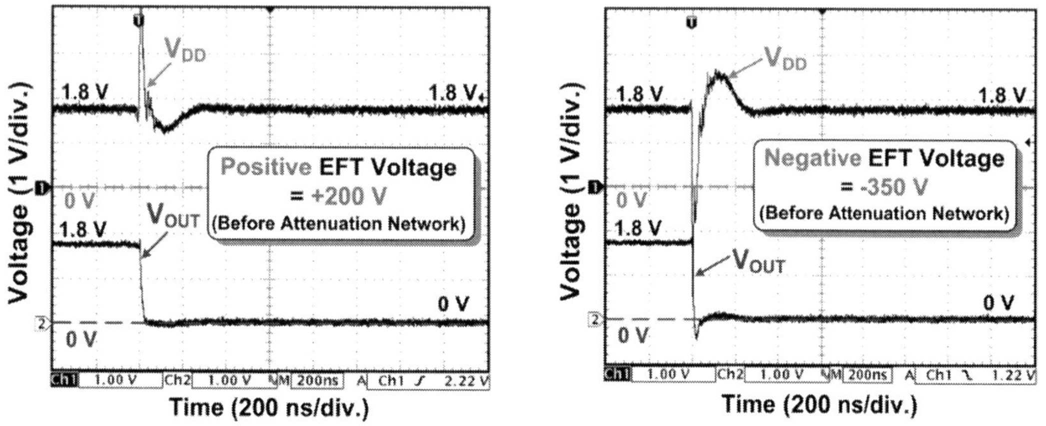

(a) (b)

Figure 4. Measured V_{DD} and V_{OUT} waveforms on the proposed on-chip detection circuit under EFT tests with (a) +200-V, and (b) -350-V, EFT voltages combined with attenuation network.

4. Application in Display Systems

The high-voltage electrical fast transients often cause CMOS ICs inside the equipment under test (EUT) to be upset or frozen after the system-level electrical transient disturbance. It has been reported that electrical transient disturbance coupled to power and ground pins of a super twisted nematic (STN) liquid crystal display (LCD) driver circuit can cause abnormal function of the LCD panel. The transient disturbance energy can couple through the indium tin oxide (ITO) film from the LCD panel into the CMOS driver ICs to cause display malfunction.

In this work, the proposed on-chip detection circuit can cooperate with firmware operation to provide hardware/firmware co-design for improving the robustness of the display products against EFT-induced stresses. The detection results can be used as firmware index to execute an automatic recovery procedure when display products are disturbed with EFT events. The detection circuit can connect with the 1.8-V power line to detect and memorize the occurrence of EFT-induced transient disturbances coupled on the digital subsystem of display products. In the beginning, the output (V_{OUT}) state of the detection circuit is set to logic "1" by the power-on procedure. When the electrical transients happen, the output state (V_{OUT}) of the detection circuit will transit from 1.8 V to 0 V. Simultaneously, firmware recovery procedure can be started to restore the display system to a known stable state as soon as possible. After the recovery procedure, the output level of the detection circuit and the firmware index are both re-set to logic "1" again for detecting the next EFT-induced electrical transient disturbance event.

5. Conclusions

An on-chip detection circuit to memorize EFT events has been proposed and successfully verified in a CMOS 0.13-μm process. The circuit performance under different positive and negative EFT-induced transient disturbance has been investigated by HSPICE simulation. The experimental results in silicon chip have confirmed that the proposed detection circuit can successfully detect and memorize the occurrence of fast electrical transients during EFT tests. The proposed detection circuit can be further combined with firmware design to provide an effective solution against the frozen or locked states caused by EFT-induced transient disturbance events in microelectronic display products.

Acknowledgments

This work was partially supported by National Science Council, Taiwan, under Contract NSC 99-2811-E-009-051; and by Himax Technologies Inc., Taiwan.

References

[1] A. Wallash and V. Kraz, "Measurement, simulation and reduction of EOS damage by electrical fast transients on AC power," in *Proc. EOS/ESD Symp.*, 2010, pp. 59–64.
[2] M.-D. Ker and S.-F. Hsu, *Transient-Induced Latchup in CMOS Integrated Circuits*, John Wiley & Sons, 2009.
[3] IEC 61000-4-4 Standard, "*EMC – Part 4-4: Testing and Measurement Techniques – Electrostatic Fast Transient/Burst Immunity Test*," IEC, 2004.
[4] W. Huang, J. Dunnihoo, and D. Pommerenke, "Effects of TVS integration on system level ESD robustness," in *Proc. EOS/ESD Symp.*, 2010, pp. 145–149.
[5] K. Muhonen, P. Erie, N. Peachey, and A. Testin, "Human metal model (HMM) testing, challenges to using ESD guns," in *Proc. EOS/ESD Symp.*, 2009, pp. 387–395.
[6] E. Grund, K. Muhonen, P. Erie, and N. Peachey, "Delivering IEC 61000-4-2 current pulses through transmission lines at 100 and 330 ohm system impedances," in *Proc. EOS/ESD Symp.*, 2008, pp. 132–141.
[7] M. Honda, "Measurement of ESD-gun radiated fields," in *Proc. EOS/ESD Symp.*, 2007, pp. 323–327.
[8] T. Smedes, J. Zwol, G. Raad, T. Brodbeck, and H. Wolf, "Relations between system level ESD and (vf-) TLP," in *Proc. EOS/ESD Symp.*, 2006, pp. 136–143.

Journal of Physics: Conference Series **301** (2011) 012054 doi:10.1088/1742-6596/301/1/012054

Influence of corona charging in cellular polyethylene film.

Gustavo Ortega Braña[a], Pedro Llovera Segovia[b], Francisco Magraner[a], Alfredo Quijano[a].

[a] Instituto Tecnológico de la Energía (ITE), Av. Juan de la Cierva 24, Parque Tecnológico de Valencia, 46980 Paterna-Valencia (Spain).

[b] Instituto de Tecnología Eléctrica - Universitat Politècnica de Valencia, Camino de Vera s/n 46022-Valencia (Spain).

Author's email adress: gustavo.ortega@ite.es

Abstract. Cellular polymers have recently attracted attention for their property of exhibiting a piezoelectric constant when they are electrically charged. The electrostatic charge generated in the voids by the internal discharges creates and internal macrodipole which is responsible for the piezoelectric effect. Charging by corona discharge is the most used method for cellular polymers. Many works has been published on polypropylene and polyethylene films mainly focused on the required expansion process or on the results obtained for raw cellular materials electrically activated. Our work is based on commercial polyethylene cellular films which have been physically characterized and electrically activated. The effect of thermal treatment, physical uniaxial or biaxial stretching and corona charging was investigated. The new method of corona charging improved the piezoelectric constant under other activation conditions.

1. Introduction.

Piezoelectricity in polymers based on their oriented crystalline structure is known at least since 1950 [1]. Recently some new piezoelectric polymers have been developed from polymer foams [2][3]. They are not based on the same physical property since electrical dipoles in polymer foams are created at the micron scale whereas in conventional piezoelectric polymers dipoles are due to the molecular structure [4][5]. Polymer foams are not initially piezoelectric but after a corona charging treatment piezoelectricity is developed in the material from a macroscopic point of view. Corona charging is applied on the surface of a flat polymer foam sample creating a high internal electric field [6]. This induces air breakdown inside the polymer foam cells and an electrostatic charge appears in the inner surface of the cells generating a macroscopic dipole at each cell where the discharge occurs [7]. Piezoelectric properties are then related to both polymer electrostatic properties and corona discharge physics.

The electric model of the charge polymer foam can be simplified into two layers of polymer and air, where σ is the equivalent surface charge density, ε is the permittivity of the material, S1 is the thickness of polymer and S2 is the thickness of air and Y the Young modulus of the polymer foam [8]:

$$d_{33} = \frac{\varepsilon\sigma}{Y} \cdot \frac{1+(S_2/S_1)}{(1+\varepsilon(S_2/S_1))^2} \qquad (1)$$

In this work we present results obtained on commercial polymer foams thermally modified and charged by means of different corona charging methods. The new contribution is the use of cellular polyethylene instead of cellular polypropylene and the reversal of the corona-poling polarity to achieve higher effects. The d_{33} static constant measurements show the influence of material modification and corona charging system on piezoelectric properties.

2. Experimental.

Samples were cut from commercial closed cell polyethylene foams 0.48mm in thickness (Alveolit TA 0.48 density 180kg/m^3). Thermal stretching is then applied by heating the sample and pulling it in particular directions. Uniaxial stretching is performed in a single step and biaxial stretching is carried out by means of two perpendicular stretches. The heating temperature is 100°C and the stretching ratio for unixially stretched samples was 50% and 30% for biaxially stretched samples (in both directions). Although thickness is reduced to 60μm in the case of uniaxial stretching and to 50μm for the case of biaxial stretching, cell shape is different in both cases and that gives very different results. After thermal stretching, samples are metalized by silver sputtering on the grounded side, then corona discharge is applied and finally the other side is also metalized.

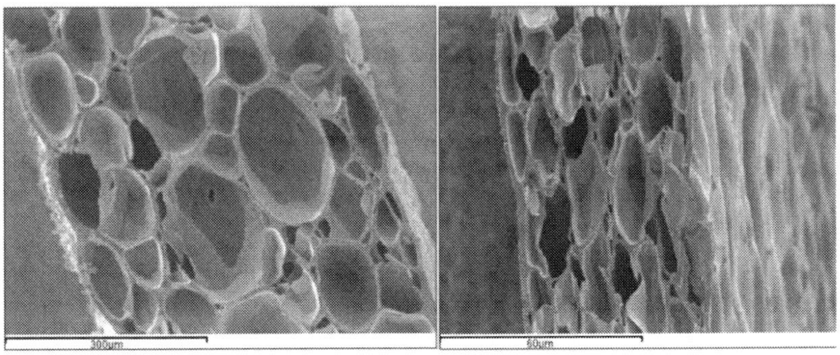

Figure 1. *Foam before thermal stretching (left) and after uniaxial thermal stretching (right).*

To determine the influence of the corona charging method, samples were not thermally treated, thickness was 0.48mm and 12kV positive corona discharge was applied for 5 min.

3. Results and discussion.

3.1. Influence of thermal stretching

Thermal stretching improves the piezoelectric constant of the material. As can be seen in figure 2, the piezoelectric constant d_{33} of biaxially stretched samples is up to 5 times higher than unixially stretched samples, although thickness is similar in both cases (170 pc/N vs. 25 pc/N). However, it may be noticed that internal charge stability is shorter than in the case of non-thermally treated samples probably because of the morphological changes induced in the material during heating. Reaching the glass transition temperature (50°C) may lead to conductivity modifications and it can be seen that piezoelectric constant decreases with time faster than in the case of non-treated samples. Besides, in the thermally stretched samples it can be expected that the internal electric field is higher and some charge injection and neutralization phenomena happen helping to dissipate static charge.

The increase of d_{33} constant may be due both to the increase of the internal poling field because of the sample thickness reduction and the modification of the cell shape. The poling field is nearly one order of magnitude higher for the stretched samples but the breakdown happens at relatively close values of corona discharge voltage for both kind of samples ($\approx$15kV corona discharge for the non-stretched samples and $\approx$14kV for the stretched samples). This is probably due to the irregular internal field distribution on the samples. Thus it is not possible to apply the same high poling field for both samples. But, on the other hand, the modification of the cells shape induces important mechanical and electrical changes in the material and this is probably the main reason for d_{33} constant increase [9]. From the internal discharges point of view, flatter cells with closer walls allow to a higher electrical charge to be stored since several breakdowns can occur inside the same cell. (In spherical cells one discharge may prevent further discharges from happening since an opposite electric field is generated by the static charge). Charge density is increased and, according to equation (1), the piezoelectric constant increases. From a mechanical point of view, the Young modulus is reduced thus improving the d_{33} coefficient [7].

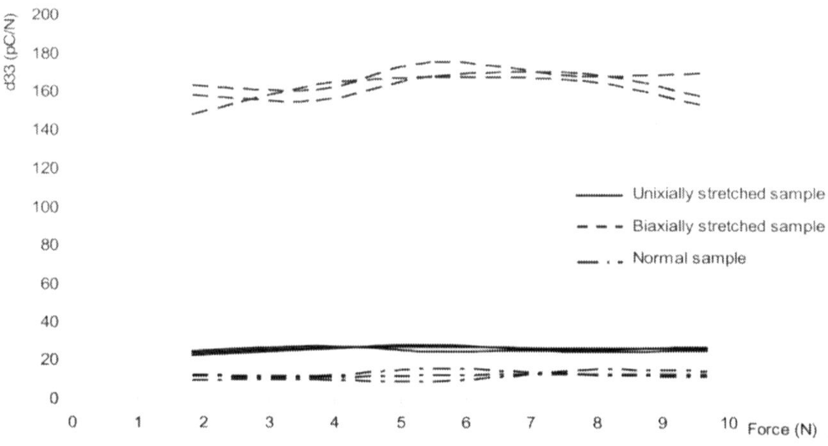

Figure 2. *Piezoelectric constant for uniaxial or biaxially stretched samples.*

3.2. Influence of corona charging system

The corona charging system is based on the well known point-to-plane arrangement but applying a new methodology. Samples are charged twice reversing their position under the corona discharge. A first corona discharge is applied for 5 minutes and the exposed side to the corona discharge is metallized. This operation also neutralizes surface charges but preserves the internal charges. Then the sample is placed again upside down under a corona discharge of the same polarity and intensity, that is, the firstly exposed and metallized side is now the grounded side of the sample. This reverses the internal electric field in the cells which now is added to the external applied electric field. (Reversing the polarity of the corona discharge instead of the sample orientation could have led probably to the same results).

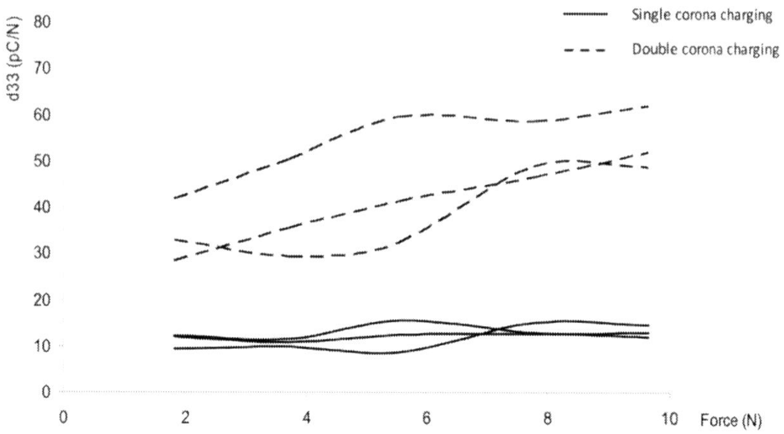

Figure 3. *Effect of the corona charging method on the piezoelectric constant.*

This new charging method provides good results as can be seen in figure 3 with no need to apply thermal stretching to the samples. This method simplifies the development of polymer foams for their application as piezoelectric materials without modifying charge stability as happen when thermal treatment is applied. Although thicker samples are used and cell geometry is more spherical, electric breakdown inside the cell is enhanced by the internal electric field reversing. In the spherical cells previous discharges will probably prevent further discharges from happening in the same cell. Of course, a flat cell will give better results and more charge density storage but thermal treatment in our samples reduced charge stability. It can be noted that longer application of the corona discharge does not provide further improvements in the piezoelectric constant.

4. Conclusions.

Piezoelectric properties from available polymer foams treated by corona discharge have been investigated. Two main factors have been considered: thermal stretching and corona discharge application methodology. The piezoelectric d_{33} constant obtained directly from corona discharge application and sample metallization can be improved significantly (up to 170 pC/N) by thermal biaxial stretching but charge stability drops. A reversal, two-step application of corona discharge also improves the piezoelectric constant (70 pc/N) while keeping charge stability. This is probably due to modification of the air breakdown dynamics in the cells of the material inducing stronger internal discharges and higher surface charge. This charging method could be easily applied to the production of larger samples.

5. Acknowledgements.

This work has been developed under the project "Intelligent Materials with Mechanical and Electrical Properties interaction (E-MAT)" which has been submitted for funding with reference number IMDEEA/2011/13 to the call of Technological Centers of IMPIVA Network 2011, Strategic development program (action 1, R+D projects) financed by the Generalitat Valenciana through the Instituto de la Mediana y Pequeña Empresa Valenciana (IMPIVA) and the European Regional Development Fund (ERDF).

References

[1] E. Fukada. "History and recent progress in piezoelectric polymers", IEEE Trans. Ultrason. Ferroelectr. Freq. Control **7** 1277-1290 (2000).
[2] G. M. Sessler and J. Hillenbrand, "Electromechanical response of cellular electret films", Appl. Phys. Lett. **75** 3405-3407 (1999).
[3] J. Hillenbrand and G. M. Sessler, "Piezoelectricity in Cellular Electret Films", IEEE Trans. Dielectr. Electr. Insul. **7** 537-542 (2000).
[4] M. Paajanen, H. Välimäki, and J. Lekkala, "Modelling the electromechanical film (EMFi)," J. Electrostatics **48** 193-204 (2000).
[5] R. Gerhard-Multhaupt, "Less can be more: Holes in polymers lead to a new paradigm of piezoelectric materials for electret transducers", IEEE Trans. Dielectr.Electr. Insul. **9** 850-859 (2002).
[6] M. Wegener and S. Bauer, "Microstorms in cellular polymers: A route to soft piezo-electric transducer materials with engineered macroscopic-dipoles", Chem. Phys. Chem. **6**, 1014-1025 (2005).
[7] J. Hillenbrand, N. Behrendt, N. Mohmeyer, V. Altstadt, H. Schmidt, G. Sessler. "Charge retention in biaxial-oriented polypropylene films containing various additives". Electrets 2005 12th International Symposium on Electrets, 0-7803-9116-0 pp 276-279 (2005).
[8] X Zhang, G. Sessler, J. Hilllenbrand "Optimization of Piezoelectric Properties of Cellular Polypropylene Films by Repeated Expansion". Electrets, 2005. ISE-12. 2005 12 Int. Symposium on Electrets, pp. 51-54, (2005).
[9] D. Pan, X. Zhang, X. Wang, G. Cao, S. Zhang, H. Li, Z. Xia. "Piezoelectricity of cross-linked polypropylene films treated by hot-stretching". Proc. 9th Int. Conf. on Properties and Applications of Dielectric Materials, 978-1-4244-4368-0/09 pp 1166-1169 (2009).

Organic electronic memory based on a ferroelectric polymer

R. Kalbitz[1], P. Frübing[1], R. Gerhard[1] and D.M.Taylor[2]

[1] Department of Physics and Astronomy, University of Potsdam, Karl-Liebknecht_Str., 24-25, 14476 Potsdam, Germany
[2] School of Electronic Engineering, Bangor University, Dean Street, Bangor, Gwynedd LL57 1UT, UK

d.m.taylor@bangor.ac.uk

Abstract. Measurements of the capacitance of metal-insulator-semiconductor capacitors and the output characteristics of thin film transistors based on poly(3-hexylthiophene) as the active semiconductor and poly(vinylidenefluoride-trifluoroethylene) as the gate insulator show that ferroelectric polarisation in the insulator is stable but that its effect when poled by depletion voltages is partially neutralised by trapping of electrons at or near the semiconductor interface. Nevertheless, the combination of materials is capable of providing an adequate memory function.

1. Introduction

Organic thin film transistors (TFT) are an attractive option for low cost electronic applications and much progress has been made in fabricating prototype circuits for active matrix displays and for RFID applications [1,2]. To extend the range of applications there is a need to develop and optimise the performance of non-volatile memory devices that are compatible with the solution-processing fabrication procedures used in Plastic Electronics. A possible candidate is an organic TFT incorporating the ferroelectric co-polymer poly(vinylidenefluoride-trifluoroethylene), P(VDF-TrFE) as the gate insulator.

2. Device concept

Although TFTs are generally considered to be dynamic devices the basic operating mechanism relies on electrostatic attraction as the means for switching on the device as seen in Fig.1(a). Here a negative voltage, $-V_G$, applied to the gate electrode attracts holes to the interface of the p-type organic semiconductor poly(3-hexylthiophene), P3HT, with the insulating P(VDF-TrFE) thus creating an accumulation layer. When a negative voltage, $-V_D$, is applied to the drain electrode then a hole current, I_D, flows through this layer from source to drain which, in the linear regime $V_D<(V_G-V_T)$ is given by

$$I_D = \frac{W}{L}\mu C_i(V_G - V_T)V_D \tag{1}$$

and in the saturation regime $V_D>(V_G-V_T)$ by

$$I_D = \frac{W}{2L}\mu C_i(V_G - V_T)^2 \tag{2}$$

where W and L are the width and length of the induced channel, μ the hole mobility, C_i the capacitance per unit area of the gate insulator and V_T the threshold voltage for the device [3]. If the electric field applied to the gate insulator exceeds the coercive field of P(VDF-TrFE) then ferroelectric domains will be established whose remanent polarisation, see Fig. 1(a), will contribute additional electrostatic forces drawing additional holes into the channel. This in turn increases the channel conductance and hence the current flow between source and drain. Furthermore, if V_G is reduced to zero, the remanent polarisation will maintain the device in its 'ON' state until it is reversed, after which the device should turn 'OFF'.

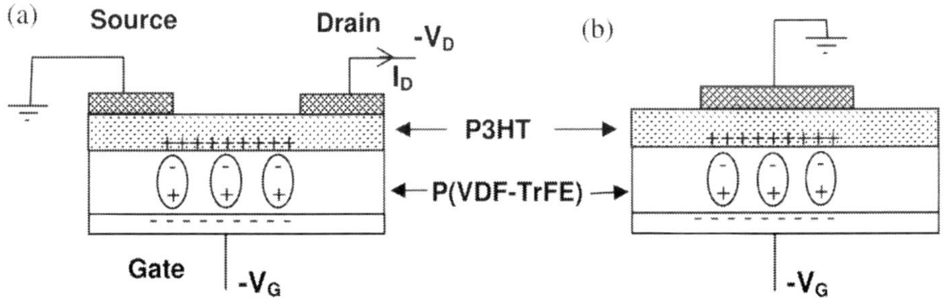

Fig. 1 *Cross-section through (a) a TFT and (b) a MIS Capacitor showing polarised ferroelectric domains in the gate insulator.*

These same processes can be observed by measuring the capacitance of a Metal-Insulator-Semiconductor (MIS) capacitor formed from the same materials (Fig. 1(b)). When the device is in accumulation, $-V_G$, the measured capacitance will be that of the gate insulator. When V_G is positive, the P3HT layer is driven into depletion and the capacitance decreases owing to the formation of an immobile space charge region at the interface which has a capacitance inversely proportional to its width. At sufficiently positive V_G the P3HT is completely depleted and the measured capacitance becomes constant and equal to the series sum of the insulator and gate layers. As in the case of the TFT if, in accumulation, the coercive field was exceeded resulting in a remanent polarisation with the orientation shown in Fig. 1(b) this will maintain the capacitor in accumulation as V_G is swept from negative to positive voltages. Only when this polarisation is reversed at sufficiently high positive gate voltages will the normal depletion behaviour be observed in the device.

These basic concepts will now be used to explain results obtained from TFTs and MIS capacitors based on P3HT and P(VDF-TrFE). It will be shown, however, that additional interfacial effects occur which limit the efficiency of such devices as memory elements.

3. Experimental

All devices were prepared in a class 1000 Clean Room by sequentially spin coating two layers of P(VDF-TrFE) (65/35 mol %) from a 2- butanone solution onto a clean glass substrate furnished with evaporated Al electrodes. The semiconducting layer was spin-coated from a P3HT/chloroform solution after drying the P(VDF-TrFE) film at 120°C for ~12 hours under vacuum. The devices were completed by vacuum evaporating top gold electrodes (capacitors) or gold source/drain interdigitated electrodes with an aspect ratio, W/L = 1485 for the TFTs. All devices were then annealed at 90°C for 12 hours under vacuum. Typical insulator and semiconductor layer thicknesses were 0.3 μm and 0.05-0.08 μm as measured on reference samples with the Dektak 3 ST surface profiler. Dielectric measurements were performed with the Novocontrol Alpha A impedance analyzer in a temperature controlled nitrogen atmosphere with bias voltages applied to the Al electrode. TFT characteristics were measured under nitrogen using an Agilent 4155C Semiconductor Parameter Analyzer.

4. Results and Discussion

Fig. 2 show the capacitance-voltage plots obtained for the MIS capacitor using the programmed voltage cycle shown in the inset. Utilising this combination of uni- and bi-polar voltage sweeps Dickens *et al* [4] and Bauer [5] were able to decouple the underlying current-voltage characteristic of their ferroelectric capacitors from effects associated with the reversal of ferroelectric polarisation. Similarly, because the measured capacitance per unit area, C_m, can be expressed as

$$C_m = C_i + \frac{dP}{dV} \quad (3)$$

where dP/dV represents a voltage-induced change in the remanent polarisation, employing such a voltage sweep sequence also allows purely capacitive effects to be distinguished from the reversal of remanent polarisation as seen in Fig. 2. On the initial sweep to positive or negative voltages, the capacitance plots exhibit broad maxima. The return sweep and subsequent unipolar sweeps (dotted curves) show only the underlying behaviour expected of a MIS capacitor based on a p-type semiconductor i.e. essentially constant capacitance in accumulation, $-V_G$, but decreasing to a lower value as the semiconductor becomes fully depleted at positive values of V_G.

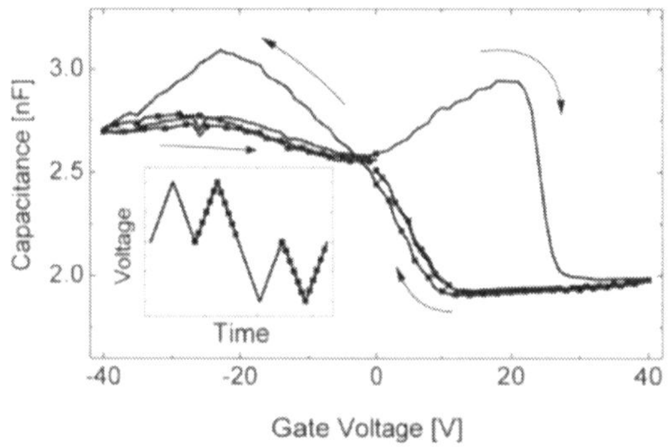

Fig. 2 *Capacitance-voltage plots obtained for the MIS capacitor using the programmed voltage sequence shown in the inset.*

We conclude, therefore, that since the maxima only occur on the first voltage sweep after polarity reversal and appear at approximately ±20 V, corresponding to the coercive field, E_C ~ 60 MV/m of P(VDF-TrFE), then they must arise from the reversal of polarisation in the ferroelectric insulator.

Interestingly, after reversing the polarisation during the positive voltage sweep, the device was expected to remain fully depleted until the subsequent negative voltage sweep had reached -20V when the polarisation reversed again. The rise in capacitance at low positive voltages on the return voltage sweep has been taken by some authors [6] as evidence that remanent polarisation is not stable in such devices. On the contrary, our measurements confirm that it is stable and is reversed only when the coercive field is exceeded. We have argued that after polarisation reversal in the depletion regime, $+V_G$, electrons must drift from the depleted semiconductor into trap states at the surface of the P(VDF-TrFE) [7].

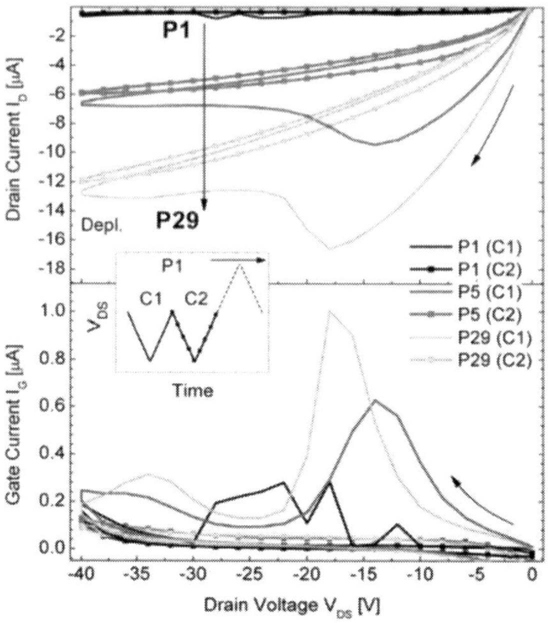

Fig. 3 *Drain current, I_D, (upper) and gate leakage current, I_G, (lower) observed in the TFT using the programmed V_D sequence shown in the inset and for $V_G = 0$ V.*

Similar effects are observed in the output characteristics of the TFT shown in Fig.3. Here the device was subjected to a series of programmed sweeps of the drain voltage, V_D, this time in sequences of two unipolar negative cycles in odd numbered programmes, P1, P3, etc. (see inset of Fig. 3) followed by two unipolar positive cycles during the even numbered programmes P2, P4, etc. In every cycle the coercive field of the ferroelectric was exceeded so that the remanent polarisation was reversed many times. Starting with programme P1 prior to any poling of the ferroelectric it is seen that, with $V_G = 0$ V, the device was clearly OFF with no evidence for the enhancement of conductivity by remanent polarisation. However, sequential sweeps to ±40 V, caused I_D to increase so that by the first cycle of programme P5, I_D had reached a maximum of about -10 μA at -15V and continued to increase thereafter, eventually saturating at about -17 μA at -20 V. The current maxima, observed during the first negative cycle, occurred therefore at voltages corresponding to the coercive field and coincided with peaks in the gate current caused by polarisation reversal, dP/dt in the ferroelectric. The maxima in I_D, however, were much greater than those for I_G and must arise from the effect of polarisation on I_D rather than from gate current flowing to the drain. This being the case, then after reversing the remanent polrisation, I_D should have decreased to the low values corresponding to the initial OFF state observed in P1.

Two interesting points emerge from the data. Firstly, ferroelectric polarisation builds up over many cycles. Such a phenomenon has been reported by others in PVDF [8] and P(VDF-TrFE) [9] capacitors and has been shown to be related to microstructural changes in the polymer. Secondly, I_D did not turn off after polarisation reversal but decreased to lower values and then followed a common curve both on the return voltage sweep and in the subsequent unipolar sweep. This is consistent with the C-V plots in Fig. 2 which show that after poling with depletion voltages, the device returns to accumulation, the condition required for the flow of source-drain current in the TFT, at $V_G \sim 0$ V. The transistor output characteristics confirm, therefore, that although polarisation reversal occurs at negative V_D, sufficient negative charges must also become trapped at the semiconductor/insulator interface to maintain a partially conducting hole channel. To neutralise the interface positive charges of the remanent polarisation as well as the shift in threshold voltage to ~0 V, requires that in excess of $\sim 10^{13}$ electrons/cm^2 must be trapped close to the interface [7]. While there is sufficient discrimination between the values of I_D for the two polarisation states, the effectiveness of the memory would be greatly improved if the number of trapping sites could be reduced. An alternative approach may be to block the injection of electrons from the semiconductor by introducing an electron blocking layer at the interface.

5. Conclusions

The results discussed above provide clear evidence that, contrary to previous reports, the remanent polarisation in organic TFTs based on the ferroelectric polymer P(VDF-TrFE) is stable but when poled with depletion voltages electrons are injected from the semiconductor into trap states in or at the insulator/semiconductor interface. A possible approach to reducing this effect is suggested.

Acknowledgements

This project was initiated by grants from the British Council (ARC 1294) and DAAD (D/07/09993).

References

[1] Bao Z and Locklin J (Editors) 2007 *Organic Field Effect Transistors* (CRC Press: Boca Raton).
[2] Klauk H (Editor) 2006 *Organic Electronics: Materials, Manufacturing and Applications*, (Wiley-VCH Verlag: Weinham).
[3] Sze S M 1981 *Physics of Semiconductor Devices: 2nd Ed.* (Wiley, New York).
[4] Dickens B, Balizer E, DeReggi A S and Roth S C 1992 *J Appl Phys* **72** 4258.
[5] Bauer S 1983 *Ferroelectrics* **49** 231.
[6] Naber R C G, Asadi K, Blom P W M, de Leeuw D M and de Boer B 2009 *Adv Mater* **21** 1.
[7] Kalbitz R, Frübing P, Gerhard R and Taylor D M 2011 *Appl Phys Lett* **98** 033303.
[8] Guy I L and Unsworth J 1987 *J Appl Phys* **61** 5374.
[9] Reynolds N M, Kim K J, Chang C and Hsu S L 1989 *Macromolecules* **22** 1092.

Advanced MEMS systems for optical communication and imaging

M N Horenstein[1], J B Stewart[2], S Cornelissen[2], R Sumner[1], D S Freedman[1], M Datta[1], N Kani[1], and P Miller[1]

[1]Boston University, 8 Saint Mary's St., Boston MA 02215, USA mnh@bu.edu
[2]Boston Micromachines, 30 Spinelli Place, Cambridge, MA 02138 USA

Abstract. Optical communication and adaptive optics have emerged as two important uses of micro-electromechanical (MEMS) devices based on electrostatic actuation. Each application uses a mirror whose surface is altered by applying voltages of up to 300 V. Previous generations of adaptive-optic mirrors were large (~1 m) and required the use of piezoelectric transducers. Beginning in the mid-1990s, a new class of small MEMS mirrors (~1 cm) were developed. These mirrors are now a commercially available, mature technology. This paper describes three advanced applications of MEMS mirrors. The first is a mirror used for coronagraphic imaging, whereby an interferometric telescope blocks the direct light from a distant star so that nearby objects such as planets can be seen. We have developed a key component of the system: a 144-channel, fully-scalable, high-voltage multiplexer that reduces power consumption to only a few hundred milliwatts. In a second application, a MEMS mirror comprises part of a two-way optical communication system in which only one node emits a laser beam. The other node is passive, incorporating a retro-reflective, electrostatic MEMS mirror that digitally encodes the reflected beam. In a third application, the short (~100-ns) pulses of a commercially-available laser rangefinder are returned by the MEMS mirror as a digital data stream. Suitable low-power drive systems comprise part of the system design.

1. Introduction

Over the past few decades, micro-electromechanical systems (MEMS) have become a mature technology found in cell phones, automobiles, cameras, medical devices, and optical imaging systems. MEMS deformable mirrors (DMs), the topic of this paper, are thin-film devices that shape and selectively reflect light. While the numerous capacitive MEMS actuators inside a DM require almost no power, the driving circuit can represent a substantial power demand for a DM optical system. This paper discusses several applications of MEMS mirrors and some new power-saving driver circuits.

2. Structure, Fabrication, and Actuation

Our mirrors, reported elsewhere [1,2], are fabricated by the following or similar process: Thin films of phosphosilicate glass (PSG) and silicon nitride are deposited on a silicon substrate, followed by 90-nm of gold. The gold is patterned and etched with 4-μm holes; these holes serve as a mask for reactive-ion etching to make holes in the silicon nitride. Unused gold is removed and electrostatic actuators that support the mirror are formed by under-etching the PSG with acid (HF) through the holes, leaving support anchors. This process yields mirrors, depicted in figure 1, that are optically flat to within 6 nm.

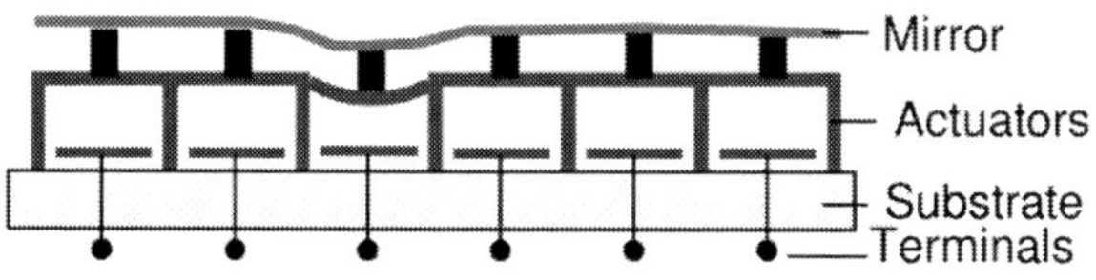

Figure 1. DM array structure.

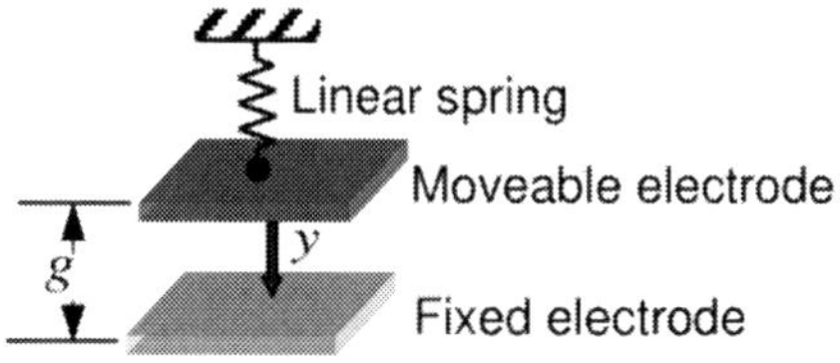

Figure 2. Simplified spring-capacitor actuator.

Each supporting actuator can be modeled, to first order, as a two-electrode capacitor of area A and spacing g, as in figure 2. A voltage V causes deflection y of the upper electrode. The balance of electrostatic and $-ky$ mechanical forces yields the well-known deflection versus voltage equation [3]:

$$y(g-y)^2 = V^2\, \varepsilon_0 A/2k \qquad (1)$$

Tests of our mirrors at 300 V show deflection of at least 200 nm, more than an optical wavelength. The need for driving signals as high as 300 V puts an extra burden on the driver circuitry.

3. Space-Based Interferometry

We have built several interesting optical systems using our electrostatic mirrors. In one system, the surface profile of a DM is continuously altered to clear up blurry or distorted images. Such DMs will play an important role in high-resolution telescope systems being developed by NASA for its exoplanet mission – the search for life on other planets [4,5]. Light from a distant star is canceled by interferometric mirrors, leaving only the images of planets in the star's halo. These optical systems must orbit the earth to avoid atmospheric distortion, hence they can consume only very small amounts of power. At the same time, however, they must have extraordinary precision approaching the picometer range. They therefore may contain up to 10^4 closely spaced actuators having picometer resolution.

Presently, all commercial analog DM-drivers use a one-amplifier-per-actuator architecture, in which analog signal processing and multiplexing are performed at low voltage. Signals are then fed to a large array of high-voltage (HV) amplifiers. This approach is necessary because, to date, no integrated circuits exist that are capable of multiplexing HV analog signals. With conventional approaches, the very large number of power-consuming HV amplifiers needed for a 10^4-element array would require many large circuit boards and tens-of-watts of quiescent power. For NASA's planned and future space-based missions, a better approach is needed. We have developed a high-voltage multiplexing circuit (MUX) that distributes the output of a *single* HV amplifier to each actuator in an array. Over a "raster" cycle, a given actuator is momentarily connected and energized by the amplifier. The MUX then holds the actuator's voltage until the next refresh event. With only one HV amplifier used, total system power, size, and cost are dramatically reduced. Our current prototype drives 144 actuators with 0-300 V HV analog signals over a 10-ms refresh cycle. Total system power consumption is about 400 mW, and the architecture is scaleable to an arbitrarily large number of actuators. The circuit configuration is also compatible with existing, commercial HV CMOS (integrated-circuit) fabrication processes. The focus of our current efforts is a 10^3-actuator mirror driver fabricated on a single high-voltage integrated circuit.

The architecture of our system is shown in figure 3. A logic-level digital bus feeds a low-voltage digital-to-analog (D/A) converter which, in turn, drives the HV amplifier. Within the MUX, each actuator has its own transistor-based driving cell (figure 4) that, when addressed by logic-level row and column lines, momentarily connects the actuator to the HV-amplifier output. A 100-nF "holding" capacitor is connected in parallel with the actuator's tiny 150-fF capacitance to enhance holding time.

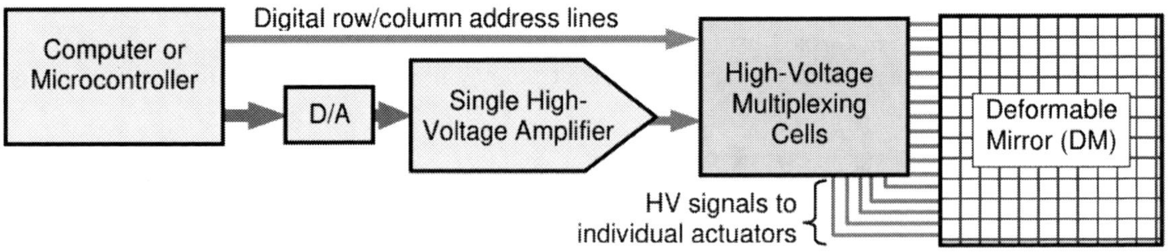

Figure 3. Block diagram of our high-voltage mutliplexer configuration.

One anomaly of this scheme is that the voltage of a cell's holding capacitor will, at times, be higher than the output of the HV amplifier (V_{CA} positive). Conventional transistor circuit design relies on the upper voltage bus being the highest in the system. Thus a key feature of the circuit of figure 4 is the emitter-coupled BJT pair Q_1-Q_2 which can maintain a high-impedance for either polarity of V_{CA}.

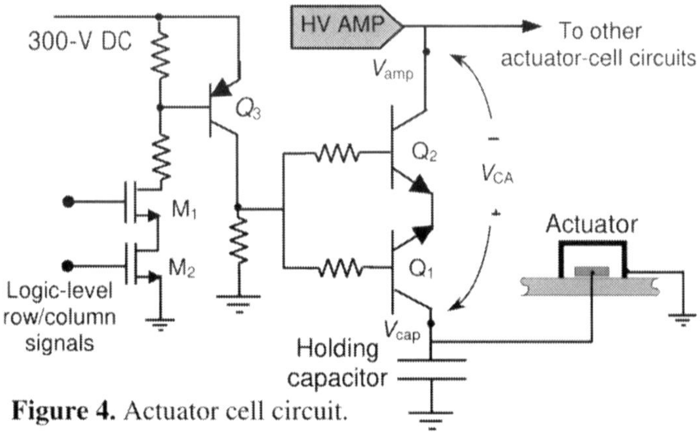

Figure 4. Actuator cell circuit.

Circuit performance was verified by measuring actuator position drift over a 10-s interval using an optical surface-mapping profilometer accurate to 1 nm. Data were then interpolated to find the actual droop over the system's 10-ms refresh cycle. The results show drift rates between 0.32 and 4.39 pm/ms (1 pm = 10^{-12} m). Holding-capacitor voltage droop over a refresh cycle was typically less than 3 mV for a full 300-V signal (i.e., one part in 2^{16}, or one LSB of a 16-bit D/A conversion). This drift is due to charge leakage over undesirable impedance paths to ground, which we minimize in our design.

4. Laser Communication via MEMS Retroreflective Mirror

Our second application involves point-to-point, stealth data communication via laser [6]. In many military and sensing applications, radio is not practical because transmission may be disrupted, unavailable, or easily detected. Also, both communication nodes must radiate power. In our alternative system, one node is active (emits laser power), but the other is passive (only reflects power). As shown in figure 5, the active node beams a constant (CW) laser toward the passive node which contains a MEMS DM retro-reflector. The mirror is digitally (on/off) modulated at high data rates to either reflect the beam (logic **1**) or scatter it (logic **0**).

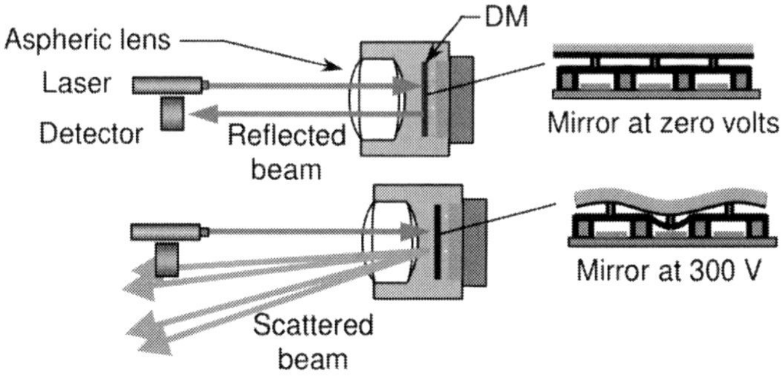

Figure 5. Laser beam is reflected (logic **1**) or scattered (logic **0**) to send back digital data.

Power consumption at the passive node is dramatically reduced via the boost-converter circuit of figure 6. Conventional MEMS switching circuits waste energy by discarding stored capacitor charge after each switching operation. Our system, on the other hand, reuses energy from cycle to cycle. The HV logic-**0** pulses that drive the DM are produced as follows: Energy is first stored in the inductor via the MOSFET switch, then transferred to the capacitor to produce a half-resonant sinusoidal pulse. The energy is recaptured by the inductor, then returned to the battery, ready for the next cycle. Equating the peak inductor energy ½ LI_p^2 with the peak transferred capacitor energy ½ CV_p^2 yields:

$$V_p = I_p\sqrt{L/C} = V_o T/\sqrt{LC} \tag{2}$$

for an inductor charging time T. In this way, 300-V-peak pulses can be produced when V_o is less than 10 V. The communication system operates at a predetermined bit rate. A voltage pulse disrupts mirror reflection, yielding a logic-**0**. The absence of an expected pulse, allowing laser reflection, sends a logic-**1**. Our goal is mirror driver that operates from a single, standard, 9-V battery for up to 24 hours.

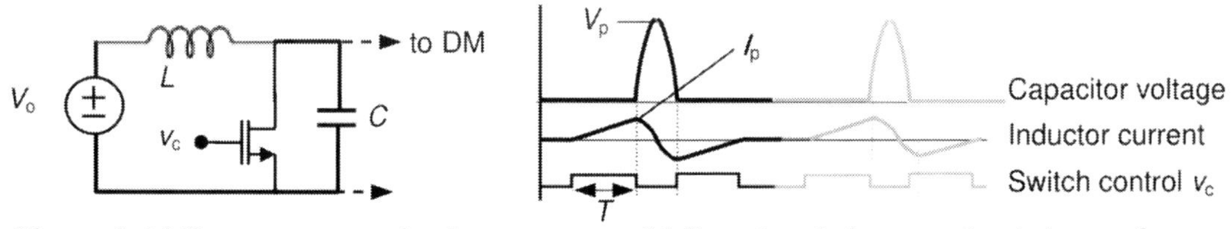

Figure 6. (a) Boost converter circuit; (b) Capacitor, inductor and switch waveforms.

5. Laser-Rangefinder Communicator

Another optical communicator builds upon a standard-issue laser ranging device [7]. For distance measuring purposes, a rangefinder measures the return time of very short (~100 ns) laser pulses emitted at about a 20-kHz rate. In our modified system, shown in figure 7, a receiving node first times the rangefinder pulses to determine their periodicity. It then synchronizes the reflective state of the mirror with the pulses to send back either a logic-**1** or logic-**0** data stream on a pulse-by-pulse basis. Because the pulses are of such miniscule duration, as is required of most rangefinders, unique timing circuitry at the passive node is required. As before, the passive node emits no radiation, hence it can operate at very low power from a single 9-V battery for long periods of time.

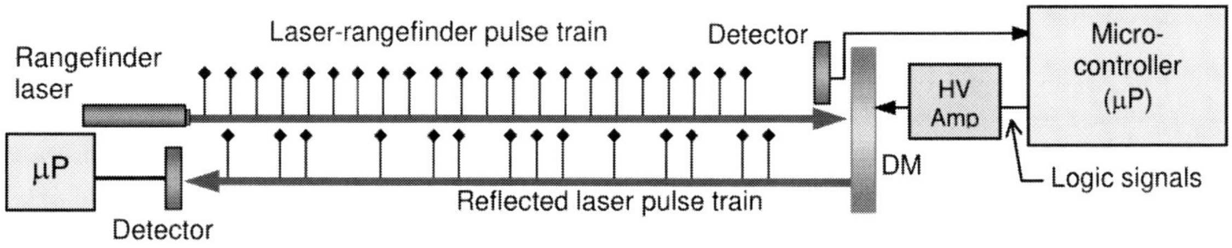

Figure 7. MEMS DM selectively returns 100 ns pulses from the laser rangefinder.

6. Conclusion

We have demonstrated the use of MEMS deformable mirrors for optical communication and imaging, and we have disclosed several techniques for reducing power consumption of the requisite driving electronics. These devices and techniques show promise for a variety of future planned systems.

7. Acknowledgements

These projects were funded by NASA SBIR NNX-10 CE08P and US-ARMY STTR W911NF-08-C-006.

8. References

[1] Bifano TG, Mali RK et al 1997 *Optical Eng* **36** 1354
[2] Cornelissen SA, Bierden PA, Bifano TG and Lam CV 2009 *J Micro/Nanolithography, MEMS and MOEMS* **8** 031308
[3] Horenstein MN, Bifano TG, Mali RK, and Vandelli N 1997 *J Electrostatics* **42** 69
[5] *http://planetquest.jpl.nasa.gov*
[4] Earth-like exoplanets: the science of NASA's navigator program 2006 *JPL Publication 06-5RevA*
[6] Stewart JB, Freedman D, et al 2010 *Proc SPIE* **7595** 759505
[7] Ziph-Schatzberg L, Bifano T et al 2009 *Proc SPIE* **7209** 72090C

Microfluidic actuation of insulating liquid droplets in a parallel-plate device

W Wang and TB Jones[1]

Department of Electrical and Computer Engineering, University of Rochester, Rochester, New York 14627 USA

E-mail: jones@ece.rochester.edu

Abstract. In droplet-based microfluidics, the simultaneous movement and manipulation of dielectric and aqueous droplets on a single platform is important. The actuation forces on both dielectric and aqueous droplets can be calculated with an electromechanical model using an equivalent RC circuit. This model predicts that dielectric droplet actuation can be made compatible with electrowetting-based water droplet manipulation if the oil droplet is immersed in water. Operations such as transporting, splitting, merging, and dispensing of dielectric droplets at voltages less than 100 V are demonstrated in a parallel-plate structure. Such capability opens the way to fully automated assembly line formation of single-emulsion droplets.

1. Introduction

Among methods to move and manipulate liquid microdroplets, electrical actuation has many advantages, principally voltage-driven process control [1]. For the parallel-plate configuration shown in Fig. 1, electric actuation is attributed to electrowetting (EWOD) for conductive droplets and dielectrophoresis (DEP) for dielectric droplets. Both mechanisms drive droplets toward an activated electrode. There is a frequency-dependent relationship between EWOD and DEP [2]. For a liquid with finite electrical conductivity and dielectric constant, a critical frequency separates conductive (EWOD) and dielectric (DEP) limits [2-4].

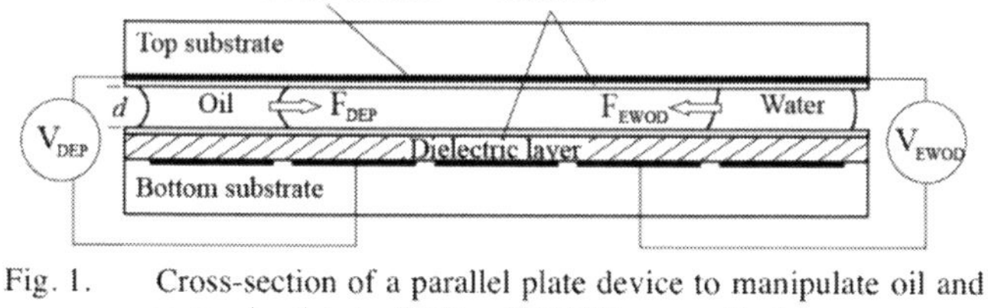

Fig. 1. Cross-section of a parallel plate device to manipulate oil and water droplets by DEP and EWOD, respectively.

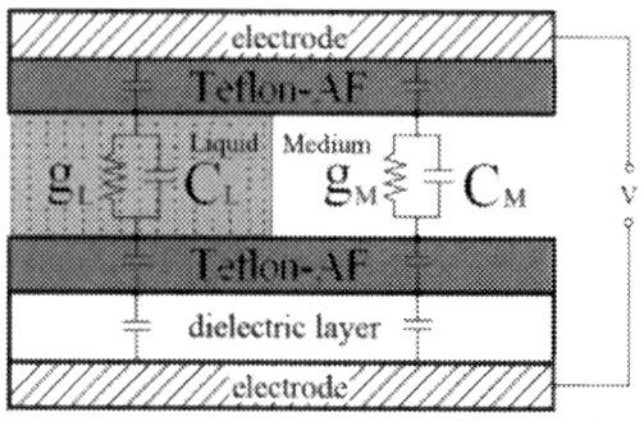

Fig. 2. Equivalent circuit model for the parallel-plate microfluidic device.

Integration of EWOD and DEP-based droplet microfluidics is an essential step to forming double-emulsion droplets on a chip [5,6]. Unfortunately, DEP usually requires much higher voltage than EWOD. Low-frequency AC in the range of 11 to 100 V is sufficient for EWOD [7-9], but ~200 V typically is needed for DEP microfluidic actuation [3,5]. Here, we propose a low-voltage droplet manipulation scheme that uses the negative DEP effect to move dielectric droplets. A general

[1] To whom any correspondence should be addressed.

Published under licence by IOP Publishing Ltd

electromechanical model predicts that dielectric droplet actuation can be made compatible with electrowetting-mediated water droplet manipulation. Companion experiments demonstrate the electric actuation of dielectric droplets, including droplet transport, splitting, merging, and dispensing at less than 100 V.

2. Modeling

Fig. 1 shows the cross-section of a typical microfluidic system consisting of two parallel glass plates. This structure is modeled by the equivalent RC circuits shown in Fig. 2. This RC circuit is similar to those of Jones [2-4, 10], Zeng [11], and Chatterjee [12]. Previous models describe the ambient medium with pure capacitance because the medium is usually air or oil. Here, however, the medium might also be conductive, so resistance is needed in parallel with the capacitance. The electromechanical force parallel to the electrodes is evaluated using coenergy $W_e^{'}=C(x)V^2/2$ [13].

$$f^e = \left.\frac{\partial W_e^{'}}{\partial x}\right|_V = \frac{1}{2}V^2\frac{\partial C(x)}{\partial x} \tag{1}$$

Fig. 3 plots calculated forces for three cases: water droplet in air, mineral oil droplet in water, and mineral oil droplet in air, all at 65 V with substrate spacing of 170 μm. The electric field distribution in the liquid depends on frequency. In fact, for the water-in-air and oil-in-water cases there exists a critical value f_c that defines the transition from effectively conductive to dielectric behavior. See Fig. 3. There is no frequency dependence for the oil-in-air case because neither is electrically conductive.

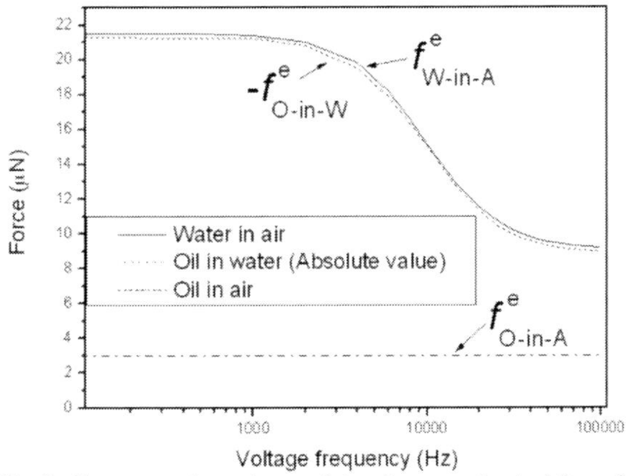

Fig. 3. Frequency dependency of the electromechanical force for three liquid combinations in the parallel-plate geometry: water droplet in air, mineral oil droplet in water medium, and mineral oil droplet in air medium. The parameters used for the calculation correspond to those of the experiments.

The force acting on an oil droplet in air is much smaller than the force experienced by a water droplet at the same applied voltage. It is for this reason that oil droplet actuation usually requires higher voltages. On the other hand, an oil droplet in water medium experiences a relatively large electric force, comparable in fact to a water droplet in air. Note, however, that this force acts in the opposite direction, that is, from high to low field intensity. This is an example of negative DEP. The liquid droplet has a dielectric constant smaller than that of the surrounding medium. Though this force acts in the opposite direction, we can still achieve low voltage oil actuation by effectively pushing the droplet away from activated electrodes. Continuous movement becomes achievable by sequential activation and de-activation of electrodes in the array.

3. Fabrication and experimental setup

The basic features of the droplet-based microfluidic system are illustrated in Fig. 1. The bottom glass substrate is evaporatively coated with 100 nm aluminum, and then photolithographically patterned into a two dimensional array of individually addressable electrodes. The plate is then spin-coated with 0.5 μm of spin-on-glass (SOG) (Futurrex IC1-200) as the dielectric layer and 1 μm of amorphous fluoropolymer (DuPont Teflon-AF) as a hydrophobic coating. The upper plate is glass coated with transparent ITO (Indium-tin oxide), which forms the ground electrode, and top-coated with 1 μm Teflon-AF as a hydrophobic coating. The top substrate is positioned above the bottom substrate using spacers. Voltages applied to individual electrodes are controlled by a LabView-based controller.

4. Results and discussion

The layout of the basic design of a transport array for oil droplets in water is illustrated in Fig. 4. The side electrodes help contain the droplets on the track. To perform an experiment, an aqueous solution is first introduced into the gap between the top and bottom substrates. The surfactant Silwet L-77 (0.0625 wt%) is used in the water phase to reduce surface tension and stabilize the mineral oil droplet in the water medium. The oil droplet is then injected.

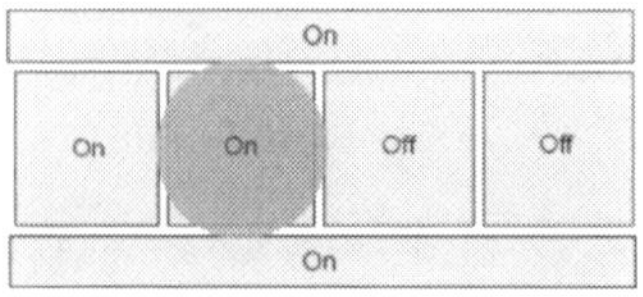

Fig. 4. Layout of the transport electrodes for moving oil droplets in water medium. When energized, the electrodes on the sides keep the dielectric droplet centered on the track.

Oil droplet transport Fig. 5 shows video frames demonstrating operation of the oil droplet transport scheme in water. The droplet initially sits on electrode 1, as shown in Fig. 5 (a). All the electrodes are initially off. When electrode 1 and the side electrodes are activated, the oil drop is pushed to the left. See Fig. 5 (b). After the droplet reaches electrode 2, this electrode is now activated forcing the oil droplet to continue moving left to the next electrode. In this way, continuous movement is easily achieved.

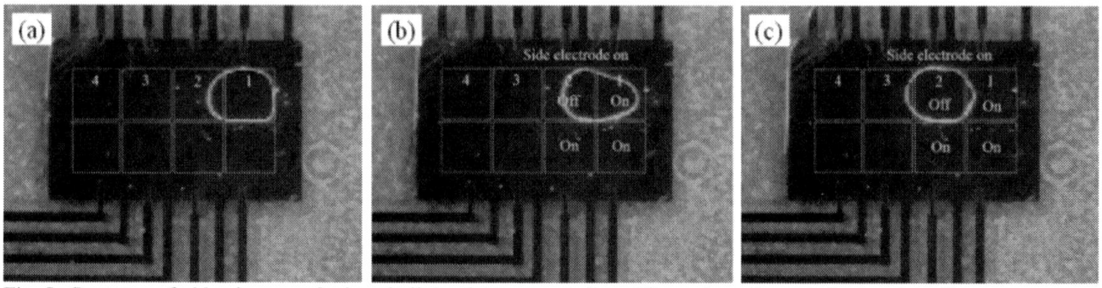

Fig. 5. Sequence of video images of mineral oil droplet transport viewed from above through the top plate. Transport electrodes are 1 mm square, the spacing between the top and bottom plates is 170 μm, and the applied voltage is 50 V-rms at 100k Hz AC.

Oil droplet splitting and merging After shown in Fig. 6, droplet division is achieved by activating the electrode on which the droplet resides while deactivating its two neighbors. In Fig. 6 (a), a large oil droplet is initially sitting on electrode 2 as shown. By activating electrode 2, the oil drop starts to move outwards onto electrode 1 and 3. See Fig. 6 (b). Eventually the oil drop splits in two. Merging is achieved simply by moving two oil droplets together. When electrodes 1 and 3 are activated, the two small oil drops are pushed back together onto electrode 2 where they quickly coalesce.

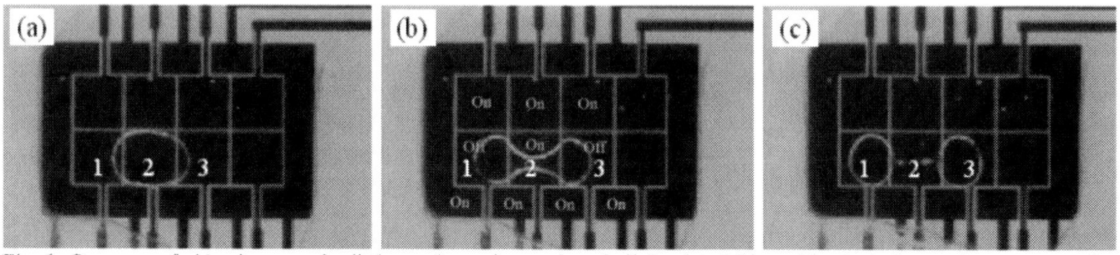

Fig. 6. Sequence of video images of splitting and merging a mineral oil droplet. Addressable electrodes are 1 mm square, the spacing between the top and bottom plates is 170 μm and the applied voltage is 70 V-rms at 100k Hz AC.

Oil droplet dispensing Fig. 7 provides a demonstration of oil droplet dispensing from a large reservoir. The oil initially covers the large reservoir electrode. When this electrode is activated, with the side electrodes also turned on, oil is drawn outward onto the transport array to form a liquid column. See Fig. 7 (b). Liquid pinch-off is achieved by activating the electrode next to the reservoir and deactivating the reservoir electrode. See Fig. 7 (c). The portion of the liquid to the left pinches off into a droplet which can then be transported away. The process may be repeated to dispense a stream of droplets. Note that sometimes a very thin film of oil connecting the dispensed droplet to the reservoir forms for a time. This film eventually ruptures.

5. Conclusions

The general electromechanical model developed here can be used to calculate electric actuation forces on both dielectric (that is, insulating) and aqueous (conductive) droplets in parallel-plate microfluidic structures. The model explains why oil droplet actuation requires higher voltage than does an aqueous droplet. It also confirms the possibility of low voltage, negative DEP actuation of oil when the droplet is in a water medium. These modeling predictions are corroborated by experiments that demonstrate some essential droplet-based microfluidic operations, including oil droplet transport, splitting, merging, and dispensing. These operations can be monolithically integrated on a single microfluidic chip to provide a versatile platform for the simultaneous processing of many different types of liquids at relatively low voltages. One particular application of interest for the assembly-line fabrication of the targets for laser fusion experiments is an automated double-emulsion droplet generator.

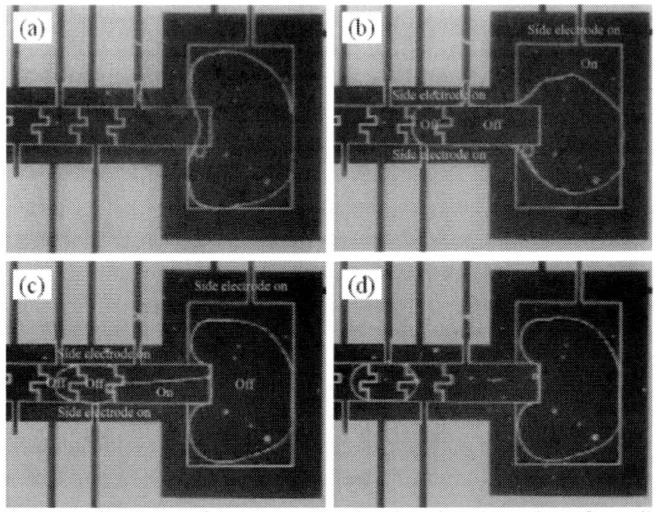

Fig. 7. Sequence of video images showing the dispensing of an oil droplet of ~230 nanoliter. The spacing between the top and bottom plates is 170 μm, and the applied voltage is 70 V-rms 100k Hz AC. Note the thin film of oil in (c) that for a short time connects the dispensed droplet to the much larger reservoir volume.

Acknowledgment

This work was supported by the U.S. Department of Energy, Office of Inertial Confinement Fusion under Cooperative Agreement No. DE-FC52-08NA28302, the University of Rochester, and the New York State Energy Research and Development Authority.

References

[1] Yang C G, Xu Z R and Wang J H, 2010 *TrAC Trends Anal. Chem.* **29** 141
[2] Jones T B, 2002 *Langmuir* **18** 4437
[3] Jones T B, Wang K L and Yao D J, 2004 *Langmuir* **20** 2813
[4] Wang K L and Jones T B, 2004 *J. Micromech. Microeng.* **14** 761
[5] Fan S K, Hsieh T H and Lin D Y, 2009 *Lab Chip* **9** 1236
[6] Wang W, Jones T B and Harding D R, 2011 *Fusion Sci. Technol.* **59** 240
[7] Moon H, Cho S K, Garrell R L and Kim C J, 2002 *J. Appl. Phys.* **92** 4080
[8] Li Y, Parkes W, Haworth L I, Ross A W S, Stevenson J T M and Walton A J, 2008 *J. Microelectromech. Syst.* **17** 1481
[9] Lin Y Y, Evans R D, Welch E, Hsu B N, Madison A C and Fair R B, 2010 *Sens. Actuators* B *Chem.* **150** 465
[10] Jones T B, Wang W and Bailey A G, 2008 Frequency-addressable apparatus and methods for actuation of liquids, US patent US20080169195
[11] Zeng J and Korsmeyer T, 2004 *Lab Chip* **4** 265
[12] Chatterjee D, Shepherd H and Garrell R L, 2009 *Lab Chip* **9** 1219
[13] Woodson H H and Melcher J R, 1968 *Electromechanical Dynamics* (New York: Wiley) part 1 chapter 3

Journal of Physics: Conference Series **301** (2011) 012058 doi:10.1088/1742-6596/301/1/012058

A numerical model of Trichel pulses in air; the effect of pressure

P Sattari, G S P Castle and K Adamiak
Dept. of Electrical and Computer Engineering, University of Western Ontario, London, Ontario, Canada N6A 5B9

E-mail: psattari@uwo.ca

Abstract. In this paper a three-species two-dimensional numerical model is used for the simulation of the Trichel pulse regime of corona discharge in air in a point-plane configuration. The effect of air pressure on Trichel pulse characteristics is investigated and the numerical results are compared with the available experimental data. The effect of corona electrode radius of curvature on charge per pulse and the relation between charge per pulse and the average charge per pulse is also studied, and compared with the experimental data reported by Atten et al.

1. Introduction

If a negative DC voltage above the corona onset level is applied to a sharp electrode in an electronegative gas, a pulsating current will appear in the circuit. These pulses are called Trichel pulses [1] after Trichel, who first discovered them and did extensive research on their physics. At voltages close to the corona onset level, Trichel pulses are irregular; however, with a small increase in voltage they become very regular. Due to the many ionic species and reactions involved in this process, Trichel pulse regime of corona discharge is an interesting yet a complicated phenomenon.

Most of the models presented so far for modelling the corona discharge are either single-species (only one charge carrier is assumed in the air gap and other species are neglected), steady-state or one-dimensional (1D). A single-species model might be useful for some engineering applications where an approximate estimate of average current and space charge density suffices. However, it does not help with understanding the detailed distributions of charge carriers during corona discharge. Moreover, all of these models fail if a good estimate of corona discharge characteristics is required. Therefore, presenting a model, which can accurately predict the Trichel pulse behavior during corona discharge has still been missing. In addition, most of the experimental and theoretical studies on corona discharge are performed at normal pressure (1 atm). Investigating the effect of pressure on the Trichel pulses is of interest and has not been widely discussed.

In this paper, a two-dimensional (2D) dynamic model, which includes three ionic species (positive and negative oxygen ions, and electrons), is used for modelling Trichel pulses in air. The effect of pressure on Trichel pulse characteristics is studied and compared with the available experimental data [2]. The ionization layer profiles in the air gap during one Trichel pulse are shown and the ionization layer thickness along the axis versus time is presented. The effect of tip radius on the charge per pulse is also investigated and the relation between charge per pulse and the average charge per pulse is studied. The numerical results are compared with the available experimental data wherever possible [2].

2. Mathematical model and numerical algorithm

The corona discharge is modelled in a point-plane configuration (figure 1). A hyperboloidal conducting needle with a tip radius curvature of r_p and length L is placed at a distance d from the infinitely large grounded conducting plate. A negative DC potential V is applied to the needle. An external (ballast) resistance R_{ext} is connected in series between the needle and the voltage source. The ambient gas is air. This system is assumed to be axisymmetric, which means that the equations should be solved in 2D using variables x and r, where x is the axial distance and r indicates the radial distance.

For the simulation of the corona discharge, both the electric field and charge density distributions need to be calculated. The electric field is governed by the Poisson equation (1), while the charge density distribution can be determined by solving charge transport equations for all three ionic species considered in the model [3]. A generic form of the charge transport equation is shown in Equation (2). These four equations are combined with the Kirchhoff voltage equation for the electric circuit, where the total current in the circuit is obtained using Sato's formula [4]. Details of the equations, reaction coefficients and the boundary conditions can be found in [3].

$$\nabla \cdot \varepsilon \mathbf{E} = N_{total} \tag{1}$$

$$\frac{\partial N}{\partial t} = S - \nabla \cdot (N\mathbf{W}) + \nabla \cdot (D \nabla N) \tag{2}$$

where ε is the gas permittivity, **E** is the electric field, N is the charge density of the ionic species, **W** is the drift velocity, D is the diffusion coefficient and S is the source term.

To investigate the effect of pressure, it should be noted that the number of oxygen molecules (NO_2) and mobility of species are affected by pressure such that NO_2 is proportional to P and μ is inversely proportional to P.

The Finite Element Method (FEM) was used to solve the Poisson equation. For the charge continuity equation, the Flux Corrected Transport (FCT) technique was added to FEM to prevent the FEM results from diverging [3, 5].

3. Results

The numerical model presented in [5] was first used to produce a series of Trichel pulses for the applied voltage of V = -9 kV, assuming an external resistance of R_{ext} = 100 kΩ and normal pressure of 1 atm (figure 2). In this case, the ionization layer thickness along the axis of symmetry versus time and 2D profiles of this layer in the air gap were shown during one Trichel pulse. The effect of pressure on the Trichel pulse frequency was investigated for a constant applied voltage. The ratio between charge per pulse and average charge per pulse as calculated by Atten et al. [2] is also calculated and discussed in detail.

Ionization layer

The ionization layer thickness along the axis is defined as the distance from the corona electrode to the point at which the ionization coefficient is equal to the attachment coefficient. Figures 3 and 4 show the thickness of this layer along the symmetry axis and electric field on the corona electrode tip versus time, respectively. The ionization coefficient is related to the electric field, therefore, the variation of the electric field and ionization layer thickness has similar profiles as these figures show.

Figure 5 shows the ionization layer profile at different instants of one Trichel pulse: half pulse rising point (1), maximum current point (2), and two points in the falling section of the pulse (points 3 and 4). At points 1 and 2, the ionization layer thickness is the largest. However at point 2, the ionization stops on the tip of the corona electrode. The diminishing of this layer gradually continues with time along the electrode surface until ionization completely stops in the entire air gap.

Effect of pressure on Trichel pulse characteristics
As reported in [6], for a given needle voltage the Trichel pulse frequency has an inverse relation with pressure. At higher pressures, there is a larger number of oxygen molecules per cubic meter and this produces an inverse relation between mobility and pressure, since ions have more frequent collisions with neutral molecules. Therefore, in these gases the Trichel pulse period increases and the frequency decreases. On the other hand, experiments show that pulse duration has an inverse relation with pressure [2], which is confirmed by our numerical results. Table 1 show the variation of Trichel pulse period and pulse duration versus gas pressure.

Comparison and discussion
Comparing numerical results with the experimental data [2] results in the following conclusions:

1) The mean charge per pulse increases with corona electrode radius: As table 2 shows, for the same applied voltage (V = -7kV), reducing the radius of the corona electrode decreases the generated charge per pulse. It can be explained by the fact that increasing the radius of the corona electrode enlarges the region involved in the ionization. Therefore, the generated charge per pulse increases. As the Q_i increases, the Trichel pulse period decreases, because larger space charge needs more time to travel. Therefore, it takes more time for the electric field to increase to a value at which a new Trichel pulse can be generated.

2) The ratio Q_i/Q (charge per pulse divided by mean charge per pulse) is always larger than 0.5. Charge per pulse, Q_i, was calculated by integrating the corona current in time intervals at which the corona current is above 10% of the maximum current. In Atten's paper [2], it is reported that in high pressure air and with no external resistance, Q_i/Q is always smaller than 0.5. Since our numerical results were for normal pressure and for non-zero external resistances this was not confirmed by our model, but from figure 6 it is obvious that decreasing the external resistance from 100 kΩ to 10 kΩ decreases this ratio significantly. Therefore, it is reasonable to extrapolate and expect that for zero external resistance, this ratio can be below 0.5 as reported in [2].

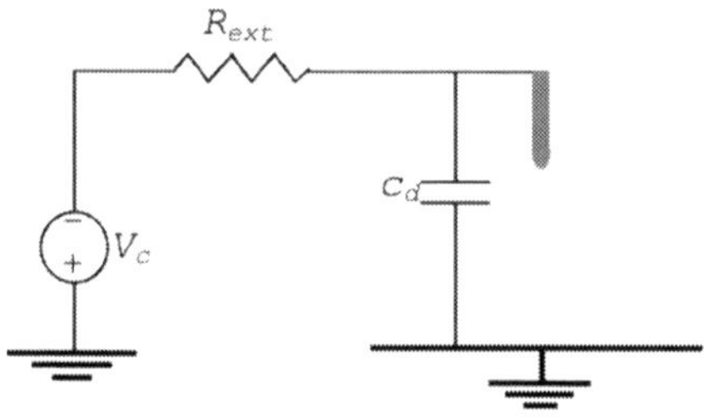

Figure 1. Circuit model of the system

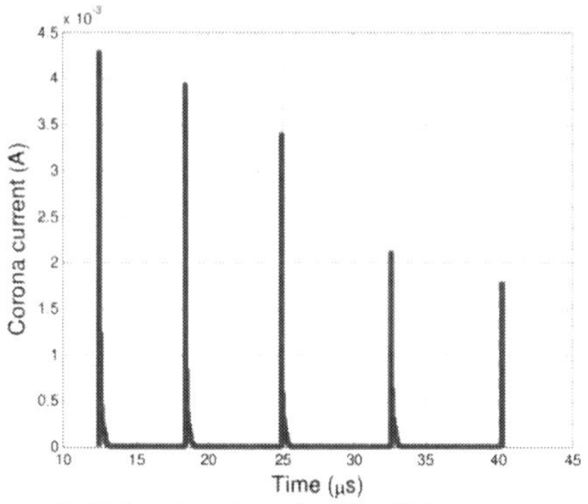

Figure 2. Trichel pulses for V=-9kV

4. Conclusions
In this paper, a 2D numerical model, which incorporates three ionic species (electrons, positive and negative oxygen ions) was used for simulating the negative corona discharge in point-plane configuration in air and Trichel pulses were obtained at different air pressures. Moreover, the variation of the ionization layer in the air gap during one Trichel pulse and the ionization layer thickness along the axis versus time were shown. The effects of pressure on Trichel pulse characteristics were investigated and it was confirmed that for a given needle voltage, the Trichel pulse frequency, the average corona current and the pulse duration are inversely proportional with pressure. It has been shown that the mean charge per pulse has a direct relation with r_p. Contrary to the experimental data presented in [2], the numerical simulation predicts the ratio Q_i/Q is larger than 0.5. It is suggested that this discrepancy is due to the presence of the external surge resistor.

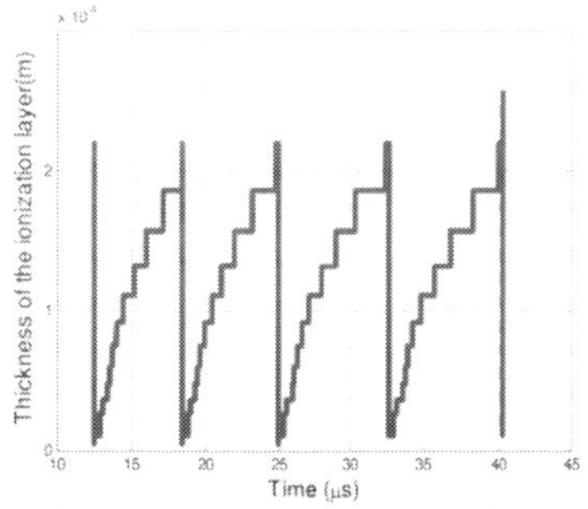

Figure 3. The thickness of the ionization layer

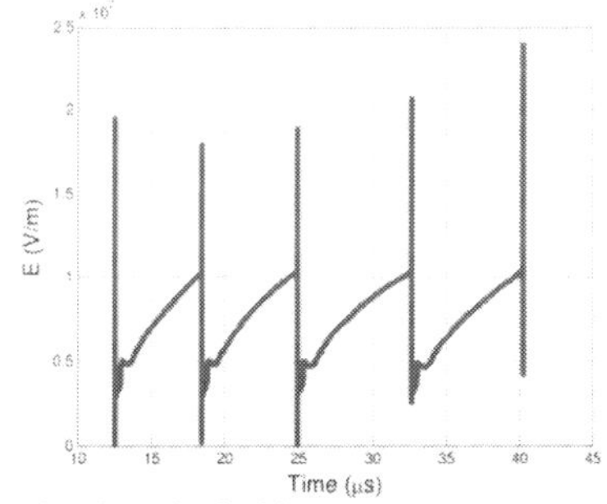

Figure 4. Electric field on the corona electrode

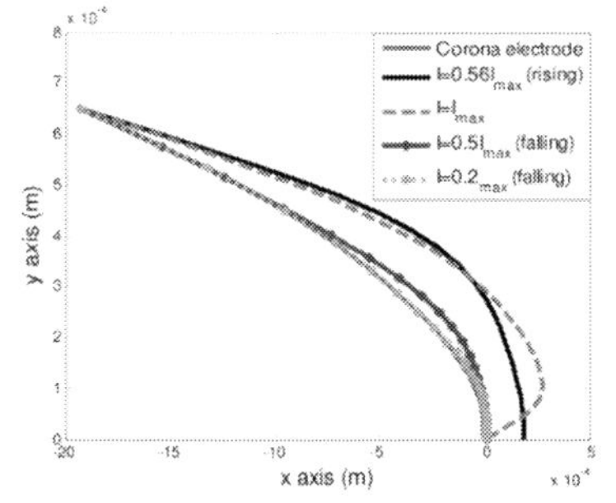

Figure 5. Ionization profile at different time instants

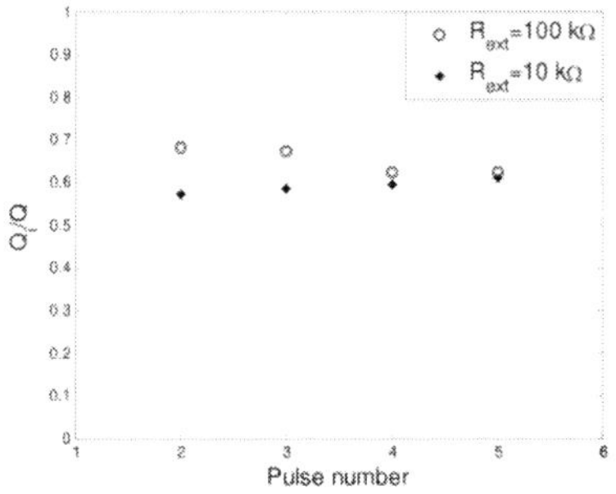

Figure 6. Q_i/Q for $V = $ -8kV

Table 1. Trichel pulse period and pulse duration versus gas pressure (V=-9kV, R_{ext}=100μΩ)

Air pressure (atm)	Trichel pulse period (μs)	Pulse duration (μs)
0.8	4.88	0.30
1	6.61	0.28
1.2	11.07	0.19

Table 2. Charge per pulse (in pico coulombs) for two configurations with different radii

Pulse number	2	3	4
Radius = 100 μm	115.61	114.17	111.49
Radius = 10 μm	23.781	21.906	21.415

Acknowledgments
The authors would like to acknowledge the helpful discussions and advice of Dr. P. Atten during the course of this work. This project was in part financially supported by the Natural Sciences and Engineering Research Council of Canada (NSERC). This work was made possible by the facilities of the Shared Hierarchical Academic Research Computing Network (SHARCNET: www.sharcnet.ca).

5. References

[1] Trichel G W 1938 The mechanism of the negative point-to-plane corona near onset *J. Phys. Rev.* **54** 1078–84
[2] Haidara M, Denat A and Atten P 1997 Corona discharges in high pressure air *J. Electrostatic* **40-41** 61–6
[3] Sattari P, Castle G S P and Adamiak K 2010 FEM-FCT based 2D simulation of Trichel pulses in air *Proc. of IEEE/IAS Annual Meeting, Houston, TX USA*, Oct 3-7, 2010
[4] Sato N 1980 Discharge current induced by the motion of charged particles *J. Phys. D* **13** L3–5
[5] Sattari P, Castle G S P and Adamiak K 2010 FEM-FCT based dynamic simulation of corona discharge in point-plane configuration *IEEE Trans. on Ind. Appl.* **46** 1699–706
[6] Lama W L and Gallo C F 1974 Systematic study of the electrical characteristics of the Trichel current pulses from negative needle-to-plane coronas *J. Appl. Phys.* **45** 103–13

A 3D numerical model of the electrostatic coating process for moving targets

N Toljic[1], G S P Castle[1], K Adamiak[1], H H Kuo[2] and H T Fan[2]

[1]Department of Electrical and Computer Engineering, University of Western Ontario, London, ON, Canada

[2]GM Global Research & Development, Warren, MI, USA

ntoljic@uwo.ca

Abstract. A 3D realistic numerical model of the electrostatic coating process has been developed including all relevant mechanical and electrical phenomena. The commercial computational fluid dynamics (CFD) software FLUENT was used to simulate the mechanical portion of the process. The electrical phenomena including space charge fields due to ions and poly-disperse particle charge distributions were added to FLUENT through the User Defined Function (UDF) capabilities. Selected example results are shown to demonstrate the effect of particle charge to mass ratio distributions, electrohydrodynamic flow effects and corona currents on particle trajectories and deposition efficiency. These numerical results correlate well with the theoretical expectations for static targets. In addition, an efficient method for handling 3D cases with moving flat parts is presented using the superposition principle. A procedure for extending the model for the general case of 3D moving targets is suggested.

1. Introduction

Advancements in the capability of computational tools introduced numerical modeling as a preferred way for enhancing numerous industrial applications. Electrostatic coating is one of the applications which benefits significantly from numerical optimization as it involves many interdependent variables. However, due to the complexity of the problem, certain challenging tasks still remain unsolved. Namely, the electrostatic coating process can be regarded as a composition of intricate mechanical and electrical phenomena, which are not entirely understood. This is especially true for the electrical portion of the process. Different particle charge and charge to mass ratio distributions, corona current effects and electrohydrodynamic flow are still a subject of ongoing theoretical, numerical and experimental studies.

The complex nature of the charging process limits our ability to assess the value of charge to mass ratio (*q/m*) of individual particles in a cloud of particles when they are electrically charged by induction or conduction. All the measurement techniques for particle charge can be classified in two major groups: static and dynamic [1-2]. In the static methods particle charge is measured directly, whereas in the dynamic methods it is calculated indirectly from the measurement of the particle mobility. Static methods are crude and simple, but they can be easily implemented. Dynamic methods are more reliable, but they require added complexity in the measurement system. It has been shown in [2] that considerable conflicting data exists for the dependence of *q/m* on particle size and it has been suggested that this area needs further research.

Also, as a result of the application of high voltage and the specific configuration of conductors, corona discharge is usually an inevitable part of the electrostatic coating process. A full mathematical model of the corona discharge is practically impossible to implement [3]. One of the least known effects of the corona discharge is the secondary electrohydrodynamic (EHD) flow, which is produced by transferring of momentum from high-speed drifting ions to surrounding air molecules. Incorporation of EHD makes numerical modeling much more computationally demanding.

In this paper, an efficient 3D model of the electrostatic coating process is presented. The mechanical part of the coating process is directly modeled with the computational fluid dynamics software FLUENT. Its User-Defined Functions (UDF) are used to solve the Poisson equation by incorporating it into the general scalar transport equations. Obtained numerical results are presented. In addition, an efficient method for handling 3D cases with moving flat parts is presented using the superposition principle and a procedure for extending the model for the general case of 3D moving targets is suggested.

2. Description of the computational domain and numerical procedure

The computational three-dimensional domain encompasses a simplified representation of the rotary bell type atomizer and a flat target of the radius 0.5m situated 0.25m away from the atomizer. The geometry of the computational domain was created by the GAMBIT software package [4] and meshed with hexagonal cells. The mesh between the atomizer and the target was refined compared to the rest of the computational domain. The numerical algorithm includes solving the gas phase of the shaping air, the discrete phase of the charged particles, the electrostatic field and coupling between them to get a self-consistent solution [5].

The shaping air flow in FLUENT was considered as incompressible, steady and viscous turbulent. The RNG (renormalization group) k-ε turbulent model was used in the modeling. The method used for simulation of the particle discrete phase in FLUENT is solved by the Lagrangian approach, in which the particles are tracked by the stochastic tracking (random walk) model in turbulent gas flow [6]. Coupling between the air flow and the particle discrete phase is included via source terms of mass and momentum. The particles are ejected with the predefined size and charge distribution and directed to the target by the shaping air and electric forces.

The electric field generated by the voltage applied to the bell cup and the space charge formed by the charged particles and ions is governed by Poisson's equation. In order to take into account the space charge density due to corona discharge into the numerical analysis, a new scalar Ψ is defined:

$$\mathbf{J} = -\nabla \Psi \tag{1}$$

where $\mathbf{J}$ is the ionic current density. The equation can be solved by the FLUENT solver as user-defined scalar transport equation. It is assumed that the total corona current at the electrodes is known and is zero at the other boundaries of the computational model domain.

In the numerical model, the shaping air flow rate was set to be 0.02 kg/s. The downdraft air acts in the negative y direction and its magnitude is 0.5m/s. The atomizer voltage was set to be -90kV, whereas the target was grounded. Diameter of particles is between 10µ and 80µm and the value of the charge to mass ratio is set to be inversely proportional to the particle radius and equal to -1mC/kg for the particle with 35µm diameter. The corona current was obtained experimentally and equal to 10µA.

3. Numerical results

Numerical results show that the air flow forms a conical shape as it moves towards the target. When the air hits the target it spreads evenly in the tangential direction. The maximum air velocity is at the tip of the atomizer and it gradually reduces as the axial position is increased. The cone is tilted in the direction of downdraft air. In the region between the atomizer and the target, the particles follow almost linear patterns. The majority of particles end at the target directly below the atomizer. The particles that escape this initial impact with the target follow the air pattern that directs them towards the edge of the target. A certain number of these particles deposit at the distal parts of the target and the rest of them travel beyond the target to the edge of the computational domain. It can be noticed that the turbulence treatment of particle movement in FLUENT results in characteristic piecewise linear trajectories of the particles.

The numerical results for the deposition profiles along the radius of the target are depicted in Figure 1. The value of the accumulation rate is very low in the centre of the target; it increases very quickly as the radius is increased. After the local maximum is reached, the accumulation rate decreases as the radial position is further increased. The particle accumulation along the target radius in the x direction exhibits local symmetry. For the particle accumulation rate along the target radius in the y direction, the influence of downdraft can be observed. The local maximum that is closer to the source of the downdraft air is lower than the local maximum that is further away from the source of the downdraft air. Transfer efficiency is calculated as a ratio of the input flow rate and the deposition mass flow rate and its value is 90%. This is relatively high value, but can be explained by the large size of the target and its flat surface, perpendicular to the droplet trajectories.

The electrical potential contours are dense in the area between the atomizer and the target, and sparse elsewhere. In the immediate vicinity of the atomizer tip the electric potential values are very high. The electric potential contours in the vicinity of the atomizer body are perpendicular to the atomizer surface. This is a result of the assumed boundary conditions for the atomizer body. Areas close to target have value of the electric potential close to zero as the target is grounded.

It should be noted that EHD flow had no significant impact on the continuous phase results and the trajectories of the charged particles. This is because the EHD flow is very weak compared to the shaping air. Its influence can, therefore, be neglected for this particular studied case.

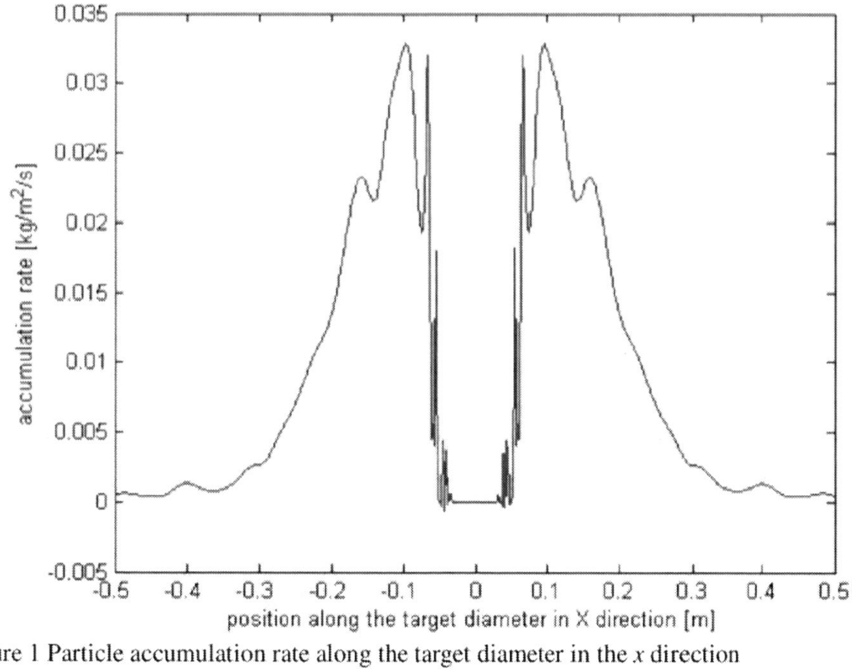

Figure 1 Particle accumulation rate along the target diameter in the x direction

4. Simulating a moving target

To simulate a moving target, we considered a simplified case where the target is flat and very wide. It was further assumed that the gun moves relatively slowly. If these conditions are fulfilled, stationary deposition patterns for different gun positions will be practically identical and all calculations can be based only on the deposition pattern for one stationary target. The deposition pattern for a moving target can be derived by superposition of the incrementally shifted deposition patterns for the stationary target. FLUENT software is required only in the initial step to obtain the stationary deposition pattern. The post processing of the data was done in a mathematical software package MATLAB.

The overlapped deposition profiles during coating period are shown in Figure 2. Higher values of deposition profile correspond to the larger increment in time. The greater value of reversal position and higher moving velocity result in more uniform coverage. In this manner the mass build up on portions of the target is avoided.

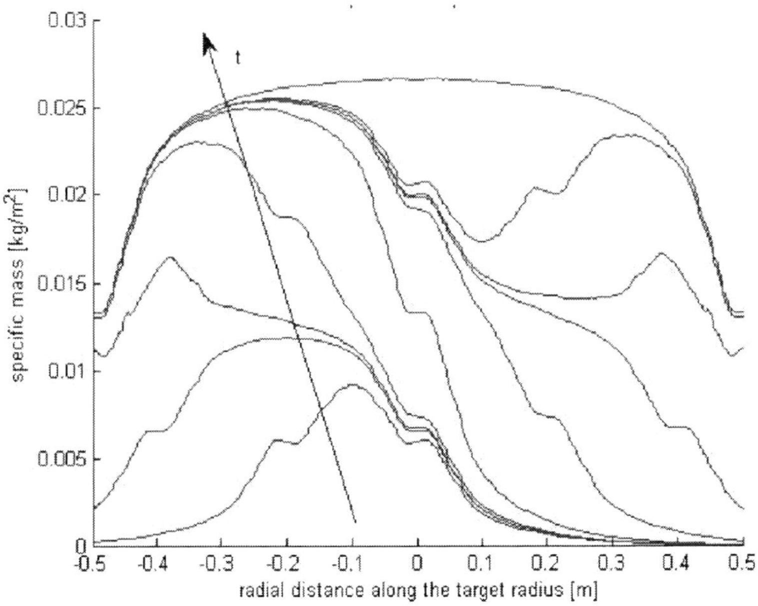

Figure 2 Overlapped graphs of specific mass profiles for 1s time increments, reversal position of 0.5m and target velocity of 0.2m/s.

5. Conclusion

In this paper, a numerical model of the electrostatic coating process has been developed including all relevant mechanical and electrical phenomena. The numerical simulation was performed for the full three-dimensional model of the problem and the results of the simulations were presented. In addition, the deposition patterns for the moving target were estimated by superposition of the incrementally shifted deposition patterns for the stationary target.

References

[1] Brown C 1997 Tutorial review: Simultaneous measurement of particle size and particle charge *J. Aerosol Sci.* 98 pp 1373–1391

[2] Toljic N, Adamiak K and Castle G S P 2009 Charge to radius dependency for conductive particles charged by induction *Journal of Electrostatics* 68 pp 57-63

[3] Zhang J, Adamiak K and Castle G S P 2007 Numerical modeling of negative-corona discharge in oxygen under different pressures *Journal of Electrostatics* 65 pp 174-181

[4] *FLUENT 6.2 User's guide.* 2005 FLUENT Inc

[5] Zhao S Adamiak K and Castle G S P 2007 The implementation of poisson field analysis within FLUENT to model electrostatic liquid spraying *Proc. IEEE Canadian Conference on Electrical and Computer Engineering* Vancouver

[6] *FLUENT 6.2 UDF manual* 2005 FLUENT Inc.

Modelling of Complex Physical Processes in Electrostatic Precipitators

KISS István[1], IVÁNCSY Tamás[1], SUDA Jenő[2], BERTA István[1]
Budapest University of Technology and Economics,
[1]Department of Electric Power Engineering, Egry J. u. 18, Budapest H-1111, Hungary
[2]Department of Fluid Mechanics, Bertalan Lajos 4–6., Budapest H-1111, Hungary
E-mail: kiss.istvan@vet.bme.hu

Abstract. Electrostatic precipitator (ESP) models have improved significantly in the past years. The dramatic development of the capacity of computers made it possible to increase the complexity of ESP models. Recently the different interactions between the gas, the electric field with ion space charge and the charged particles to be precipitated can be described more accurately by the newly developed complex approach. However even some of the newest computer models are limited; they are not able to follow the interactions of the complicated physical phenomena properly. For example pulse energisation of short time impulses cannot be described correctly with models assuming continuous corona current. There is another important problem, namely the examined duration of operation. Some of the models determine the trajectories of dust particles assuming that they are unchanged during the operation of an ESP. The validity of this assumption is very limited in such cases, where the development of certain phenomena is time dependent (e.g. back corona formation). In this paper the authors focus on the "long term" models, analysing such situations in which it is vital to investigate a longer period of operation of ESP-s. Using the newly developed model the effect of back corona, rapping, etc. can be analysed with higher reliability than it has been performed in previous ESP models.

1. Introduction

Numerical models are useful tools for predicting the performance of an electrostatic precipitator. There were numerous models developed by experts to predict the collection efficiency of ESP's. The difficulty in the modelling of ESP's is the large number of processes and phenomena which have influences on each other. In the modelling some of the effects are neglected because of the limited computational capacity or their minor influence on the modelled phenomena.

In the majority of the models 2D modelling is used. Both the electric and flow field can be treated as 2D, and the effect of gravity is neglected. This method provides a simple and good approximation of the processes in the electrostatic precipitator. The use of a 2D model reduces the amount of computation therefore longer sections – a whole ESP channel – and more complex processes can be modelled. 3D models also exists, but because of the high computational needs, these are just for short ESP sections (e.g. Chang[1], Adamiak[2]).

2. The modelled phenomena

In a numerical model it is necessary to include the parameters of the precipitated particles, the parameters of the energization and the gas flow. It is essential to know the *electric field* created inside the ESP channel since this is relevant for *particle charging* and the charged particles transferred in the *flow field*.

2.1. Calculating the electric field

The early models supposed a uniform electric field through the whole channel[3]. Later analytical approximation[4, 5], then the finite differences method[6], the finite elements method[7] and the combination of the latter two with the finite volume method were used. Nowadays combined methods such as the combination of finite differences method and finite elements method[8, 9], and the boundary element method combined with finite volume method are used.

It is highly important to include the effects of ionic and dust space charge in the calculation of electric field. Based upon practical experience and measurements, the dust space charge has a notable influence on the electric field and therefore on the corona current, back corona formation and on the efficiency of the precipitation.

2.2. Particle charging

For the calculation of the particle charge in numerical models there evolved two major ways. One assumes for a given particle diameter a constant saturation charge through the whole ESP. The particles are assumed charged to the saturation charge from their entry in the ESP. Numerous models use this method[10, 11]. The other possibility is to calculate the charge of the particles in a time and position dependent way, taking into account the local electric field and the charge distribution. This method was also used by P. A. Lawless and later in a Lagrangian particle movement simulation by Meroth[12] and Schmid et al.[13]. There are just a few models which calculate the dust space charge using the Lagrangian method (one example is the model from I. Gallimberti[14]). The best solution is the use of the combined charging model and the Lagrangian approach for the particle movement[15].

2.3. Flow field

Simple models assume constant gas velocity profile in the cross-section of the channel (e.g. Deutsch[16] and Ramadan & Soo[17]). Beside the laminar flow models there were numerous different turbulent flow models created, such as $k-\varepsilon$ turbulence model: Kallo & Stock[9]; DNS: Soltani et al.[10]; LES: Ullum[11]. These CFD simulations made the models more accurate, and through the determination of the diffusion coefficient (D_p) it is possible to calculate important turbulent flow parameters, for example the flow modification effect of the corona electrode geometry.

Nowadays the effect of the ionic wind on the turbulent gas flow can be better analysed. The differences in the EHD simulations are in the details of how they take into account the charge densities (ρ_i and ρ_d). One example of a model where the space charges and the ionic wind are also taken into account is the LES model created by Ullum[11]. Because of the large computational demand it just calculates an ESP section with 2 corona electrodes.

3. Long-time modelling

In a numerical ESP model the following effects play a role in the efficiency of precipitation:

- electric field modified by space charges (ionic space charge and dust space charge)
- ionization (the generated free charges on the corona electrode and the ionic wind)
- particle charging, saturation charge and charging process (diffusion, field and mixed)
- gas flow (turbulent and boundary flow)
- dust collection, dust layer expansion
- dust re entrainment, back corona

To have an accurate model it is necessary to model these phenomena as accurately as possible and holding the computational needs of the simulation as low as possible. For the long time models it is necessary to generate a model, which calculates the processes in a non stationary manner i.e. it takes the short time (fast) changes and also the long time (slow) changes into account, especially in the case of pulse energization.

If there is a change in the parameters of the incoming dust (e.g. by fuel changes) the collection efficiency changes. In such cases the relative permittivity, particle size and resistivity of the dust changes, which influences the saturation charge, cohesion of the collected dust and back corona formation. Also, rapping has an influence on the dust emission. Closer to the inlet, larger particles are collected while closer to the outlet small particles[18], so the collected dust layer is different.

The different dust layer formation needs a different rapping cycle. The rapping of the sections closer to the outlet has greater influence on the dust emission, since the re entering dust particles are closer to the outlet and have smaller particle size which needs longer to collect. However the rapping cycle has a longer time period, because less dust reaches the last sections of the ESPs.

Also the back corona formation depends highly on the collected dust layer thickness, therefore the back corona occurrence is supposed closer to the inlet.

There are dust deposits on the corona electrodes which should be handled in the long-time modelling also. These dust deposits can be handled as changes in the corona wire diameter.

4. Conclusions

In the long-time modelling it is essential to track the amount of dust collected on the collecting electrode and modelling the effect of rapping. Practical experience shows that, the changes in the dust properties can generate lower efficiency periods for a while. The model should be able to handle dust mixtures with different electric properties and different concentrations. These property changes should be tracked on the collected dust layer also, to get reliable results.

References

[1] Brocilo D, Podlinski J, Chang J S, Mizeraczyk J and Findlay R D 2008 *Journal of Physics: Conference series* **142** 012032
[2] Farnoosh N, Adamiak K and Castle G S P 2011 *IEEE Transactions on Dielectrics and Electrical Insulation* **18** 211–220 ISSN 1070-9878
[3] Williams J C and Jackson R 1962 *The motion of solid particles in an electrostatic precipitator, interaction between fluids and particles* (London, UK: Institution of Chemical Engineers) pp 282–288
[4] Cooperman P 1960 *AIEE Transaction* **79** 47–50
[5] jr Oglesby S and Nickols G B 1978 *Electrostatic precipitators* (New York: Marcel Dekker Inc.)
[6] McDonald J R, Smith W B, Spencer III H W and Sparks L E 1977 *Journal of Applied Physics* **48** 2231–2243
[7] Egli W, Kogelschatz U, Gerteisen E and Gruber R 1997 *Journal of Electrostatics* **40-41** 425–430
[8] Kallio G A and Stock D E 1985 *Journal of Applied Physics* **59** 999–1005
[9] Kallio G A and Stock D E 1992 *Journal of Fluid Mechanics* **240** 133–166
[10] Soltani M, Ahmadi G, Ounis H and McLaughlin J B 1998 *Journal of Multiphase Flow* **24** 77–92
[11] Ullum T 2003 *Investigations on turbulence and flow structures in electrostatic precipitator* Ph.D. thesis Technical University of Denmark Lyngby, Denmark
[12] Meroth A M 1997 *Numerical electrohydrodynamics in electrostatic precipitators* Ph.D. thesis University of Karlsruhe
[13] Schmid H J, Stolz S and Buggish H 2002 *Applied Scientific Research, Flow, Turbulence and Combustion* **68** 63–89.
[14] Gallimberti I 1997 *VIII International Conference on Electrostatics* pp 1–30
[15] Suda J M 2007 *Modeling of two-phase flow in electrostatic precipitators* Ph.D. thesis BME
[16] Deutsch W 1922 *Annalen der Physik* **68** 335–344
[17] Ramadan O E and Soo S L 1969 *Physics of Fluids Research Notes* **12** 1943–1945
[18] Grass N 2002 *Industry Applications, IEEE Transactions on* **38** 1190–1195 ISSN 0093-9994

Table 1. The main parameters of Electrostatic Precipitator models (numerical simulation)[15]

	Modelling of electric field				Charging of particles		Modelling of flow field					Modelling of two-phase flow	
	Uniform, constant electric field intensity	Analitical approximation	Numerical simulation	Including of ion- and particle space charge	Determination of saturation chgarge	Modelling of charging process	Constant velocity profile	Analitical velocity profile	CFD	Including the ionic-wind	Including discharge electrode geometry in simulation	Eulerian modelling $0 < D_p < \infty$	Lagrangian modelling
Literature 1922–2007													
Deutsch (1922)												∞	
Anderson (1924)							□					∞	
Williams & Jackson (1962)												Const.	
White (1963)	□						□					Const.	
Ramadan & Soo (1969)				I+P						*			
Feldman (1975)												∞	
Cooperman, P. (1971, 1977)	□											Const.	
McDonald et al. (1977)			FD / FD			F + D							
Yabe et al. (1978)			FD / FD	I					Lam.	*			
Bernstein & Crowe (1979)			FD / FD	I					Lam.	*			
Leonard et al. (1980, 1982, 1983)												Const.	
Yamamoto & Velkoff (1981)			FD / FD	I					Lam.	*			
Yamamoto (1984, 1989)			FD / FD	I					Lam.	*		0	
Berta (1984)			FD / FD	I					Lam.	*			PTM
Self et al. (1984)												Const.	
Shaughnessy et al. (1985)			FD / FD	I					TH	*			
Davidson & Shaughnessy (1986)			FD / FD	I					TH	*			
Kihm et al. (1985, 1987)												Const.	
Petroll & Födisch (1988)						F + D						Const.	
Watanabe (1989)				I									PTM
Zamany (1992)			FD / FD			C			$k-\varepsilon$			Const.	
Riehle (1992)	□			I		C						Const.	SRF
Zhibin et al. (1992, 1993, 1994)			FD / FD	I		C	□			TH		Const.	
Kallio & Stock (1992)			FD / FE	I					$k-\varepsilon$	*	⊙		
Riehle & Löffler (1993)	□			I		C						Const.	
Liang & Lin (1994)			FD / FD			C			$k-\varepsilon$			Loc.	
Bai et al. (1995)						F			Lam.			0	
Cañadas et al. (1995)				I + P		F + D							
Lami et al. (1995)			FD / FD	I + P						*			
Riehle & Löffler (1995)	□			I		C						Const.	SRF
Palmer (1996)			FD / FE	I					$k-\varepsilon$	*	⊙		
Khare & Sinha				I		F							
Lawless (1993, 1996)			FD / FD	I	□	FDM						Const.	
Schmid & Schmidt (1996)			FD / FD			C						∞	
Riehle (1996b)	□			I		C						Const.	SRF
Goo & Lee (1996, 1997)			FD / FD	I		F + D			$k-\varepsilon$		⊙		CRW
Choi & Fletcher (1997)			FD / FV	I + P		F			$k-\varepsilon$		⊙		EL
Egli et al. (1997)			FE / FE	I		F			Lam.	*			PTM
Meroth (1997)			FE / FV	I		FMD			$k-\varepsilon$				EL
Gallimberti (1997)			FD / FD	I + P		FMD			$k-\varepsilon$	*			PTM
Suda (1997) /M.Sc./	□			I + P	□	C			BL			Loc.	
Medlin (1998)			FD / FV	I + P		C			$k-\varepsilon$	*	⊙	Const.	
Lu & Huang (1998)			FD / FD	I		C						∞	
Soltani et al. (1998)									DNS				PTM
Soldati et al. ('93, '97, '98, 2000, 2002)			FD / FD	I					DNS				CRW
Parasram (2001)			FD / FE	I		C			$k-\varepsilon$				CRW
Kim et al. (2001)			FD / FD	I		C						Loc.	
Böttner & Sommerfeld (2001)			FV	I		F			$k-\varepsilon$				PTM
Park & Chun (2002)						C						Const.	
Varonos et al. (2002)			FD			C			$k-\varepsilon$				PTM
Schmid et al. (2002)			FE / FV	I		FMD			$k-\varepsilon$	*	⊙	Loc.	CRW
Böttner (2002, 2003)			FV	I		C			$k-\varepsilon$	*			PTM
Ullum (2003)			FV	I + P		C			LES	*			PTM
Lind et al. (2004)			FD / FD	I + P		F + D			$k-\varepsilon$	*		Const.	
Fujishima et al. (2004)			FD / FD	I + P		F + D			$k-\varepsilon$	*			PTM
Nikas et al (2005)			FD / FD	I		F + D			$k-\varepsilon$	*			PTM
Zhang et al. (2005)			FV	I		F + D			$k-\varepsilon$	*			CRW
Talaie et al. (2001, 2005)			FD	I + P		FMD			$k-\varepsilon$	*			CRW
Skodras et al. (2006)			FV	I + P		C			$k-\varepsilon$	*	⊙		PTM
Kiss & Suda (2007)			BEM / FV	I + P		C			$k-\varepsilon$	*	⊙	Loc.	(PTM)

Legend:

(shaded) model by the referenced author; (shaded) □ parameter of the previous version

Modelling of electric field
- FD Finite differences method
- FE Finite elements method
- FV Finite volume method
- BEM Boundary element method
- I Including ionic space charge
- P Including dust space charge

Modelling of particle charging
- F Field charging, impact ionization
- D Diffusion charging
- C Field and diffusion charging
- FMD Field modified diffusion charging theory
- FCT Flux Corrected Transport method

Modelling of flow field
- ⊙ Corona electrode geometry modelling
- * Including ionic wind
- TH Theoretical examination
- Lam. Laminar model
- $k-\varepsilon$ $k-\varepsilon$ Turbulence model
- BL Boundary layer equation
- DNS Direct numerical simulation
- LES Large eddy simulation
- EL Eddy lifetime

Modelling of two-phase flow

Eulerian modelling
- 0 / ∞ D_p diffusion coefficient
- Const. Constant D_p value
- Loc. Local D_p value

Lagrangian modelling
- CRW Continuous random walk model
- SRF Simulated particle flow with random generated fluctuating velocity
- PTM Particle trajectory model

Higher order dielectrophoretic force characterisation of non-spherical particles

H Nili[1], T Sun and N G Green
School of Electronics and Computer Science, University of Southampton, Southampton SO17 1BJ, UK

E-mail: hn07r@ecs.soton.ac.uk

Abstract. A numerical method is presented for determination of higher-order dielectrophoretic forces on non-spherical particles. The method is applied to ellipsoidal particles of different dimensions subjected to electric fields with different degrees of non-uniformity. The results are compared against net force calculations using the Maxwell stress tensor method. The contribution from higher-order forces is shown to exceed the dielectrophoretic force based only on the dipole approximation when particle dimensions are comparable to the length scale of electric field non-uniformity. It is also shown that individual contributions from quadrupolar and octupolar dielectrophoretic forces strongly depend on particle aspect ratio.

1. Introduction

Dielectrophoresis (DEP) [1] is one of the most commonly used techniques for manipulation and characterisation of micrometre and sub-micrometre particles suspended in fluid media [2-4]. The DEP force on particles arises from the interaction of a non-uniform electric field and polarisation charges at the particle/electrolyte interface. The dipole approximation that ignores all but the first-order of force terms is commonly invoked for determining the DEP force on spherical and non-spherical particles.

While competent in many circumstances, the dipole approximation is expected to provide an accurate account of the DEP force only when particle dimensions are much smaller than a characteristic length scale of electric field non-uniformity [5]. With the current trend towards micro-electrode geometry, there are an increasing number of instances where particle dimensions are indeed comparable to field non-uniformity length scales. Furthermore, most biological particles are non-spherical in which case shape-dependent polarisation gives rise to the induction of higher-order moments.

This paper presents a numerical method for determining higher-order DEP forces on particles of arbitrary shape. The method is applied to spherical and ellipsoidal particles of different dimensions subjected to electric fields with different degrees of non-uniformity to quantify the significance of higher-order forces in different cases of particle and field geometry. The results are compared against total force calculations made using the rigorous Maxwell stress tensor method both for verification and for determination of the contribution of individual force terms to the total DEP force.

2. Background and theory

Based on the dipole approximation, the dielectrophoretic force is given by [1]:

[1] To whom any correspondence should be addressed.

$$\mathbf{F}_{DEP} = \mathbf{p} \cdot \nabla \mathbf{E} \tag{1}$$

where $\mathbf{p}$ is the effective dipole moment and $\nabla \mathbf{E}$ the gradient of the applied electric field $\mathbf{E}$. In general, the DEP force consists of an infinite number of terms with the nth-order term given by [6]:

$$\mathbf{F}^{(n)} = \frac{1}{n!} \mathbf{p}^{(n)} [\cdot]^{(n)} \nabla^{(n)} \mathbf{E} \tag{2}$$

where $\mathbf{p}^{(n)}$ is the nth-order effective moment, $\nabla^{(n)}\mathbf{E}$ is the nth gradient of the electric field and $[\cdot]^{(n)}$ is the nth-order dot product. There is no analytic method for deriving higher-order moments of non-spherical particles. However, in the case of axial symmetry, where the effective multipoles are linear, the moments can be found numerically by an integration of the potential due to the particles over a spherical enclosing surface [7]:

$$p^{(n)} = 4\pi\varepsilon_m R_t^{n+1} \int_0^{\pi} \varphi_{R_t} P_n(\cos\theta) \sin\theta d\theta \tag{3}$$

where ε_m is the dielectric constant of the suspending medium, R_t is the radius of the spherical integration surface, and $P_n(\cos\theta)$ are the Legendre polynomials. The following expressions, consistent with the charges on and positions of the poles of the first three linear multipoles, represented in figure 1, were used to calculate the field derivatives:

$$\nabla^{(1)} E = \frac{d}{dz} E(z) = \frac{E(z + \frac{d_1}{2}) - E(z - \frac{d_1}{2})}{d_1} \tag{4a}$$

$$\nabla^{(2)} E = \frac{d^2}{dz^2} E(z) = \frac{E(z + d_2) - 2E(z) + E(z - d_2)}{d_2^2} \tag{4b}$$

$$\nabla^{(3)} E = \frac{d^3}{dz^3} E(z) = \frac{E(z) - 3E(z - d_3) + 3E(z - 2d_3) - E(z - 3d_3)}{d_3^3} \, . \tag{4c}$$

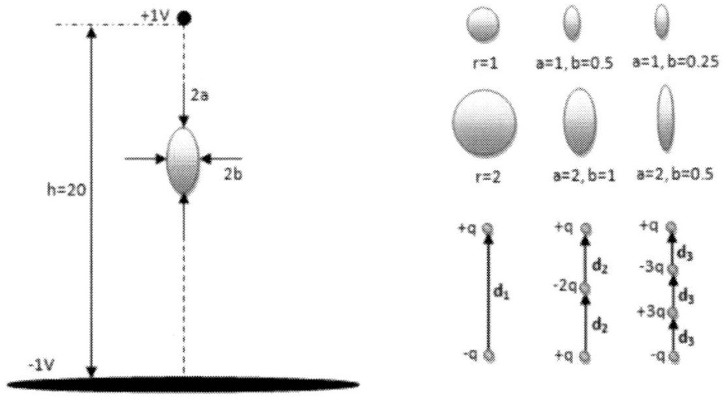

Figure 1. The electrode geometry (point-plane) and particle shapes and dimensions analysed in this work, and a representation of the first three linear multipoles

Using values of the first three effective moments obtained from equation (3) and values of the field derivatives obtained from equations (4a)-(4c), the first three terms of the DEP force on the particles were calculated from equation (2). Calculation was also made in each case of the total DEP force on the particles by integrating the Maxwell stress tensor over the surface of the particle [8]:

$$\mathbf{F}_{DEP} = \frac{1}{2}\varepsilon_m \oint E_m^2 \cdot \mathbf{n}dA \,. \quad (5)$$

In equation (5), E_m is the field magnitude inside the medium and **n** is the vector normal to the integration surface A. The Maxwell stress tensor (MST) method is known as the most unassailable in calculating field-induced force.

Figure 1 shows the particle shapes studied in this work and the way they are positioned within the electrode geometry. The particle and the medium are assigned dielectric constants of 3 and 80, respectively. With no loss of generality, the simulations were performed without considering units for the dimensions. Force values were calculated at fifteen different particle centre heights above the plane electrode: 2, 3, ..., 17 with higher points corresponding to a higher degree of field non-uniformity. We used FlexPDE (FlexPDE Inc) to numerically solve the electric field in each case.

3. Results and discussion

Figures 2 and 3 show the results for spherical and ellipsoidal particles, respectively. Contributions to the total DEP force from the first three force terms when the particles are at h = 17 (highest point above the plane electrode where the field gradient is strongest) are summarised in Table 1.

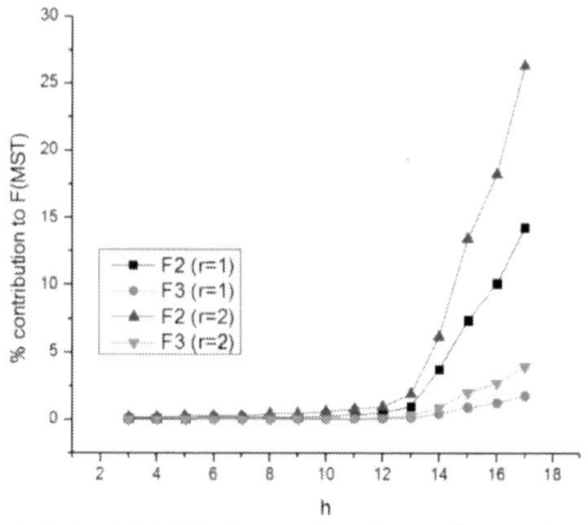

Figure 2. Variations with particle height (h) above the plane electrode of the contributions from the first three DEP force terms to the total DEP force on spherical particles of radii r = 1 and r = 2

Table 1. Contribution from the first three force terms to the total DEP force on the particles in figure 1 when they are at h = 17 where the field gradient is strongest

Particle	$F^{(1)}/F^{(MST)}$	$F^{(2)}/F^{(MST)}$	$F^{(3)}/F^{(MST)}$	$F^{(2)}+F^{(3)}/F^{(MST)}$
r = 1	83.5%	14.3%	1.8%	16.1%
r = 2	69.3%	26.4%	3.9%	30.3%
a = 1, b = 0.5	65.2%	24.9%	9.2%	34.1%
a= 2, b = 1	39.1%	43.6%	16.9%	60.5%
a = 1, b = 0.25	66.0%	20.2%	13.5%	33.7%
a = 2, b = 0.5	38.8%	35.9%	25.0%	60.9%

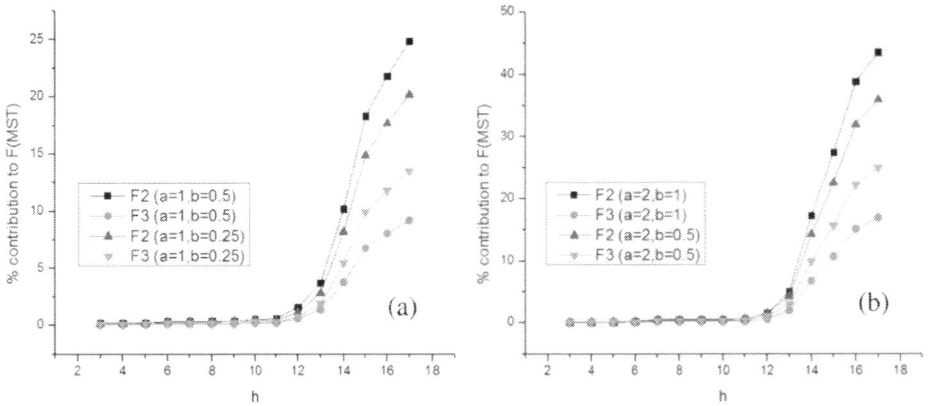

Figure 3. Variations with particle height (h) above the plane electrode of the contributions from the first three DEP force terms to the total DEP force on ellipsoidal particles with dimensions along the symmetry axis equal to (a) a = 1, and (b) a = 2

As suggested by the plots and the table, higher-order forces are less significant for spheres compared to ellipsoids of similar dimensions. The data confirms that higher-order forces become increasingly significant as particle dimensions become comparable to the length scale of electric field non-uniformity. At h = 17 where the field is most strongly divergent, higher-order contribution to ellipsoids of a = 1 is almost equal to 34% - interestingly, this is the same for both ellipsoids of a = 1 indicating that the parameter controlling the significance of higher-order DEP forces is the particle dimension along the axis of symmetry – whereas for ellipsoids of a = 2, higher-order contribution is a significant 61% exceeding the dielectrophoretic force based only on the dipole approximation.

It may also be seen from the data that although the overall contribution from higher-order forces is similar – to within an error of <1% - for particles with equal dimensions along the symmetry axis, individual contributions from second- and third-order force terms strongly depend on particle aspect ratio. The quadrupolar and octupolar DEP forces show inverse and direct relationships with particle aspect ratio, respectively.

4. Conclusion

A method was presented for determining higher-order dielectrophoretic forces on particles of arbitrary shape subjected to an axially symmetric electric field. The method was applied to spherical and ellipsoidal particles of different dimensions in electric fields with varying gradients. It was shown that higher-order forces are more significant for ellipsoidal particles and that it is the particle dimension along the axis of symmetry that controls the significance of higher-order contribution to the total force. It was also shown that unlike the overall higher-order contribution, quadrupolar and octupolar forces are strongly dependent on particle aspect ratio with the quadrupolar force decreasing and the octupolar force increasing with increasing aspect ratio. The results were validated against total force calculations using the Maxwell stress tensor method.

5. References

[1] Pohl H A 1951 *J. Appl. Phys.* **22** 869
[2] Morgan H, Hughes M P and Green N G 1999 *Biophys. J.* **77** 516
[3] Markx G H, Talary M S and Pethig R 1994 *J. Biotechnol.* **32** 29
[4] Krupke R, Hennrich F, Lohneysen H V and Kappes M M 2003 *Science* **301** 5631
[5] Jones T B 1985 *IEEE Trans. Ind. Appl.* **IA-21** 930
[6] Jones T B and Washizu M 1996 *J. Electrostat.* **37** 121
[7] Green N G and Jones T B 2007 *J. Phys. D: Appl. Phys.* **40** 78
[8] Wang X, Wang X B and Gascoyne P R C 1997 *J. Electrostat.* **39** 277

Computational strategies for 3D computation of the electric field in a vibratory tribocharger for mixed granular solids

J-C Laurentie, Ph Traoré, J Wu and L Dascalescu

PPRIME Institute, UPR 3346 CNRS – University of Poitiers – ENSMA, France

E-mail: jean.charles.laurentie@univ-poitiers.fr

Abstract. In a previous paper, the authors introduced an original model for the computation of the triboelectric charges due to particle-to-particle collisions in a vibrated bed. The evaluation of the electric field inside the particle bed is a prerequisite for the accurate modelling of the tribochaging process, as it determines the trajectory of the particles as well as the charge swapping during collision. Two different approaches are presented and compared in terms of accuracy and computational time consumption. One consisted of computing the electric field as the superposition of the contributions of each point charge. The other, which proved to be much more effective, implied solving the Poisson equation for the potential generated by a given spatial distribution of charge density.

1. Introduction

Electric field separation of insulating materials and charging of toner particles for photo-copiers are two electrostatic processes that make use of the triboelectric effect [1 - 3]. Tribocharging has also disastrous effects, such as explosions in flour mills, cereal silos and supertankers, due to electrical discharges occurring between charged granules [4]. Triboelectricity is also at the origin of the electrostatic adhesion forces between charged powders and transport pipes in the chemical or pharmaceutical industries, and may cause a severe reduction of the flow-rates at which such materials are processed.

Most of the models elaborated for simulating the tribocharging effect are based on the colliding surface area and surface state [5 – 8]. In a previous paper [9], the authors have proposed an original model that takes into account the various tribo-charging mechanisms, as well as the electrical and mechanical forces to which the granules are subjected. The tribo-charging model presented in that paper is mainly based on the work done by Ali et al. [5], while the granular dynamics is modelled with the Distinct Element Method (DEM) [10].

The aim of the present work is to validate a numerical method for the computation of the electric field appropriate for the simulation of tribocharging processes in vertically-vibrated granular beds in relation to possible application in the electrostatic separation of plastics in the recycling industry.

2. Mechanical and electrical models

2.1. Tribo-charging model

The particles motion is governed by Newton's second law

$$m_i \frac{d\vec{V}_i}{dt} = \vec{F}_i \qquad (1)$$
$$I_i \vec{\varpi}_i = \vec{T}_i$$

where m_i and I_i stand for mass and moment of inertia of particle i respectively, $\vec{V}_i$ and $\vec{\varpi}_i$ are the velocity and angular velocity. $\vec{F}_i$ and $\vec{T}_i$ are respectively the set of forces and torque acting on the particle i. In our context the force $\vec{F}_i$ is only limited to the sum of the gravity force $m_i\vec{g}$, the contact force applied to particle i: $\vec{f}_i^{\,contact}$ obtained by the Distinct Element method [10], and the Coulomb force $\vec{f}_i^{\,coulomb} = q_i\vec{E}_i$ where q_i and $\vec{E}_i$ are respectively the charge of particle i and the electric field at the location of particle i. The charge of particle i at time step n during its collision with $ncol$ other particles is given by:

$$q_i^n = q_i^{n-1} + \sum_{j=1}^{ncol} \delta_{ij}^n \qquad (2)$$

where δ_{ij}^n is the charge exchanged per unit area during the collision between the two spherical insulating particles i and j. This charge exchanged during the collision between particles i and j is computed thanks to the tribo-charging model of [5] and modified to account for the contact surface A_{ij} which appears during the impact of two particles [9] may be expressed as follows:

$$\delta_{ij}^n = \frac{\varepsilon_0}{\delta e}(\Phi_i - \Phi_j - \vec{E}_{ij} \cdot \frac{\vec{d}_{ij}}{\left\|\vec{d}_{ij}\right\|}\delta e) \qquad (3)$$

where:
ε_0, the permittivity of free space;
δ, the cutoff distance of the charge transfer (typically 500 nm);
e, the electron charge;
Φ, the surface work function, representative of the energy needed to extract an electron from the surface of a particle;
$\vec{E}_{ij}$, the electric field at the point of impact of the two particles.
$\vec{d}_{ij} = \vec{X}_j - \vec{X}_i$ is the distance vector between particle i and j.

2.2. Computation of the Electric Field.
Both this tribo-charging model and the estimation of the Coulomb force involved in Newton's second law require the effective computation of $\vec{E}_i$. Indeed both the charge exchanged between the colliding particles and the electric forces acting on them are functions of the local intensity of the electric field. In the case of an electrostatic separator of granular materials, there are two sources of electric field: the electric field due to the charge carried by the particles themselves and the external field generated by an electrode system energized from a high-voltage supply.

To compute the electric field due to the particles, the charge is assumed to be homogeneously distributed on the surface of each granule. Consequently, the total charge of a particle is supposed to be concentrated at its center. Although prima facie this seems to be a very strong hypothesis for the insulators considered in the present study, in the conditions of a very good mixing rate of the granular materials, the contact with other particles is equiprobable for each point on the surface of a granule. This authorizes the hypothesis of a homogeneous charge distribution on particle surface.

The total value of the electric field at the center of a particle may be computed with two different strategies. The first one used the principle of superposition:

$$\vec{E}_i = \vec{E}^{ext} + \sum_{k=1}^{N} \frac{1}{4\pi\varepsilon_0} \frac{q_k}{\left\|\vec{d}_{ik}\right\|^3} \vec{d}_{ij} \tag{4}$$

where N is the total number of charged particles involved in the simulation.
The second one is based on the Poisson equation together with the definition of the electric field with regard to electrical potential:

$$\Delta V = -\frac{\rho}{\varepsilon_0} \text{ on } \Omega \tag{5}$$

$$V = V_0 \text{ on } \Gamma \tag{6}$$

$$\vec{E} = -\nabla V \tag{7}$$

ρ is the volume charge density, computed as the sum all the charge of each particles located in the computational cell; Ω is the computational domain having Γ as boundary. This Poisson equation is discretized with a second order finite volume method [10] and the respective system of linear equations is solved with the iterative Stone method [11].

3. Numerical results

Several numerical tests have been undertaken to compare the performance of the two above-described strategies of electric field computation in terms of accuracy and CPU time consumption. The 3-D computational domain has been divided into cells each containing a charged particle. Two tests have been performed with two sets of 9260 and 10^6 particles respectively. In each node of the domain the programme computes the electric field generated by the presence of the particles located in each cell. The iso-contours of the electric potential in the x middle plane computed with the superposition law and the Poisson equation are displayed in figure 1.

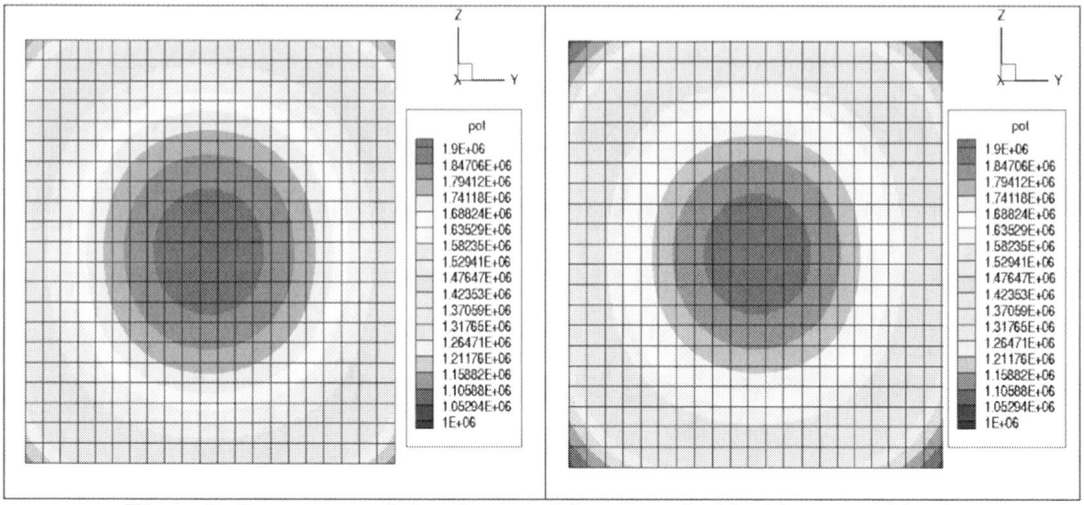

Figure 1. Isocontours of electric potential computed with: (a) superposition law; (b) Poisson equation, for uniformly charged particles (1.10^{-10} C)

Compared to the exact solution given by the superposition law, the accuracy of the results obtained with the Poisson equation is remarkable. However, the main difficulty with solving the Poisson equation is related to the boundary conditions. With the exception of the case of metallic walls where electric potential could be set to a fixed value (zero if the wall is grounded), the specification of boundary conditions in some other cases could be very complicated. A good alternative could be to impose on the boundary nodes the values obtained by the superposition law.

$$V_i = \sum_{k=1}^{N} \frac{1}{4\pi\varepsilon_0} \frac{q_k}{\left\|\vec{d}_{ik}\right\|} \tag{8}$$

where i holds for a point which belongs to the boundary. The boundary values can be conveniently computed before calling the Poisson solver.

The main drawback of the superposition law is the overall computational time which could be dramatically long as it is directly linked to the number of charged particles in the domain. Table 1 indicates the CPU time consumption for the computation of the electric field by the superposition law and by the Poisson equation for different sets of particles. For less that 10000 particles the two computational times are slightly comparable. But as the number of particles increases, solving the Poisson equation becomes largely more efficient. As seen on table 1, the computation of 10^6 is definitely not feasible for a transient computation of several thousand of time steps.

Table 1. CPU time consumption for the two computational strategies and for different sets of particles on a 3 GHz workstation

CPU time	$N_1 = 9260$ particles	$N_2 = 10^6$ particles
Superposition law	1.8 s	10.45 h
Poisson equation	1.07 s	34.3 s

4. Conclusions

(1) Accurate computation of the electric field is a prerequisite for modelling the behaviour of granular mixtures of insulating materials in vibrated or fluidized beds. Both numerical methods of field computation analyzed in this paper ensure similar accuracy. However, the finite volume method of solving Poisson equation is characterized by a much shorter computational time (less than 1 min, for 10^6 particles) as compared to the one based on the superposition method (more than 10 h, for the same number of particles).

(2) All the computations have been performed on a single processor since the superposition law has very bad scalability, contrary to the Poisson solver which has been efficiently parallelized. This is another argument in favour of the Poisson equation approach for the computation of the electric field, as it reduces by several orders of magnitude the overall CPU time.

References

[1] Dascalescu L, Morar R., Iuga A, Samuila A and Neamtu V 1998 *Part. Sci. & Technol.* **16** 25
[2] Das S, Medles K, Younes M, Dragan C and Dascalescu L 2007 *IEEE Trans. Ind. Appl.* **43** 1137
[3] Chang J-S, Kelly A J and Crowley J M 1995 *Handbook of Electrostatic Processes* (New York: Marcel Dekker)
[4] Glor M 2005 *J. Electrostat.* **63** 447
[5] Ali F S, Ali M A, Castle G S P and Inculet I I 1998 *Conf. Rec. IEEE IAS Ann. Meet.* vol. 3, pp. 1884-1890
[6] Watano S, Saito S and Suzuki T 2003 *Powder Technol.* **135-136** 112
[7] Matsusyama T and Yamamoto H 2006 *Chem. Eng. Sci.* **61** 2230
[8] Cundall P A and Strack O D L 1979 *Geotechnique* **29** 47
[9] Laurentie J C, Traoré P, Dragan C and Dascalescu L 2010 *Conf. Rec. IEEE IAS Ann. Meet.*, DOI: 10.1109/IAS.2010.5615669, pp 1-6
[10] Patankar S V 1980 *Numerical Heat Transfer Fluid Flow* (Whashington, DC: Hemisphere)
[11] Stone H L 1968 *SIAM J. Numer. Anal.* **5** 530

Lightning protection: challenges, solutions and questionable steps in the 21st century

István BERTA

1526 Budapest, HUNGARY

Budapest University of Technology and Economics

Department of Electric Power Engineering, High Voltage Laboratory

berta@shock.hu

Abstract. Besides the special primary lightning protection of extremely high towers, huge office and governmental buildings, large industrial plants and resident parks most of the **challenges** were connected to the secondary lightning protection of sensitive devices in Information and Communication Technology. The 70 year history of Budapest School of Lightning Protection plays an important role in the research and education of lightning and development of lightning protection. Among results and **solutions** the Rolling Sphere designing method (RS) and the Probability Modulated Attraction Space (PMAS) theory are detailed. As a new field Preventive Lightning Protection (PLP) has been introduced. The PLP method means the use of special preventive actions only for the duration of the thunderstorm. Recently several non-conventional lightning protection techniques have appeared as competitors of the air termination systems formed of conventional Franklin rods. The questionable steps, non-conventional lightning protection systems reported in the literature are the radioactive lightning rods, Early Streamer Emission (ESE) rods and Dissipation Arrays (sometimes called Charge Transfer Systems).

1. Lightning protection – past, present, future

The general problem in the research of electrostatics in complicated natural and industrial systems is basically the approximation of complicated multivariable functions of input and output variables. If speaking generally about the electrostatic hazards there are two main traditional fields considering the source of the dangerous situation, namely the **atmospheric electrostatics** (lightning phenomenon, primary, secondary and preventive lightning protection) and **industrial electrostatics** (passive electrostatics – technological problems caused by static charging, fires, explosions and active electrostatics – the industrial use of electrostatic charging of materials).

The approximately 100 years research and education of **Budapest School of Electrostatics** were always important parts of the 228 year old history of Budapest University of Technology and Economics [15]. From the foundation of the Department of Electrical Engineering by Károly Zipernowsky (in 1893), through the research and educational work of László Verebélÿ, continuing the theoretical and practical results of Tibor Horváth [8,9], static electricity was and will remain in the focus.

The first records of lightning started with the early observations at the beginning of the 18th century. The first academic results were published in Buda at the end of the 18th century in German and Latin and only later in Hungarian. The 19th century brought only a few results in lightning research and lightning protection all over the world. Lightning research was started in Budapest by

László Verebélÿ. He published the first results in the proceedings of the Hungarian Electrotechnical Association (1940) and in his inaugural at the Hungarian Academy of Sciences (1947).

The results of Tibor Horváth were summarized in Dresden (1963) and also published at the 7th International Conference on Lightning Protection (ICLP) in Arhem (1969, The Netherlands) [7]. These were the first steps of the so called probabilistic approach using the attraction volume (space) theory. The sudden use of the theory in practical life is proved by the MOSz 274 Hungarian standard of lightning protection (1952) closely followed by the next generation of the Hungarian standards on lightning protection MSZ 274 (in 1962). These periodically renewed guidelines helped the generations of lightning experts both in design and building of air termination systems. These publications back to the 60s already detailed the **probability modulated attraction space theory** (PMAS) and the **method of rolling spheres** (RS).

Generations after generations of former BSc, MSc and PhD students went on with research and education in lightning protection. Nowadays 4 generations of Budapest alma mater deliver papers on their results in international lightning protection conferences.

2. Challenges

Lightning was, is and will be one of the important challenges for mankind. Already in the early days people were faced with the problem of lightning and thunder. Explanation and ways of the protection were always important questions. One can hardly find any mythology, religion and art not trying to handle these both terrible and beautiful natural phenomena. The history of mankind is closely connected to lightning, one of the most dangerous "Acts of God".

From the very beginning any idea of protection was extremely important. It was Benjamin Franklin, who presented our world with the so called Franklin rod (1752). He invented a fantastic tool, but the explanation of how it works was questionable [16]. (His original thought was that the lightning rod would silently discharge the thundercloud.) This problematic explanation has been the main reason of several misunderstandings and misleading steps to date.

The centuries of lightning research has always been accompanied by the research and development of lightning protection, theory and practice and the protective effect of air termination systems. The first steps of primary lightning protection were closely connected to defining the protected volume and searching the appropriate protected angle based upon the statistics of the high voltage power lines. New theories (i.e. electro-geometrical theory based on the probability approach, the attractive space concept and the accepted risk concept) and new methods (i.e. calculation of equivalent area, method of rolling spheres) have been worked out for solving basic contradictions [8].

After coming up with extremely sensitive elements and systems (informatics, communication and media) to protect information and communication systems, the secondary lightning protection theory and practice have appeared (in the second half of the 20th century). During the last decades of the same century lightning detection and location systems appeared around the world. These possibilities together with the pressure of risk management have led to the preventive lightning protection theory and practice of the last some years (see Chapter 3.3).

3. Some of the steps for solutions

3.1. Probability modulated attraction space (PMAS)

The probability-based so-called electro-geometrical method was fist published in the DSc thesis of Tibor Horváth in 1972. By the help of his theory the "annual damage frequency", the "average strokeless period", the "weighting factors" and the "risk of damage during time t" can be rather easily explained [8]. By the help of his calculations the residual risk of lightning can be determined in spite of the presence of a lightning protection system.

The theory is based on the physical process of lightning formation. It is well-known that the formation of a lightning stroke starts by a downward leader, which reaches a certain point called the orientation point, where it initiates an upward leader from the objects at the ground. The lightning current starts to flow after the attachment of the leaders. The probability, that the orientation point is situated at a given distance from an object can be determined according to a distribution function [9].

Examining a certain object, there is a conditional probability, that lightning reaches the object if its distance from the orientation point has a certain value. Taking into consideration the 50% regression surface of this probability distribution, the space can be separated into two parts. The space containing those orientation points from which the lightning strike hits the object is called the attraction space. The number of expected lightning strikes can be calculated by modulating this space using the probability distribution function described in the previous paragraph.

Evaluation of the efficiency of the lightning protection system can be made by the determination of the probability modulated attraction space of the object. If having air termination system on the object, the original attraction space was reduced by the attraction space of the air termination system. From the upper part of the original attraction space all lightning strokes will not hit any more the object to be protected, but the air termination system [8, 9]. Integrating the probability distribution for the attraction space remained and multiplying it by the annual ground flash density, the expected frequency of lightning strikes hitting the building can be calculated. This is an important parameter of the risk calculation. [8,9]

The first step is to estimate the expected annual frequency of lightning that strikes the structure to be protected. This is the resulting value of two components: the equivalent area and the annual ground flash density. The equivalent area (A_{eq}) is defined as the area of such a field terrain, which would be struck by lightning with the same frequency as the structure to be protected. It is connected to the structure independent of the lightning activity in the region. On the other hand, the ground flash density (N_g) characterises the territory of the structure as a meteorological datum. As a result, the structure will be struck by N_F number of flashes per year:

$$N_F = N_g \times A_{eq}.$$

When a lightning stroke causes any damage, the consequences can be very divergent. These circumstances can be taken into account with weighting factors. The consequences of damage are also divergent, which can be expressed by weighting factors. Occurrence and weighting finally conclude at the weighted damage related to each type of damage. This involves the frequency of a strike, the probability for causing damage and the weighting of the consequences. The result of this calculation is the expected annual frequency of weighted damage (D).

The resulting frequency of weighted damage D is an average value that expresses the danger of lightning. When there is a lightning protection system installed, its value is usually very low and cannot be easily interpreted. It is better to understand the period T. This is an average value, which gives no information on what should be expected in the next years.

$$T = 1/D \text{ [years]}$$

Where T is the average interval between two total weighted damages caused by lightning. This is usually too long a time, and can cause a false conclusion concerning the safety. A more comprehensive view can be obtained with further evaluation.

The lightning events are analyzed according to the Poisson process. The Poisson relations express the probability of a given number of events occurring during a given time. This calculation results in the probability called "Risk of damage". According to the Poisson process the risk that the weighted damage occurs during time t can be calculated. The simplified form of the Poisson relations expresses the probability that total damage occurs during time t when T is the average period of damage [9].

There is a rather new tool for solving the aforesaid problems, namely the use of soft computing methods: fuzzy logic, neural networks and genetic algorithms. The new approaches can be effectively used for both the risk evaluation of working systems (checking and auditing) and by using the predictive power of the methods for designing and planning activities both in atmospheric and industrial electrostatics. These methods were efficiently used for the handling of uncertainties in the determination of value D [2].

Unlike general practice, handling of electrostatic hazards should be carried out through a specific risk management strategy. This strategy is specific not only for the examined arrangement but also for the risk taker.

Protection methods may include all the tools which are effective for tuning the level of risk taken (i.e. insurance, use of financial savings, changing the site). If the aim is to determine the level of risk of a given system the suggested method is to use a fault diagnostic system (observations, and interventions). Coming closer to the risk level, one has to determine the damage frequency. The risk caused by lightning can be calculated.

3.2. Rolling sphere method (RS)

The aim of the method of the rolling spheres is to design an air termination system in case of complex geometrical conditions. However the method is usually rather simple (especially while using computers). It is important to advise procedures for designing with the rolling sphere method in the case of rather complicated geometrical conditions to avoid the occurrence of false applications.

Applying this method, the positioning of the air termination system is adequate if it can prevent a fictitious sphere from intersecting the structure to be protected when approaching it from any possible direction. In extreme fall, the sphere may contact the structure to be protected only at a point or along a line, while rolling anyway on or around the air-termination system.

Although the radius of the sphere can be in connection with a lightning current, occurring with known probability, the radius is not really related to the interception efficiency of the air termination system. The probability of occurrence of a lightning current only expresses that a stroke of higher lightning current has to be excluded.

Of course, this does not mean that all lightning of lower current will strike the structure in the protected volume constructed with the given radius of the rolling sphere. So the probability of interception failures may be much lower (may be one or more magnitudes) than the probability related to the lightning current [11,13].

3.3. Preventive lightning protection

The original definition of preventive lightning protection [1] – as it was introduced on the 28th ICLP (Kanazava, Japan, 2006) by the researchers of the Budapest School of Lightning Protection – is as follows.

"The **preventive lightning protection** method means avoiding damage of a lightning strike with special preventive actions. The preventive actions can be of various types, and the primary goal of preventive lightning protection is to decrease the risk of damage due to lightning for the duration of the thunderstorm." [3,4]

This novel approach in lightning protection uses forecasting in conjunction with preventive actions, plan both of them according to the other. This type of lightning protection, is extremely fruitful in the case of objects having a changing level of risk in time and in space (e.g. extra high towers and high buildings, airports, live line maintenance work on high voltage power lines, open air mass performances [14]).

The key to preventive lightning protection is using different forecasting governing the preventive action. The most important types of PLP systems are:

- Preventive Lightning Protection using Local Detectors (LDPLP), [5]
- Zonal Preventive Lightning Protection (ZPLP), [6]
- High Reliability Preventive Lightning Protection (HRPLP), [6]
- Fuzzy Preventive Lightning Protection (FPLP), [14].

4. Questionable steps

As it has been described in the history on lightning protection the use of the **conventional** lightning protection systems (air termination systems consisting of the so called **Franklin rods**, down conductors and grounding system) go back to the 18th century [16].

Later in the 20th century and even nowadays several types of the so called **non-conventional** systems appeared. Many of the lightning experts remember the "glory and fall" of the **radioactive lightning rod**. In his early publications of one of the "fathers of lightning protection" D. Müller-Hillebrand already criticized the use of radioactive lightning rods, and the heavy discussion at 12th ICLP, Portoroz, Yugoslavia (1973) led by another "giant of lightning research" C. Berger showed both the theoretical and practical problems. Recently several non-conventional lightning protection systems have appeared again both in the scientific and business world. The non-conventional lightning protection solutions are the **Early Streamer Emission (ESE) rod** and **Dissipation Arrays** (sometimes called Charge Transfer Systems). There was and even now there is pressure at the international level (scientific conferences, national and international standards, www.intlpa.org) to use these systems in many parts of the word.

The competitor of the conventional lightning protection systems (air termination system built of several Franklin rods positioned on the basis of theoretical considerations) is usually a single non-conventional protector (chosen and built on the idea of its improved protected area). The problem is that this rivalry was often based only on economic considerations instead of theoretical discussions and world wide validation of performance and application in practice for several decades.

The International Conference on Lightning Protection (Cagliary, 2010, www.iclp-certre.org) paid special attention to the non-conventional lightning protection systems. One of the invited lectures focused on this topic [2] and a special Session was also organized. As the speaker Vernon Cooray summarized recently, the experts have had difficult discussions on the theory and practical operation and the comparison of the efficiency of the conventional and non-conventional lightning protection methods. In the above mentioned invited paper both the scientific basis of these systems and their performance were reported following the authors own results and the scientific literature among other publications that of CIGRE WG 405 of SC C4.

To tell the truth several problems have appeared.

- The discussions turned away from scientific questions.
- Instead of basic physical phenomena often only the protected volume is the topic of the debate.
- Based upon the suggestion of the users of ESE the efficiency of one single ESE rod is compared with the air termination system built of several Franklin rods (because of economic reasons).

However the performance of the Franklin rods has been validated in a large number of studies conducted around the world for many decades, the non-conventional systems have been introduced into several lightning protection standards without testing them over a long period of time in the field to validate their performances.

5. Discussion

The original idea that a protected space exists around air termination systems has been disproved. Nowadays the world wide accepted opinion is that the protected effect of an air termination has to be treated as a probability problem. Instead of searching the protected volume, scientists want to define the attractive volume and based upon the probability approach the actual risk has to be compared with the accepted risk defined for the object [16].

The Probability Modulated Attractive Space theory (see Chapter 3.1.) and the designing method, namely the Method of Rolling Spheres (see Chapter 3.2.) is widely accepted and used nowadays all over the world. Even the international (IEC) and European standards (EN) adopted both of them (IEC CENELEC EN 62 305, Protection against Lightning, 2006) however the practical use of these new regulations are interfered with by several misunderstandings and even some failures [10, 11, 12, 13].

Preventive Lightning Protection uses forecasting in conjunction with preventive actions. This newly developed method is extremely fruitful in case of objects having changing level of risk in time and in space. Based upon soft computing methods PLP can be handled (planned, built up and operated) in several practical cases.

There are several laboratory experimental results showing that ESE is an interesting idea, but the principle does not work under natural field conditions. Till now there is no justification to prove that the performance of ESE rods is better than that of the Franklin rods. The results in the literature also demonstrate that dissipation arrays cannot dissipate an approaching lightning flash either to the protected object or to the air terminal system. Contrary to the non-conventional devices, the performance of the Franklin rods has been validated for many decades. The use of protected volume leads to unlimited (uncontrolled) discussion with ESE. Problems can arise with both tall tower and flat roofs. Neither a single Franklin rod, nor a single ESE device can fulfil the requirements of air termination systems. Based upon the electro-geometrical concept and the probabilistic approach they both have to be used with well defined reduced attractive volume of the object to be protected.

6. Acknowledgement

Thanks are due to my teacher Professor Emeritus Dr Tibor Horváth, and my former students and present co-workers, especially István Kiss, PhD, associate professor, Dr. Norbert Szedenik lecturer, Bálint Németh, Head of the High Voltage Laboratory, Attila Gulyas MSc in electrical engineering.

7. References

[1] Berta, I: Static Control, 2005 Journal of Electrostatics **63** 6-10 679

[2] Cooray, V: Non Conventional Lightning Protection Systems, 30th International Conference on Lightning Protection - ICLP 2010, (Cagliari, Italy, 2010) Inv. Paper L2, p 7

[3] Gulyás, B Németh, S Szonda and I Berta, Application of Preventive Measures in Lightning Protection, Kanazawa, Japan, 2006, p 8-1.

[4] Gulyás, B Németh, I. Kiss, and I. Berta, The Theoretical Framework of Preventive Lightning Protection, Uppsala, Sweden, 2008, p. 8-6-1.

[5] Gulyás, A, Mäkelä, J, Németh, B, Cooray, V, Kiss, I and Berta, I: Local Detectors in Preventive Lightning Protection, 30th International Conference on Lightning Protection - ICLP 2010, (Cagliari, Italy - September 13th -17th, 2010) Session 8a, Paper 1105, p 10

[6] Gulyás, A: Application of preventive measures in lightning protection, PhD Th, Budapest, 2011

[7] Horváth, T: Eine neue Methode zur theoretischen Ermittlung der Wahrscheinlichkeit von Bliteinschlag sowie des Schutzeffektes. Berichte der 7. Internationalen Blitschutzkonferenz, Arhem, The Netherlands, Proc, 1969. S. 76-82.

[8] Horváth, T: Computation of Lightning Protection, 1991, Research Studies Press, Taunton, p 204

[9] Horváth, T: Understanding Lightning and Lightning Protection. 2006, Research Studies Press (Wiley & Sons) http://www.wiley.com/go/horvath

[10] Horváth, T: Interception Efficiency of Lightning Air Termination Systems Constructed with Rolling Sphere Method, Kanazawa, Japan, 2006, p. IV-5.

[11] Horváth, T: False Application of the Rolling Sphere Method in the International Standards, 29th International Conference on Lightning Protection, Uppsala, Sweden,2008, Paper 10-03

[12] Horváth, T: Problems and Solution of Utilizing the Protection Angle Method, 30th International Conference on Lightning Protection - ICLP 2010, (Cagliari, Italy - September 13th -17th, 2010) Session 4B, Paper 1018, p 7

[13] Horváth, T: Construction of the Protected Volume with Proper Application of Rolling Sphere Method, 30th International Conference on Lightning Protection - ICLP 2010, (Cagliari, Italy - September 13th -17th, 2010) Session 4B, Paper 1018, p 5

[14] Németh, B, Gulyás, A and Kiss, I: Lightning Protection of Open Air Mass Performances, Uppsala, Sweden, 2008, p. 8-9-1.

[15] Németh, J: From the BME to World Renown, BME Univ. Publishers, Budapest, 2005, p 230

[16] Rakov, V, A and Uman, M, A: Lightning, Physics and Effects, Cambridge University Press, New York, 2003, p 679

Evaluation of performance of footwear and flooring systems in combination with personnel using voltage probability analysis

Jeremy Smallwood[1], **David E. Swenson**[2].

[1]Electrostatic Solutions Ltd, U.K. [2] Affinity Static Control Consulting LLC, U.S.A

jeremys@static-sol.com

Abstract. Evaluation of electrostatic performance of footwear and flooring in combination is necessary in applications such as electrostatic discharge (ESD) control in electronics manufacture, evaluation of equipment for avoidance of factory process electrostatic ignition risks and avoidance of electrostatic shocks to personnel in working environments. Typical standards use a walking test in which the voltage produced on a subject is evaluated by identification and measurement of the magnitude of the 5 highest "peaks" and "valleys" of the recorded voltage waveform. This method does not lend itself to effective analysis of the risk that the voltage will exceed a hazard threshold. This paper shows the advantages of voltage probability analysis and recommends that the method is adopted for use in future standards.

1. Introduction

Evaluation of electrostatic performance of footwear and flooring in combination is necessary in applications such as electrostatic discharge (ESD) control in electronics manufacture [1], evaluation of equipment for avoidance of factory process electrostatic ignition risks and avoidance of electrostatic shocks to personnel in working environments. For this reason, standards such as IEC 61340-4-5 [5] have been developed. These typically use a walking test in which the voltage produced on a subject is measured. Evaluation is typically by manual identification and measurement of the magnitude of the 5 highest "peaks" and "valleys" of the recorded voltage waveform. This method does not lend itself to effective analysis of the risk that the voltage will exceed a hazard threshold, e.g. 100V body voltage for protection of electronic components or 4kV as the threshold of perceptible shocks to personnel. The measurement is typically done on one test subject and floor area only.

The method of voltage probability analysis has several advantages. Firstly, the results may be directly interpreted to find the risk of exceeding any desired hazard threshold voltage. Secondly it can be easily automated using a computer. The results of many measurements (e.g. at different floor positions, or with several subjects or shoe types) may also be combined to give greater confidence in an overall performance of the footwear-flooring system in practice.

As an example the risk of shocks to personnel in an office environment, where a number of subjects wore different shoe types of their choice, is evaluated, based on a sample of subjects selected from that population. Typically in this situation measurements on one person will not be representative of the office population, and measurements of several personnel may be required. In this case comparative measurements on different floors led to identification of a particular floor covering as a key source of electrostatic charging and shocks to personnel.

In a second example, voltage probability analysis is used to evaluate the ElectroStatic Discharge (ESD) risk from personnel within and electronics industry ESD Protected Area (EPA).

2. Methodology

The method of measurement of body voltage is well known and hardly needs description here. Various standards have slight variations on the equipment required. Typically the human subject holds a hand-held electrode which is connected to an electrostatic voltmeter. The output of the voltmeter is recorded, for example by a datalogger or digital oscilloscope. In proprietary instruments, these stages may be combined in one unit.

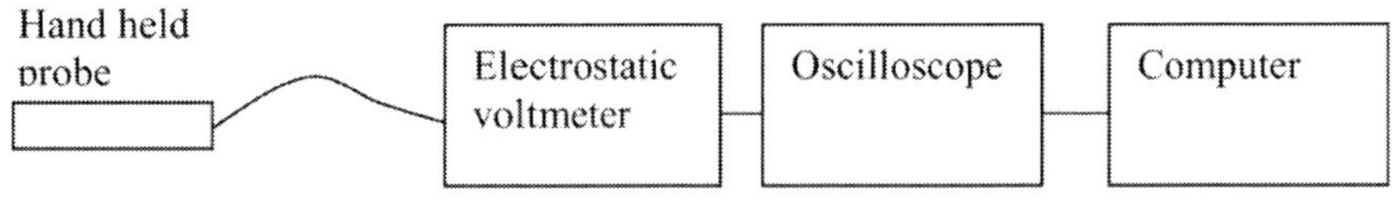

Figure 1. Typical arrangement for measurement of body voltage.

As an example, we have made measurements of body voltage using an electrostatic voltmeter connected to a software oscilloscope running on a laptop computer. The oscilloscope software recorded a voltage trace as a series of digitised voltage values. These were then saved as a comma separated value (csv) file for processing using a spreadsheet software package. The csv file is presented by the spreadsheet as a table of voltage samples at a series of time intervals, determined by the operator on setting the oscilloscope. This table may be presented conventionally as a graph or subjected to further processing and voltage probability analysis. Each time and value pair represents the body voltage at one sample instant over the trace.

One simple method of analysis is to convert the data to a voltage frequency table as follows. Parallel columns of the spreadsheet may be set up, in which the column value is set to 1 if the voltage lies between set limits (e.g. 0-50V, >50 – 100V, >100 – 150V etc). The number of 1's in each column is summed to give the number of data samples within that range. Each of these sums may then be divided by the total number of samples to give the frequency with which the voltage appears in each range. These values may then be presented conveniently as a barchart or graph.

If a simpler output evaluation is required, e.g. the percentage of samples which lie above a 100V hazard threshold value, then it is sufficient to set up 1 column (>100V). The percentage of data samples over that value is easily calculated from the number of 1's in the column divided by the total number of data samples.

3. Results and discussion

A typical body voltage waveform generated by a walking test over a 20 second period using the above method is shown in Figure 2. Table 1 shows sample data from this (Waveform F data) and 6 other similar waveforms (A, B, C, D, E and G) produced by different subjects walking on the same surface material in an office floor evaluation. The table summarises the number of data points for each waveform in the voltage ranges 0. 0.5, 1, ...5kV. It can be seen that different personnel (presumably wearing very different shoe types) could produce very different voltage data distributions.

An overall voltage distribution for the population as a whole was produced by summing all the data values for waveforms A-G, as well as the total number of data samples. The frequency of the data in each voltage range was then calculated for the population.

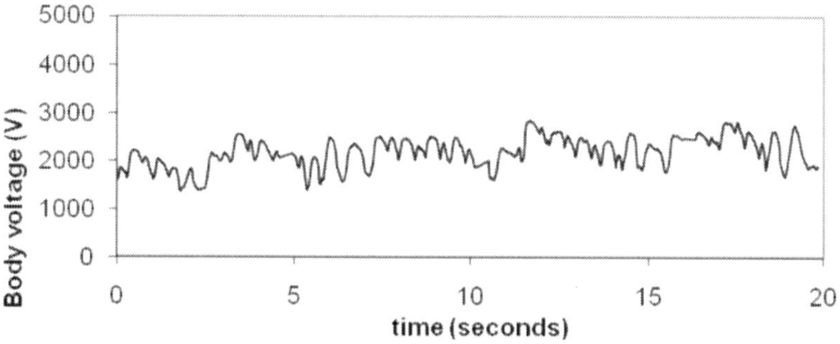

Figure 2. Typical body voltage waveform generated from walking test. This waveform gave Waveform F data shown in Table 1

		Voltage range (kV)											Total no. of samples
		0	**0.5**	**1**	**1.5**	**2**	**2.5**	**3**	**3.5**	**4**	**4.5**	**5**	
frequency of data in voltage ranges (%)		0%	1%	3%	12%	36%	27%	14%	5%	1%	0%	0%	
Sum of samples over all waveforms		0	83	251	1051	3071	2291	1175	397	97	15	0	8431
Waveform data	A	0	37	62	109	687	311	0	0	0	0	0	1206
	B	0	35	101	134	205	230	376	118	7	0	0	1206
	C	0	11	84	429	685	0	0	0	0	0	0	1209
	D	0	0	4	237	507	396	78	0	0	0	0	1222
	E	0	0	0	0	81	245	481	273	90	15	0	1185
	F	0	0	0	128	535	511	38	0	0	0	0	1212
	G	0	0	0	14	371	598	202	6	0	0	0	1191

Table 1. Table of data arising from waveforms A, B, C, D, E, F and G from walk tests with seven personnel on the same surface

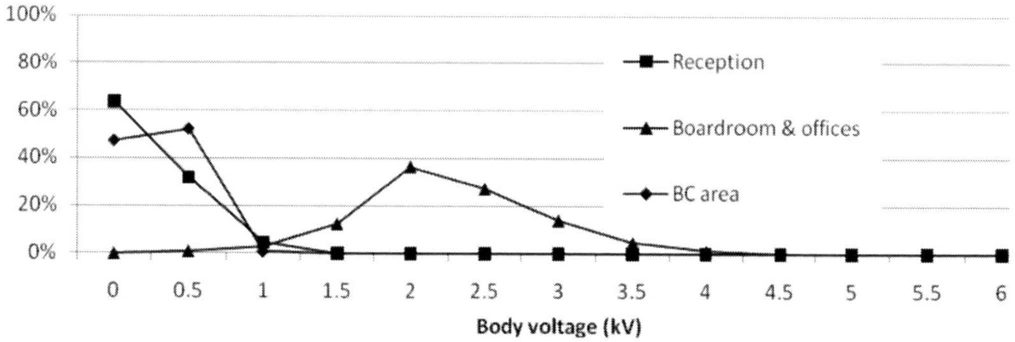

Figure 3. Comparison of body voltages experienced in 3 floor areas in an office environment

Figure 3 shows the results of similar measurements made in three floor areas in an office environment compared for body voltage where shocks had often been reported by personnel. The example data given above is for the "Boardroom and offices" areas, obtained using seven walk tests with volunteer personnel from the area. The data for the "BC area" and "Reception" were obtained from four walk tests. Some, but not all of the personnel participated in tests for all three areas.

The difference in performance of the three floor coverings can immediately be seen. The purpose of this investigation was to identify floor coverings which may contribute to shocks to personnel, with an estimated threshold of 4kV. It can immediately be seen that the Boardroom and office floor covering are the most likely to contribute to this. Under the ambient relative humidity of 36% r.h. there was already a small chance (circa 1%) that personnel may experience shocks. When humidity is reduced, the probability curve would be expected to move to the right and the frequency of shocks increase. The data shows clearly why personnel who are more sensitive to shocks would expect to feel them with increased frequency.

4. Conclusions

Simple voltage probability analysis gives a useful means of evaluating the performance of personnel in combination with their footwear and floor coverings against a hazard threshold. Furthermore, the results of many tests may easily be combined to give a greater confidence in the performance of a system, especially where many personnel or types of footwear may be present. The analysis process may easily be automated in a spreadsheet or by some other computer based method.

It is recommended that this type of analysis should be adopted by future standards as an option for evaluation of footwear and flooring in combination with personnel.

References

[1] Swenson D.E. Process Monitoring to determine electrostatic risks, J. Phys. Conf. Se. 142 (2008) 012005

[2] British Standards Institute. (2001) Electrostatics - Part5-1: Protection of electronic devices from electrostatic phenomena - General requirements. BS EN 61340-5-1: 2001. ISBN 0 580 32346 3

[3] ESD Association. (2007). ESD Association Standard for the Development of an Electrostatic Discharge Control Program for – Protection of Electrical and Electronic Parts, Assemblies and Equipment (excluding Electrically Initiated Explosive Devices). ANSI/ESD S20:20-2007. ISBN 1-58537-121-1

[4] ESD Association ESD Association standard test method for the protection of electrostatic discharge susceptible items. Floor materials and footwear – voltage measurementin combination with a person. ANSI/ESD STM97.2-2006

[5] IEC 61340-4-5 Standard test methods for specific applications – Methods for characterizing the electrostatic protection of footwear and flooring in combination with a person

Journal of Physics: Conference Series **301** (2011) 012065 doi:10.1088/1742-6596/301/1/012065

On the problems regarding the risk calculation used in IEC 62305

T. B. Gellén, N. Szedenik, I. Kiss, B. Németh

Budapest University of Technology and Economics; Egry J.u.18, Budapest Hungary

szedenik.norbert@vet.bme.hu

Abstract. The 2nd part of the international standard on lightning protection (IEC 62305) deals with risk management. The explanations of the mathematical principles and the basic terms of this part facilitate the proper application of the standard. This paper gives additional information for better understanding of the standard and highlights some issues that might occur in its practical application.

1. Introduction

The lightning flash to earth, as a random event, can be modeled using stochastic analysis. Knowing the flash density, the number of the lightning flashes to a given area can be calculated according to a Poisson process. It means, that the parameter describing the time interval between two subsequent lightning flashes has an exponential distribution, and the parameter determining the number of lightning flashes to earth follows a Poisson distribution. Knowing these parameters the probability of the expected number of lightning flashes to a given area during a given time interval can be calculated using simple equations.

2. Risk

The EN 62305-2 standard defines risk as "the value of probable average annual loss (humans and goods) due to lightning, relative to the total value (humans and goods) of the object to be protected". The level of the risk can be calculated by the following equation: $R = N \times P \times L$, where N is the number of dangerous events due to flashes, P is the probability that a dangerous event will cause damage to the structure to be protected and L is the mean amount of loss (humans and goods) consequent on a specified type of damage due to a dangerous event, relative to the value (humans and goods) of the structure to be protected. If we assume that $P \approx 1$ (i.e. a lightning strike will cause damage almost certainly), and $L = 1$ (maximum loss is caused), then $R \approx N$. In this way the risk can be interpreted as the expected annual frequency of the maximum loss in the structure. For example, if the level of this risk is 2×10^{-5} then this means that a lightning strike that destroys the structure completely can be expected in every 50 000 years.

The risk of loss of human life (R_1) in the IEC 62305-2 means the annual frequency of the death of all persons in the given building. Let us call this event *tragedy*. For example, if the value of this risk is 5×10^{-3} then one tragedy can be expected in every 200 years. But what is the probability that such an event will *not* occur in e.g. 100 years? For the answer the events caused by lightning should be modeled by a Poisson process.

3. Probabilistic description of the lightning hazards

Let us introduce the following probability variables: $\xi_1 \ldots \xi_n$, meaning the time intervals between lightning strikes. They have a "memorylessness" feature, i.e. the remaining time to the next event does not depend on the time elapsed from the previous event. Hence the $\xi_1 \ldots \xi_n$ probability variables follow an exponential distribution with parameter λ. This is shown in figure 1.

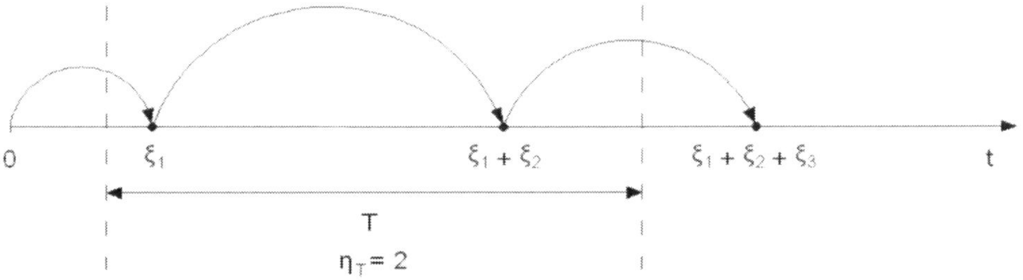

Fig.1: Illustration of the Poisson process

The points on the time axis illustrate a uniform Poisson process. Hence the number of events during any time interval ($T \geq 0$) can be described by the η_T probability variables, which follow a Poisson distribution with parameter $\lambda \times T$ according to the following equation:

$$P(\eta_T = k) = \frac{(\lambda\, T)^k}{k!} \cdot e^{-\lambda \cdot T}, \quad k = 0, 1, \ldots \tag{1}$$

where

T: is the chosen time interval;

k: is the number of events occurred during time interval T;

λ: is the parameter of the $\xi_1 \ldots \xi_n$ probability variables, i.e. the rate of the Poisson process.

The parameter of an exponential distribution is the reciprocal value of its expected value. After many independent experiments the expected value can be estimated by the mean value of the detected values. Hence the expected value $M(\xi_1 \ldots \xi_n)$ of the $\xi_1 \ldots \xi_n$ probability variables equals approximately the reciprocal value of the mean number of events detected during the given time interval. In our case it is the reciprocal value of the annually expected frequency of events. The risk is the annually expected frequency of the events caused by lightning:

$$\lambda = \frac{1}{M(\xi_1 .. \xi_n)} \approx R. \tag{2}$$

Hence the parameter of the Poisson distribution related to the probability variable η_T, which expresses the number of events occurring during the time interval $T \geq 0$, is $\lambda \times T \approx R \times T$. The probability of having no occurrence of tragedy in 100 years in a building with $R_1 = 5 \times 10^{-3}$ is: $P(k = 0) = e^{-0.5} \approx 0.61$. The probability of one single occurrence of this event under the same conditions is: $P(k = 1) = 0.5 \times e^{-0.5} \approx 0.3$.

This high value of the risk cannot be allowed. This is guaranteed by the tolerable risks specified in the standard. In the case of loss of human life its value is $R_{1T} = 10^{-5}$. This means that one tragedy in every 100,000 years in average is tolerable. Table 1 shows the probability of having at least one single occurrence of tragedy during different time intervals (T) if the tolerable risk is $R_{1T} = 10^{-5}$.

Table 1: Probability of loss of human life during different time intervals

Time interval [years]	Probability of tragedy
10	$\approx 9.99 \times 10^{-5}$
100	$\approx 9.99 \times 10^{-4}$
1 000	$\approx 9.95 \times 10^{-3}$
10 000	$\approx 9.52 \times 10^{-2}$
100 000	≈ 0.63

It can be seen that the probability of the event is not proportional to the time. It is also true when the process is evaluated for a 100-year-long time interval (T = 100) with risk values $R_1 = 10^{-2}$ and $R_1 = 2 \times 10^{-2}$ (the risk of loss of human life can be even higher without lightning protection measures). In this case the density functions of a Poisson distribution with parameters $\lambda \times T \approx 10^{-2} \times 100 = 1$ and $\lambda \times T \approx 2 \times 10^{-2} \times 100 = 2$ are shown in the following figures.

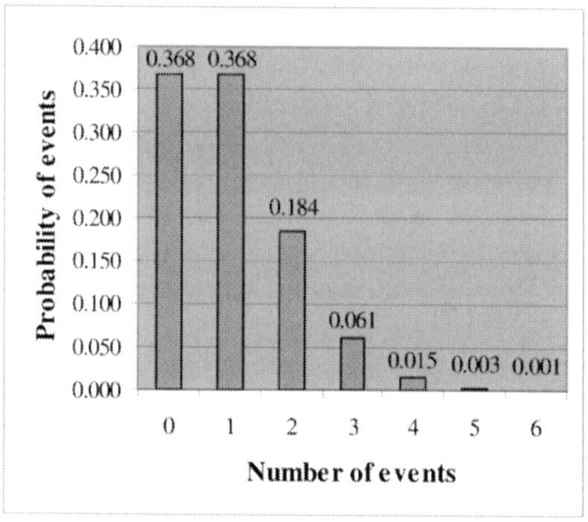

Fig. 2: Probability of tragedy in 100 years when $R_1 = 10^{-2}$

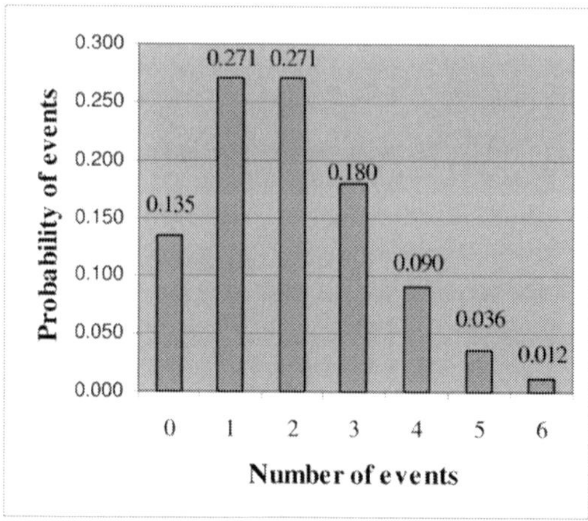

Fig. 3: Probability of tragedy in 100 years when $R_1 = 2\times10^{-2}$

In the first case (Fig. 2) the probability of the tragedy is $1 - 0.368 \approx 0.63$. In the second case (Fig. 3) this probability is $1 - 0.135 \approx 0.87$. It can be shown that the probability of the single events is not proportional to the risk. If it was so, then the probability in the second case would be $2 \times 0.63 = 1.26$, which is impossible. Hence the value of the risk itself does not give the probability of the events but is proportional to the expected value of the probability variable η_T, expressing the number of events: $M(\eta_T) = \lambda \times T \approx R \times T$. (The expected value of a Poisson distribution is equal to its parameter.) Sometimes risk defined by the expected value expresses the hazard much better than the given probability values.

4. Application of the risk analysis

At the beginning of the risk analysis it is usually assumed that the structure is not covered by any lightning protection. The value of N depends on the local lightning activity and the physical parameters of the building. The value of P can be assumed to be 1 without any protection measure. However the value of L cannot be estimated very well, since it depends e.g. on the mean values of the loss, the fire hazard and the endangered number of persons and the time per year for which the persons are present in the hazardous zone.

According to our results the accuracy of the risk calculation depends basically on the correct estimation of the losses. For example, let us suppose that we have an office building with 800 m^2 area,

25 m height and the constant presence of 200 persons. 20 persons are present in the archive of the building. Fig.4 shows the actual risk for the whole building related to the tolerable risk depending on the risk of fire in the archive.

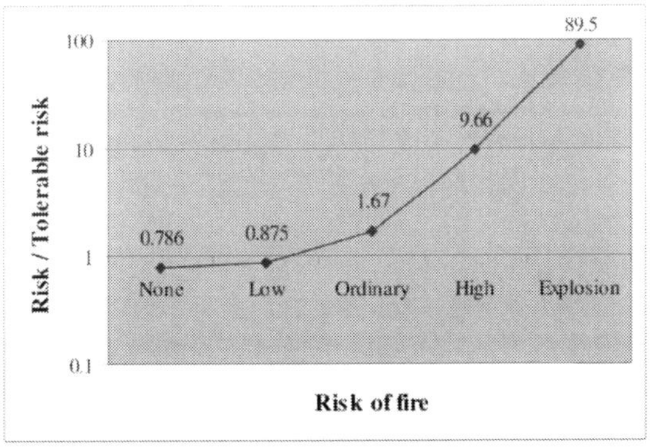

Fig.4: R_1/R_T via risk of fire of archive in a case of office building

Time / day	Number of persons				
	8	9	10	11	12
24 h	1.15	1.23	1.3	1.38	1.45
20 h	0.9	0.95	0.99	1.04	1.08
16 h	0.82	0.86	0.89	0.93	0.96
12 h	0.76	0.78	0.81	0.83	0.86
8 h	0.69	0.71	0.72	0.74	0.76

Table 2: R_1/R_T via number of persons and the time for which the persons are present in the archive

Based on our experiments we can conclude that knowing the risk of fire in each part of the structure is crucial if we want to calculate the risk accurately. In our example, the total risk of the office building related to the tolerable risk is shown in the following table as a function of the number of persons and the time per day for which the persons are present in the archive when an automatic fire extinguisher is provided.

The endangered number of persons and the time per year in which the persons are present in the hazardous zone are usually poorly estimated. However the necessity of protection measures can only be decided if we know these values. If the exact values are not available, then the worst-case method should be used. Otherwise the planned measures will not meet the requirements of the lightning protection standard.

5. Conclusions

The risk analysis described in IEC 62305-2 quantifies the effects of the lightning. However the term "risk" can have different interpretations. The paper identifies the risk used in the standard. The paper helps to clarify the mathematical basis of the risk calculation applied in the IEC 62305 international standard on lightning protection. It focuses on the critical parameters, simplifications, which can lead to false results in the risk calculation followed by improper lightning protection measures. This issue is illustrated by a case study example.

References

[1] T. Horváth: Computation of Lightning Protection, Research Study Press LTD., Taunton, Somerset, England 1991
[2] T. Horváth: A Practical Method to Consider the Accepted Risk in Standardizing of Lightning Protection, 22nd International Conference on Lightning Protection, Budapest, Hungary, 19–23 September 1994
[3] T. Horváth: Understanding Lightning and Lightning Protection, Research Study Press LTD., Taunton, Somerset, England 2006
[4] IEC 62305-1:2006, Protection against lightning - Part 1: General principles
[5] IEC 62305-2:2006, Protection against lightning - Part 2: Risk management

Journal of Physics: Conference Series **301** (2011) 012066 doi:10.1088/1742-6596/301/1/012066

Factors that influence the efficiency of a fluidized-bed-type tribo-electrostatic separator for mixed granular plastics

L Dascalescu[1]*, O. Fati[1], M. Bilici[1,2], F. Rahou[1], C. Dragan[1], A. Samuila[2] and A. Iuga[2]
[1]PPRIME Institute, UPR 3346 CNRS – University of Poitiers – ENSMA, France
[2]High-Intensity Electric Fields Lab, Technical University, Cluj-Napoca, Romania
E-mail: lucian.dascalescu@univ-poitiers.fr

Abstract. Fluidized bed devices have already been used as tribochargers for various industrial electrostatic separation processes. In the present paper, the authors investigate the behaviour of polyamide - polycarbonate granular plastic mixtures in a parallelepiped bed, the height of which is roughly 2 times its length or width, so that the collisions between granules become the prevailing tribocharging mechanism. Two of the opposite walls of the tribocharging chamber consist of metallic plates connected to two DC high-voltage supplies of opposite polarities, so that the charged particles are attracted to the electrodes and separated while still in the fluidized state. The collecting hoppers are designed as Faraday cups connected to two electrometers, thus allowing the instantaneous measurement of the charge carried by the separated particles. Experimental design methodology was employed for the optimization of the tribo-aero-electrostatic separation process, the input variables being the high-voltage applied to the electrodes and the duration of the tribocharging. Higher voltages applied to the electrode system do not necessarily lead to larger quantities of collected products but improve the purity of the concentrates. The composition of the mixture influences the outcome of the process.

1. Introduction

A wide range of applications in the chemical industry and in powder processing take advantage of the tribo-charging effects occurring in fluidized bed reactors [1-4]. The size of solid particles processed in standard reactors ranges from less than 1 μm to 6 cm [5-9]. In any fluidized bed, the granules become electrically charged by contacts between each other or with the walls of the tribocharger [1, 10].

In a previous paper, the authors have studied a cylindrical fluidized bed device, characterized by elevated air velocities and height/diameter ratios, which favor more frequent granule-to-wall collisions [11]. By choosing wall materials in accordance with the triboelectric series [12-14], the two classes of particles to be separated charge with opposite polarities [15] and are sorted by the electric forces exerted on them while falling freely in the electric field generated between vertical plate electrodes.

The present paper is aimed at experimental modelling of the tribo-electrostatic separation process in a parallelepiped fluidized bed, the height of which is roughly 2 times its length or width, so that the collisions between granules become the prevailing tribocharging mechanism. The peculiar advantage of this configuration in comparison with the "standard" cylindrical one is that an electric field can be easily generated inside the fluidized bed by the metallization of two of the opposite walls of the tribocharging chamber and connecting them to two DC high-voltage supplies of opposite polarities. The charged particles will be attracted to the electrodes and separated while still in the fluidized state.

* To whom any correspondence should be addressed.

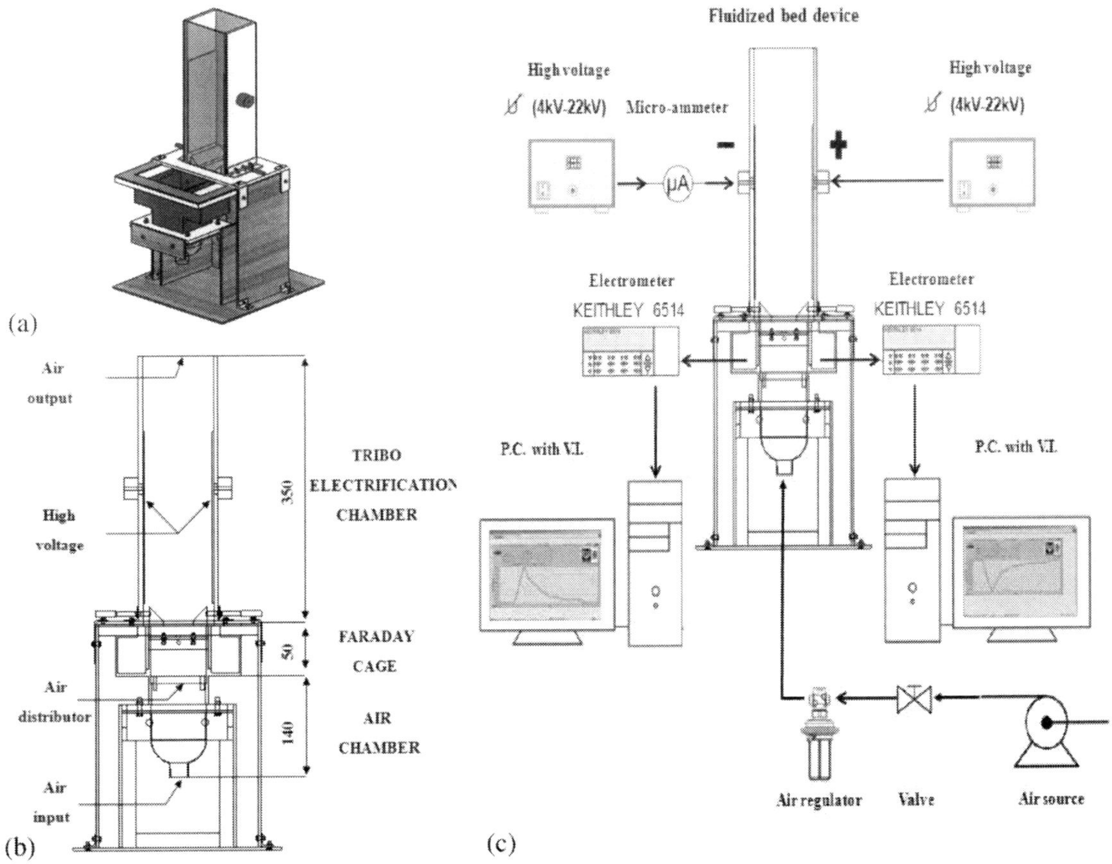

Figure 1. Experimental installation: general aspect (a) and main dimensions (b) of the tribocharging fluidized bed device; schematic representation of the experimental set-up (c).

2. Experimental set-up

The fluidized bed built for this experiment is a rectangular prism (115 mm x 85 mm x 400 mm), with two opposite vertical walls made of polycarbonate, the other two consisting of aluminium plates connected to two adjustable DC high-voltage supplies of positive and negative polarity (Fig. 1).

The granular material is deposited on the screen type air distributor and is maintained in the fluidized bed state generated by the ascending air. The charged granules are attracted to the electrodes of opposite polarity and fall into the two collecting hoppers, designed as Faraday cups connected to two electrometers that measure the charge carried by the separated particles.

3. Material and methods

The tribo-electrostatic separation experiments were performed on blue virgin polyamide (PA) and orange polycarbonate (PC) granules, used in the plastics industry (characteristic size: 2 mm). Two types of samples (mass of a sample: 100 g) were prepared as binary mixtures with different compositions. The experiments were carried out under relatively stable ambient conditions: temperature T = 17 - 22°C, relative humidity RH = 44 - 60%, and at given air velocity v = 6 m/s, measured with a miniature cup-type anemometer in the absence of particles. The charge measurement data acquired by the electrometers (model 6514, Keithley Instruments) during the steady-state operation of the tribocharger were processed by a virtual instrument (VI) developed in the LabView environment.

The high voltage U applied to the electrodes and the duration t of the tribocharging were the two process variables investigated. The effects of each variable on the output of the process (i.e. the charge/mass ratio of the granules collected at the electrodes), were evaluated for two classes of binary granular mixtures: 50% PA – 50% PC; 75% PA – 25% PC.

In order to obtain a quadratic polynomial model of the tribo-aero-electrostatic separation process, the composite factorial experimental design [34] was employed for the present study. The domain of variation of the high voltage provided by each supply (U_{min} = 6 kV ; U_{max} = 18 kV) and of the tribocharging duration (t_{min} = 30 s ; t_{max} = 120 s), were established based on the results of preliminary experiments. The data were processed using commercial software, MODDE 5.0, developed by Umetrics [17]. The program calculates the coefficients of the mathematical model, evaluates their statistical significance, as well as two statistical criteria (the goodness of fit R^2, and the goodness of prediction Q^2), and identifies the best adjustments of the input variables for optimizing the process.

4. Results and discussion

The results of the composite factorial experimental design carried out for the two classes of samples characterized by different percentages of PA and PC granules are summarized in table 1. The corresponding mathematical models calculated by MODDE 5.0 are given in table 2. The statistics R^2 and Q^2 of these models, listed in the same table, are excellent.

The equal-charge curves plotted in figure 2 for the 50 g PA + 50 g PC granular mixture illustrate the effects of the two factors under study. Optimum operation is achieved at U = 10.4 kV; t = 120 s.

Table 1. Charge of the PA and PC products obtained by tribo-aero-electrostatic separation of two classes of granular mixtures

Run No.	U [kV]	t [s]	50 g PA + 50 g		75 g PA + 25	
			Q_{PA} [nC]	Q_{PC} [nC]	Q_{PA} [nC]	Q_{PC} [nC]
1	6	30	24.2	-44.1	18.1	-31.8
2	18	30	81.7	-92.4	61.5	-82.5
3	6	120	319	-350	190	-110
4	18	120	217	-282	188	-164
5	6	75	181	-237	90	-82.7
6	18	75	174	-214	145	-156
7	12	30	91.6	-98	38.3	-57.2
8	12	120	363	-389	246	-156
9	12	75	288	-293	117	-132
10	12	75	298	-303	124	-134
11	12	75	305	-313	132	-137

Table 2. Quadratic models of Q_{PA} [nC] and Q_{PC} [nC] calculated by MODDE 5.0 from the data in Table 1, for the two classes of granular mixtures

Class	Model	Statistics
50 g PA 50 g PC	Q_{PA} = 293.5 -8.58 U^* + 105 t^* + - 111 U^{*2} - 25 t^{*2} -39.9 U^*t	R^2 =0.996 Q^2=0.952
	Q_{PC} = -297.5 +7.11U^* -131 t^* + - 63.7 U^{*2} – 45.7 t^{*2} -29.1 U^*t	R^2 =0.990 Q^2=0.919
75 g PA 25 g PC	Q_{PA} = 123.9 +31.2 U^* + 99.5 t^* -5.86 U^{*2} +18.8 t^{*2} + 11.4 U^*t	R^2 =0.995 Q^2=0.952
	Q_{PC} = -133.2 –29.7 U^* - 43 t^* - + 12.1 U^{*2} +24.9 t^{*2}	R^2 =0.984 Q^2=0.934

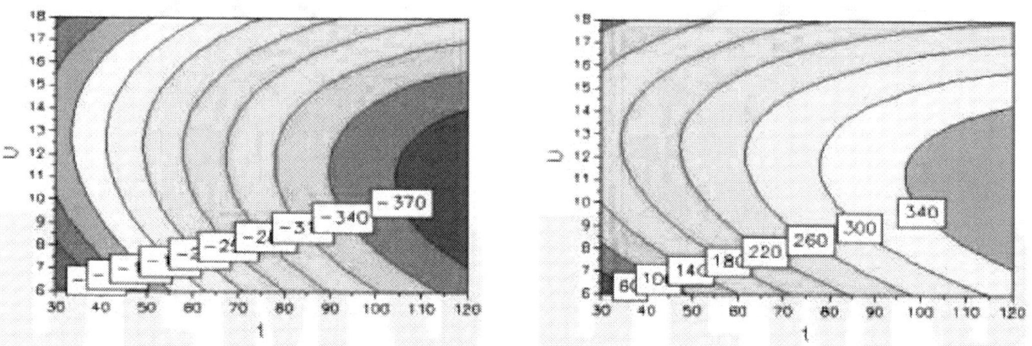

Figure 2. MODDE 5.0 predicted equal-charge q [nC] contours for PC and PA particles separated from a 50 g PA + 50 g PC granular mixture, as a function of the voltage U [kV] and the duration t [s].

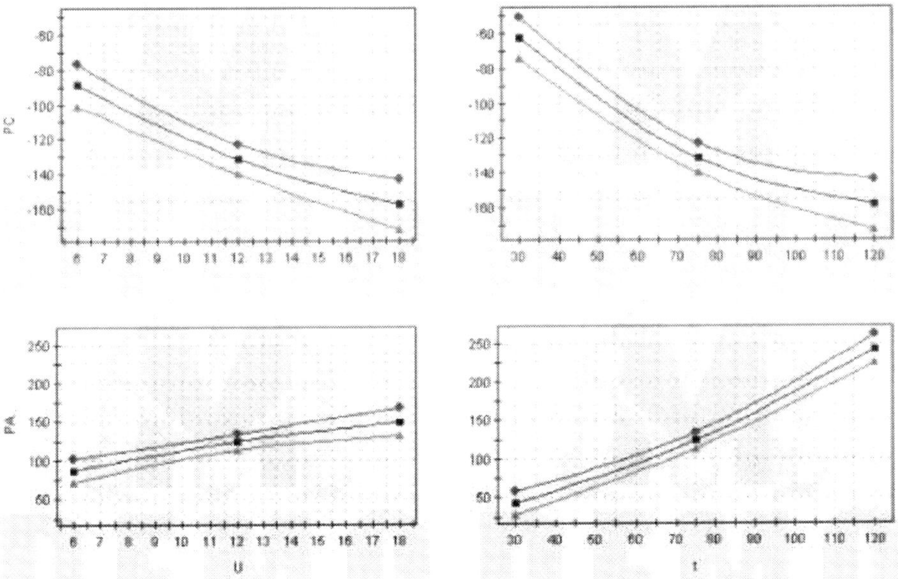

Figure 3. MODDE 5.0 predicted charge *q* [nC] of PC and PA particles separated from a 75 g PA + 25 g PC granular mixture, as function of the high-voltage *U* [kV] (a) and the tribocharging duration *t* [s] (b). The upper and the lower curves on each graph indicate the limits of the 95% confidence interval.

In the case of the 75 g PA + 25 g PC granular mixture, better results are obtained at higher applied voltages and longer tribocharging durations (figure 3).

The PA-PC granular mixtures of various compositions can be successfully tribocharged and separated in the electric field of a fluidized bed. As one sort of particles may be collected faster than the others, the composition of the mixture in the fluidized bed may vary in time. This may impose severe constraints on the integration of fluidized bed devices into a continuous industrial process.

References

[1] Ciborowski J S and Wlodarski A 1962 *Chem. Eng. Sci.* **17** 23
[2] Boland D and Geldart D 1971/72 *Powder Technol.* **5** 289
[3] Dwari R K and Rao K H 2009 *Minerals Eng.* **22** 119
[4] Rokkam R G, Fox R O and Muhle M E 2010 *Powder Technol.* **203** 109
[5] Pell M, Dunson J B and Knowlton T M 2007 *Perry's Chemical Engineers' Handbook 8/E Section 17: Gas-Solid Operations and Equipment* (Toronto: McGraw-Hill Professional)
[6] Howard J R, 1989 *Fluidised Bed Technology. Principles &Applications* (New York: A Hilger)
[7] Liu Z, Bi X T and Grace J R 2010 *J. Electrostat.* **68** 321
[8] Sowinski A, Salama P and Mehrani P 2010 *J. Electrostat.* **67** 568
[9] Guardiola V, Rojo V and Ramos G 1996 *J. Electrostat.* **37** 1
[10] Ali F S, Inculet I I and Tedoldi A 1999 *J. Electrostat.* **45** 199
[11] Calin L, Mihalcioiu A, Iuga A and Dascalescu L 2007 *Part. Sci. & Technol.* **25** 205
[12] Castle G S P 1997 *J. Electrostat.* **40&41** 13
[13] Inculet I I, Castle G S P and Brown J D 1998 *Part. Sci .Technol* **16** 91
[14] Iuga A, Calin L, Neamtu V, Mihalcioiu A and Dascalescu L 2005 *J. Electrostat.* **63** 937
[15] Calin L, Caliap L, Neamtu V, *et al* 2008, *IEEE Trans. Ind. Appl.* **44** 2008.
[16] Frigon NL and Mathews D 1996 *Practical Guide to Experimental Design* (New York: Wiley)
[17] Eriksson L, Johansson E, Kettaneh-Wold N, Wikström C and Wold S 2000 Design of Experiments. Principles and Applications (Stockholm: Learnways AB)

Optimisation of belt-type electrostatic separation of granular plastic mixtures tribocharged in a propeller-type device

M Miloudi[1], K Medles[1], A Tilmatine[1], M Brahami[1] and L Dascalescu[2]

[1]IRECOM, University Djillali Liabes, 22000 Sidi Bel Abbes, Algeria
[2]PPRIME Institute, UPR 3346 CNRS – University of Poitiers – ENSMA, France

E-mail: miloudi.mohamed67@gmail.com; lucian.dascalescu@univ-poitiers.fr

Abstract. Electrostatic separation has already proven its efficiency as a method to recover metals and plastics from granular waste electrical and electronic equipment (WEEE). Nevertheless, there still is some work to do regarding the sorting of the various sorts of plastics (ABS, ABS-PC, HIPS, PC) contained in information technology wastes. The main objective of this paper is to optimize the electrode configuration and the operation of a belt-type electrostatic separator employed in conjunction with a new tribo-aerodynamic charging device. This application is particularly designed for the recovery of granular materials issued from shredding of computer cases, which contain large quantities of ABS. The following control variables were considered for the optimization: (i) distance between the axis of the high-voltage electrode and the surface of the grounded belt electrode; (ii) angular position of the high-voltage electrode with respect to the horizontal plane; (iii) high-voltage level. Experimental design methodology was employed for determining the optimum value of each variable, using commercial software (MODDE, Umetrics, Sweden).

1. Introduction

The recycling of materials contained in waste electrical and electronic equipment (WEEE) is an important issue for the sustainable use of natural resources [1]. Electrostatic separation, which is the generic term given to sorting of granular mixtures by making use of the electrical forces acting on charged or polarized bodies[2, 3], has already proven its efficiency to recover metals and plastics from chopped WEEE [4, 5]. Nevertheless, there still is some work to do regarding the sorting of the various sorts of plastics (ABS, ABS-PC, HIPS, PC) contained in information technology wastes. In most of the existing plastic – plastic electrostatic separation technologies, the granular mixtures are first processed in a tribocharging device [6], and then introduced in the electric field generated between two vertical plate electrodes. The efficiency of tribocharging being affected by a number of uncontrolled factors (ambient temperature and relative humidity, particle surface state, etc.), the free-fall plastic/plastic tribo-electrostatic separators [7, 8] are not as effective as the role-type corona-electrostatic equipment employed for the sorting of plastic/metal mixtures.

The work presented in this paper was aimed at optimizing the electrode configuration and the operation of a belt-type electrostatic separator employed in conjunction with a new tribo-aerodynamic charging device [9]. The study focused on the recovery of granular materials issued from shredding of computer cases, which contain large quantities of ABS. Experimental design methodology was employed for the evaluation of the various factors and the estimation of their optimal values.

Published under licence by IOP Publishing Ltd

2. Experimental set-up

The experiments were carried out using a belt-type tribo-electrostatic separator (Fig. 1, a) consisting in a custom-designed tribocharging device, a vibratory feeder, a metallic conveyor (roll diameter=5cm), connected to the ground, an elliptical electrode (axis: 15 cm and 5 cm), energized from a fully-adjustable DC high voltage power supply (40 kV; 1 mA) and a three-compartment collector of particles: two boxes for the pure products and one for the middling. The granular mixture to be separated is introduced in the tribocharging device, which consists with a cylindrical tube made of PVC (diameter: 10 cm; height: 35 cm) and a coaxial plastic propeller inside it (Fig. 1, b). The optimum operation of this device, i.e. propeller speed v = 3533 RPM and duration t = 40 s of the tribocharging, has been established based on the experimental data reported in a previous paper [9].

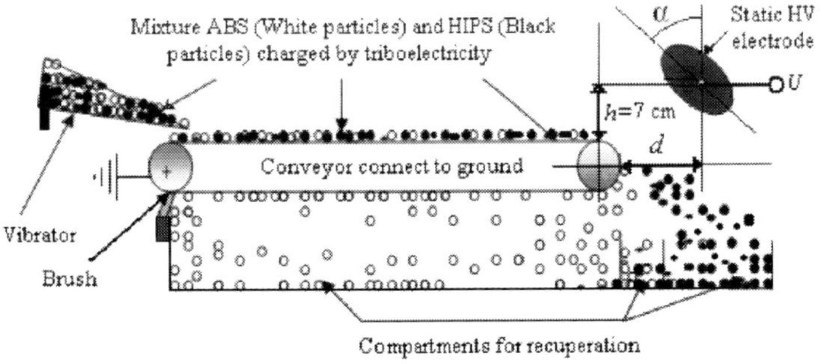

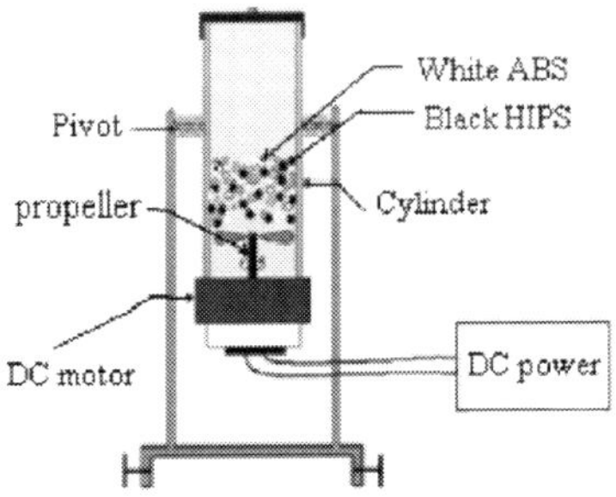

Figure 1. Schematic representation of the belt-type tribo-electrostatic separator (a) and of the propeller-type tribocharging device (b).

The tribocharged materials are then deposited by gravity on the tray of a vibratory feeder that transfers them at a constant rate on the metallic belt of a variable-speed conveyor, entrained by a DC motor. The feed-rate and the speed of the conveyor are co-related, so that the material to form a mono-layer at the surface of the belt.

The conveyor introduces the particles in the electric field zone generated between the high-voltage electrode of positive polarity and the grounded belt. The negatively-charged particles are subjected to an attraction force exerted by the electrode of opposite polarity, while the positively-charged ones are pinned to the surface of the belt, until a brush will mechanically remove them, as shown in Fig. 1, a. Based on the results of several preliminary experiments, the high voltage electrode was positioned at h = 7 cm above the plane of the belt, while the speed of the DC motor drive was adjusted at 15 RPM.

3. Materials and method

The 10 g samples used for the experiments consisted in a granular mixture of 50% white ABS and 50% black HIPS (size distribution: 1 mm to 2 mm). After each experiment, the masses of the collected products were measured with an electronic balance (resolution: 0.01 g).

Using the experimental design methodology [10], a the quadratic model was obtained for of each of the responses y_{ABS}, y_{HIPS}, $y_{Mixture}$, i.e. the mass of the three collected products, as function of the normalized centred values u_i* of each factor u_i considered in the present study, i.e. the applied voltage U, the angle α and the horizontal position d of the electrode:

$$y = a_0 + a_1 U^* + a_2 d^* + a_3\,\alpha^* + a_{11} U^{*2} + a_{22}\, d^{*2} + a_{33}\alpha^2 + a_{12}\, U^* d^* + a_{23}\, d^*\alpha^* + a_{31}\,\alpha^* U^* \quad (1)$$

$$u_i^* = (u_i - u_{ic})/\Delta u_i\,;\; u_{ic} = (u_{imax} + u_{imin})/2;\; \Delta u_i = (u_{imax} - u_{imin})/2. \quad (2)$$

In order to obtain such a quadratic model, the data of a composite factorial experiment (Table I) data were analysed with MODDE 5.0 software (Umetrics, Sweden), which calculates the coefficients of the mathematical model, draws the response contours and identifies the best adjustments of the parameters for optimizing the process.

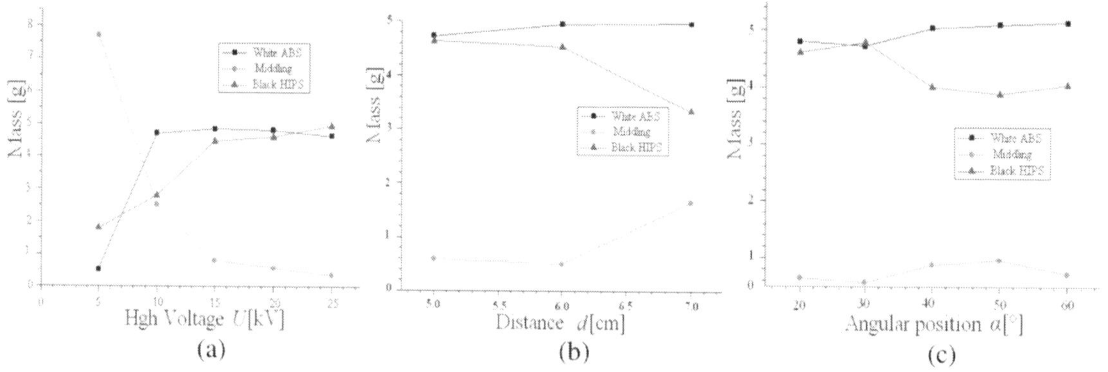

Figure 2. Mass of the products obtained from 10-g samples of 50% ABS – 50% HIPS, as function of: (a) the applied voltage U, for $\alpha = 30°$, $d = 7$ cm; (b) the horizontal position d, for $\alpha = 30°$, $U = 15$ kV; (c) the angular position α, for $U = 15$ kV; $d = 7$ cm.

4. Results and discussion

4.1. Domain of variation of the control variables

All the experiments were performed at temperatures ranging between 20.2°C and 23.3°C, while the relative humidity varied between 37.2% and 41.3%. Three sets of "one-factor-at-a-time" experiments were performed in order to establish the domain of variation of the control variables. In the first set, the applied voltage was adjusted at various values between 5 and 25 kV, while the other two variables were kept constant: $\alpha = 30°$, $d = 7$ cm. The results are illustrated by the curves in figure 3, where each point is the average value of at least three experiments. At voltages higher than 15 kV, more than 80% of the ABS and HIPS particles in the feed are recovered in the final products. The lower value of the HIPS mass at 25 kV can be explained by the strong electric field forces that made some of the particles impact the high-voltage electrode and be deviated in the ABS compartment.

For the second set of experiments $\alpha = 30°$, $U = 15$ kV and the horizontal position d of the static electrode was adjusted between 5 and 7 cm (Fig. 2, b). Slightly smaller quantities of HIPS were collected at $d = 7$ cm, but the process is quite robust with respect to this factor. This effect is due to the modification of the electric field distribution. At higher values of d, the reduction of the local strength of the electric field is accompanied by a diminution of the electric forces acting on the HIPS particles, which will detach sooner from the surface of the belt and be collected in the middling compartment. In the third set of experiments, the angular position α of the static electrode was varied between 20° and 60°, at constant $U = 15$ kV and $d = 7$ cm (Fig. 2, c). At $\alpha > 30°$, the quantity of HIPS diminishes, for reasons similar to those given for the variation of d. Based on the above data, the domain of the variables was established as follows: $U = 15$ to 25 kV; $\alpha = 20$ to 30°; $d = 5$ to 7 cm.

4.2. Modeling and optimization of the separation process

The mathematical models of the responses y_{HIPS} ($R^2 = 0.999$; $Q^2 = 0.994$) and y_{ABS} ($R^2 = 0.997$; $Q^2 = 0.983$), and were obtained with MODDE 5.0 using the data in Table 1 (the non-significant coefficients were eliminated):

$$y_{HIPS} = 5.02 + 0.045\,U^* + 0.067\,d^* + 0.134\,\alpha^* - 0.0016\,U^{*2} - 0.14\,d^{*2} + 0.104\,U^*d^* - 0.0098\,U^*\alpha^* - 0.037\,d^*\alpha^* \quad (5)$$

$$y_{ABS} = 4.58 - 0.039\,U^* - 0.139\,d^* - 0.034\,\alpha^* + 0.072\,U^{*2} + 0.107\,d^{*2} + 0.04\,\alpha^{*2} - 0.037\,U^*d^* + 0.018\,U^*\alpha^* + 0.0197\,d^*\alpha^* \quad (6)$$

The two statistical criteria computed by MODDE 5.0 were excellent (i.e., close to unity) for both models: the goodness of fit $R^2 = 0.999$ and 0.997; the goodness of prediction $Q^2 = 0.994$ and 0.983. The predicted mass of collected ABS is represented in Fig. 3. Similar curves were obtained for y_{HIPS}.

Table 1. Masses of the products obtained from a 5g HIPS + 5 g ABS granular mixture, using a composite factorial experimental design

N°	U [kV]	d [cm]	α [°]	HIPS [g]	Middling [g]	ABS [g]
1	15	5	20	4.26	0.46	5.2
2	25	5	20	4.09	0.72	5.17
3	15	7	20	4.23	0.81	4.9
4	25	7	20	4.71	0.67	4.64
5	15	5	40	4.77	0.29	5.02
6	25	5	40	4.51	0.41	5.09
7	15	7	40	4.49	0.7	4.84
8	25	7	40	4.89	0.42	4.67
9	15	6	30	4.98	0.34	4.71
10	25	6	30	5.09	0.33	4.61
11	20	5	30	4.72	0.4	4.95
12	20	7	30	4.86	0.51	4.64
13	20	6	20	4.57	0.59	4.67
14	20	6	40	4.92	0.48	4.58
15	20	6	30	5.02	0.38	4.6
16	20	6	30	4.99	0.4	4.57
17	20	6	30	5.03	0.39	4.61

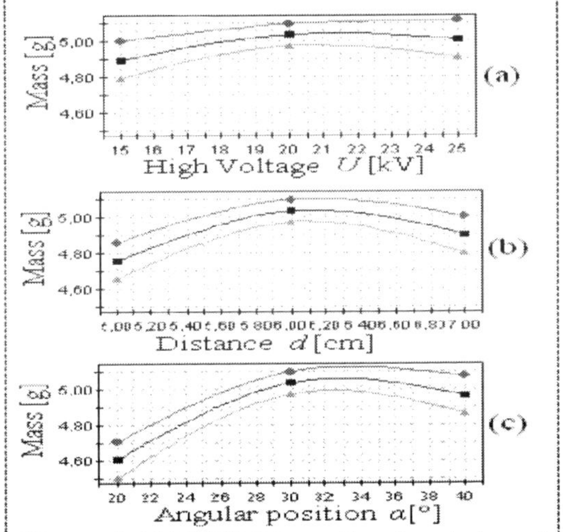

Figure 3. MODDE 5.0-predicted variation of collected masses of ABS, as function of the applied voltage (a), the horizontal position (b) and the angular position (c) of the electrode. The upper and the lower curves on each graph indicate the limits of the 95% confidence interval.

MODDE 5.0 offered also the possibility of identifying the optimal point of the process: $U = 15$ kV, $d = 5.7$ cm and $\alpha = 33°$, for which the predicted masses of collected products were: $y_{HIPS} = 5.02$ g and $y_{ABS} = 4.86$ g. The masses collected in an experiment conducted in the optimal conditions were $m_{HIPS} = 4.97$ g and $m_{HIPS} = 4.64$ g, in very good agreement with the predictions.

5. Conclusions

(1) The tribo-electrostatic process associating a propeller-type tribocharging device and a belt-type electrostatic separator is a promising technology for the recovery and recycling of the plastic materials contained in granular WEEE.

(2) Experimental design methodology proved to be an effective tool for the optimization of the novel tribo-electrostatic separation process.

References

[1] Tilmatine A, Medles M, Bendimerad S, Boukholda F and Dascalescu L 2009 *Waste Management* **29** 228

[2] Haga K, 1995 *Handbook of Electrostatic Processes* Eds. Chang J-S, Kelly AJ and Crowley JM, (New York: Marcel Dekker) pp. 365-386

[3] Dascalescu L, Morar R., Iuga A, Samuila A and Neamtu V 1998 *Part. Sci. & Technol.* **16** 25

[4] Iuga A, Samuila A, Neamtu V, Morar R, Belaca R, Das S and Dascalescu L 2011 *IEEE Trans. Ind. Appl.*. **47** 222

[5] Blajan M, Beleca R, Iuga A and Dascalescu L 2010 *IEEE Trans. Ind. Appl.*. **44** 1558

[6] Iuga A, Calin L, Neamtu V, Mihalcioiu A and Dascalescu L 2005 *J. Electrostat.* **63** 937

[7] Inculet I I, Castle G S P and Brown J D 1998 *Part. Sci. & Technol.* **16** 91

[8] Calin L, Caliap L, Neamtu V, Morar R, Iuga A, Samuila A and Dascalescu L 2008 *IEEE Trans. Ind. Appl.* **44** 1045

[9] Miloudi M, Medles K, Tilmatine A, Brahami M and Dascalescu L 2010 *Proc. 7th Conf. French Society of Electrostatics* p. 340

[10] Frigon NL and Mathews D 1996 *Practical Guide to Experimental Design* (New York: Wiley).

Comparison of the effect of grounding the column wall in gas-solid fluidized beds on electrostatic charge generation

Andrew Sowinski, Antonio Mayne, Bassam Javed, Poupak Mehrani

University of Ottawa, Chemical and Biological Engineering Department,
161 Louis Pasteur St., Ottawa, Ontario, Canada, K1N 6N5

Email: poupak.mehrani@uottawa.ca

Abstract. In gas-solid fluidized beds as particles are fluidized, they continuously come into contact with other particles, as well as the fluidization column wall. This generates electrostatic charges by means of triboelectrification and frictional charging, leading to particle agglomeration, reactor wall fouling, and eventually process downtime and large financial losses. Grounding the fluidization column has been considered as a means of helping electrostatic charge dissipation within fluidized beds; however, in industrial applications despite the process vessels being grounded, the electrostatic problem still persists. This work focused on the effect of fluidization column grounding on particle wall fouling. Experiments were conducted in an atmospheric system consist of a 0.1 m in diameter carbon steel fluidization column. The mass and charge-to-mass ratio (q/m) of the particles that remained adhered to the column wall upon the completion of one hour fluidization period were measured in an electrically isolated and grounded columns to quantitatively determine the amount of reactor wall fouling. Polyethylene particles with different particle size ranges (300- 1000 μm) were fluidized with extra dry air at 1.5 times their respective minimum fluidization velocity (u_{mf}). Results obtained in the grounded fluidization column were not significantly different from those in the isolated column for all particle size ranges tested where the particles mass collected and q/m and were found to be generally similar.

1. Introduction

Gas-solid fluidization is a process where gas travels through a bed of particles at a high enough velocity such that the solid particles behave as a fluid and are subjected to particle motion within the bed. Gas-solid fluidization is a common process with a variety of applications such as drying, coating, and gas-solid reactions. In such reactors, particles are subsequently introduced to a significant number of particle-particle and particle-column wall interactions, which ultimately generate electrostatic charges due to triboelectrification and frictional charging. Generally, these electrostatic charges are not desired as they cause particle agglomeration, particle build-up on reactor walls, and electrical discharge which could potentially result in fires and explosions. In poly-olefin production, the formation of a particle layer on the reactor wall due to bed electrification has been reported as a significant problem for many years [1]. The particle layers tend to form sheets on reactor walls that can become large enough to dislodge and plug the gas distributor plate causing immediate reactor shut-down, reduction in reactor operation and ultimately production.

It is generally believed that grounding the reactor wall is an effective method for reducing electrostatic charges within a gas solid fluidized bed [2]. However, the majority of academic works on electrostatic charge generation have focused primarily on fluidized beds composed of transparent plastic. Only a handful of works [3-8] have conducted studies in metallic fluidization columns, some of which utilized a grounded unit [7-11]. However, the differences of electrostatic charge generation between grounded and electrically isolated columns have not been exclusively investigated.

The objective of this work was to examine the effect fluidization column grounding on particle wall fouling, and to determine its application as a charge reduction technique.

2. Experimental Setup and Methodology

The experimental apparatus and methodology used for this work has been described elsewhere [2]; with the exception that this work focuses solely on the wall fouling of the fluidized bed. The fluidization column, fabricated out carbon steel, was 0.1 m in diameter. The column was normally electrically isolated from the rest of the system (i.e., distributed plate and top section of the column). For the grounding experiments, a grounding wire was attached to the column. The fluidizing particles were polyethylene resin, produced using a metallocene based catalyst in an industrial gas-phase fluidized bed reactor. The received resin was sieved into five different particle size ranges (300-425 μm, 425-500 μm, 500-600 μm, 600-710 μm, and 710-1000 μm), with a constant density of 918 kg/m^3. The bed height to column diameter ratio (L/D) was maintained at 4 for all trials. The particles were fluidized for 60 minutes and the fluidizing gas velocity was maintained at 1.5 times the minimum fluidization velocity (u_{mf}) of each particle size range. The fluidizing gas was at 23 ± 0.6 °C, and was dried to 0 ± 0.02 % relative humidity.

After the fluidization period (60 min) was completed, the fluidizing gas was turned off and the bottom of the column was opened to remove the majority of the particles [2-4]. The particles that were remained adhered to the column wall denoted wall fouling. Pictures of the wall fouling were taken from the bottom of the column. Next, a Faraday cup was placed below the column and the column wall was lightly tapped to remove as many particles as possible from the wall and to record their charge. The mass of the collected particles was then recorded and the charge-to-mass ratio (q/m) was calculated. The mass percent (m%) was calculated from the mass of the particles collected in the Faraday cup divided by the initial mass of particles placed into the fluidization column. Each experiment was repeated at least three times to ensure reproducibility of the data.

3. Results and Discussion

Figure 1 shows the m% and q/m of particles that were adhered to the column wall for all particle size ranges under electrically isolated and grounded column conditions. Results clearly indicate that regardless of the particle size, a grounded column wall was not significantly different from that of the isolated column. The collected m% remained relatively constant within each particle size range with the exception of the 600-710 μm. With this range, the opposite of the general perception that column wall grounding reduces electrostatic charges was observed since more particles were adhered to the grounded column. For all particle ranges, a similar q/m was obtained for both the isolate and grounded column. At a particle size range greater than 710 μm, the gravitational force on the particles dominated electrostatic forces causing for the least amount of mass collected from the column wall. Figure 2 shows typical images taken from the bottom of the fluidization column showing particle-reactor wall fouling. From these images, it is clear that grounding the column wall did not reduce the amount of wall coating. Grounding the column wall is believed to be a method for reducing bed electrification and column wall fouling because as particles come into contact with the grounded wall, electrons can be easily exchanged while keeping the column wall neutral and thus aid in dissipating particles charge. However, in most cases the fluidizing particles are insulators where they have less of a tendency to give up their electrostatic charge. Additionally, at any given time, only a small fraction of particles within the fluidized bed will have the opportunity to come into contact with the column wall for some of their charge to be dissipated. In the meantime, the rest of the particles will continue to generate electrostatic charge through triboelectrification by particle-particle contacts, which would result in bi-polar charging. Moreover, even if the layer of particles adhered to the column wall were able to dissipate their charge, the particles that come into contact with these adhered particles would not lose any charge due to the restricted electron flow through insulators. Overall, if grounding the column wall would have helped in charge dissipation, it would have been expected that the net q/m of the particles to approach zero and the collected mass would decrease. However, such outcomes were not observed. Another study has shown that the charge transfer and conductivity of a fluidized bed composed of copper particles was independent of particle size and dependent on the void fraction of fluidized bed and the presence of chain-like particle in contact [12]. Therefore, with the factors mentioned above and the experimental results presented here, it is evident that grounding the column wall for systems containing non-conductive particles cannot provide an adequate method for electrostatic charge reduction through charge dissipation as previously expected.

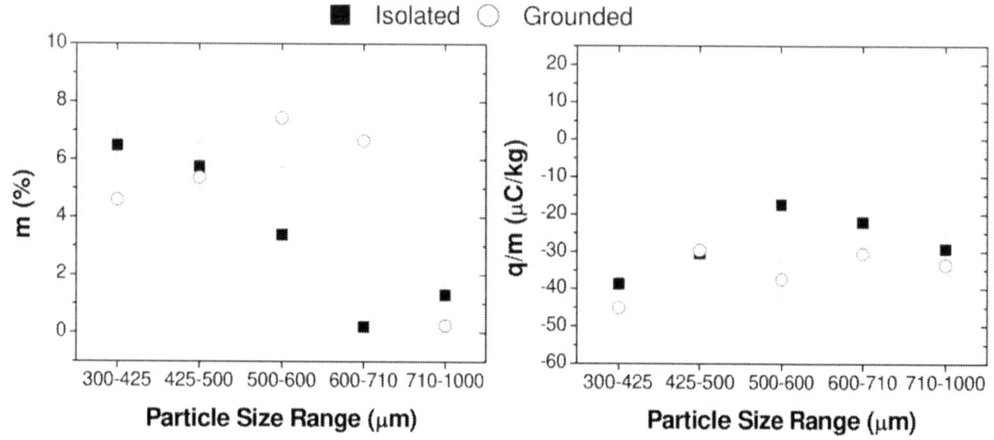

Figure 1. The m% and q/m of particles that adhered to the column wall at different size ranges.

300-425 μm

600-710 μm

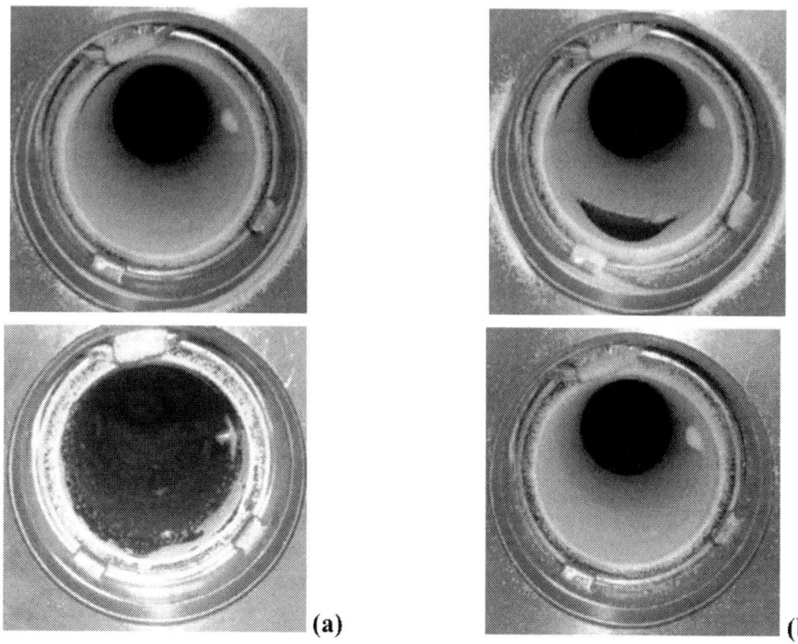

(a) **(b)**

Figure 2. Typical imaged obtained from wall fouling taken from the bottom of the column with 300-425 μm and 600-710 μm particles with the column a) Isolated; b) Grounded.

4. Conclusions

It is often believed that grounding the fluidization column wall can help in reducing electrostatic charges and ultimately wall fouling in gas-solid fluidized beds. However, it was shown that grounding had no effect on the reduction of charging and wall fouling of non-conductive particles in a pilot-scale fluidized bed, and in one case even promoted more wall fouling. This also validates that in industrial size gas-solid fluidized beds grounding the fluidization column does not provide any means of electrostatic charge reduction.

Acknowledgements

Financial assistance from the Natural Sciences and Engineering Research Council of Canada (NSERC) and input from Univation Technologies LLC (joint venture between ExxonMobil Chemical Company and The Dow Chemical Company, USA) are acknowledged with gratitude.

References

[1] Hagerty R O, Muhle M E, Agapiou A K, Kuo C-T, Goode MG, Hussein F D, Pannell R B and Szul J F 2005 Method for controlling sheeting in gas phase reactors. US Patent 2005/0148742 A1
[2] Ciborowski J and Wlodarski A 1961 On electrostatic fluidized beds *Chem. Eng. Sci.* **17** 23–32
[3] Sowinski A, Miller L and Mehrani P 2010 Investigation of electrostatic charge distribution in gas-solid fluidized beds *Chem. Eng. Sci.* **65**(9) 2771–81
[4] Sowinski A, Salama F and Mehrani P 2009 New technique for electrostatic charge measurement in gas-solid fluidized beds *J. of Electrostatics* **67**(4) 568–73
[5] Giffin A and Mehrani P 2010 Comparison of influence of fluidization time on electrostatic charge build-up in the bubbling vs. slugging flow regimes in gas-solid fluidized beds *J. Electrostatics* **68**(6) 492–502
[6] Ally M R and Klinzing G E 1985 Inter-relation of electrostatic charging and pressure drops in pneumatic transport *Powder Tech.* **44**(1) 85–8
[7] Kisel'nikov V N, Vyalkov V V and Filatov, V M 1967 On the problem of electrostatic phenomena in a fluidized bed *Int. Chem. Eng* **7**(3) 428–31
[8] Fujino M, Ogata S and Shinohara H 1985 The electric potential distribution profile in a naturally charged fluidized bed and its effects *Int. Chem. Eng.* **25**(1) 149–59
[9] Ali F S, Inculet I I and Tedoldi A 1999 Charging of polymer powder inside a metallic fluidized bed. *J. of Electrostatics* **45**(3) 199–211
[10] Moughrabiah W O, Grace J R and Bi X T 2009 Effects of pressure, temperature, and gas velocity on electrostatics in gas-solid fluidized beds *Ind. Eng. Chem. Res.* **48**(1) 320–5
[11] Boland D and Geldart D 1971/72 Electrostatic Charing in Gas Fluidised Beds *Powder Tech* **5** 289-94
[12] Sabacky B J and Evans J W 1977 The electrical conductivity of fluidized bed electrodes – its significance and some experimental measurements *Metal. Trans.* B **8** 5–13

Effect of magnetic field on the plasma generated during a sliding contact

K Nakayama[1, 2]

[1]Advanced Research Institute, Chiba Institute of Technology, 2-17-1, Tsudanuma, Narashino, Chiba 275-0016, Japan
[2]Nanosystem Research Institute, AIST, 2-12-1, Namiki, Tsukuba, Ibarakai 305-8564, Japan

E-mail: Keiji.nakayama@it-chiba.ac.jp

Abstract. The triboelectrically generated plasma discharge generated in the gap behind a sliding contact is thought to be one of the main causes of the unexplained serious tribological problems met recently in the development of advanced technologies. To solve the tribological problems arising from the triboplasma, it is essential to develop a triboplasma control technology. In this report, the effect of magnetic field on triboplasma generation is investigated as a method for developing a control technology using the Lorenz force, which acts on moving charged particles in the plasma under the influence of an applied magnetic field. Photographs of the photons emitted from the sliding contact and its vicinity were obtained with an ICCD camera while sliding a diamond pin along a sapphire disk in ambient air. By applying a magnetic field, both the distribution and intensity of the tribo-plasma changed greatly, demonstrating that a triboplasma can be controlled by applying a magnetic field.

1. Introduction

Tribological problems have long been analyzed and solved based on the temperature rise that occurs during frictional contacts [1, 2]. However, in many kinds of tribological problems, especially in the development of advanced technology, curious phenomena occur which could not be explained by traditional frictional heat theory alone [3, 4]. This suggested that some unknown high energy source is present at and near the sliding contact other than frictional heat energy. The author and his team have searched for this unknown energy source for many years and reported the discovery of a triboplasma generated in the gap of a sliding contact [5]. They showed that the triboplasma was generated by discharges in the surrounding gas due to triboelectrification (figure 1), since the photon spectrum in the UV region coincided completely with that of the plasma generated during an electrically-induced discharge in the same gas [6]. It was concluded that this triboplasma must be the unknown energy source. It has been observed that the triboplasma not only has harmful effects by decomposing lubricants [8] and accelerating wear [4] but also has beneficial effects by forming friction- and wear-reducing organic films [7]. Consequently, it is necessary to develop technologies to control triboplasma generation. It has been shown that triboplasma can be controlled by changing the electrical resistivity of the mating sliding partners [9]. One of other possible ways to control the triboplasma is by the application of a magnetic field to the triboplasma, since the direction of the moving charged particles in the triboplasma will be changed by the action of the Lorenz force. In the

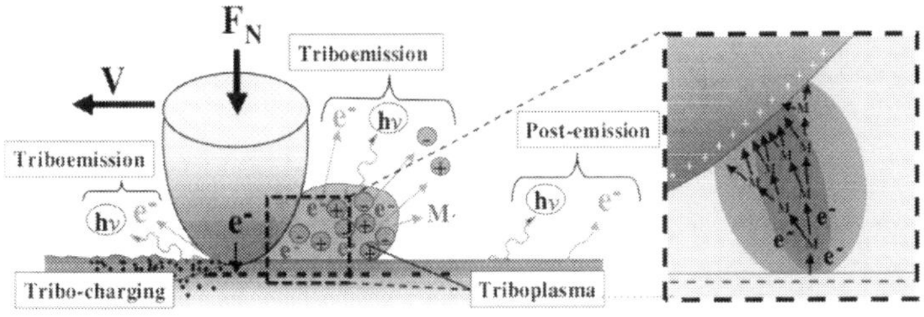

Figure 1 Model of triboplasma generation through discharging of gas due to triboelectrification.

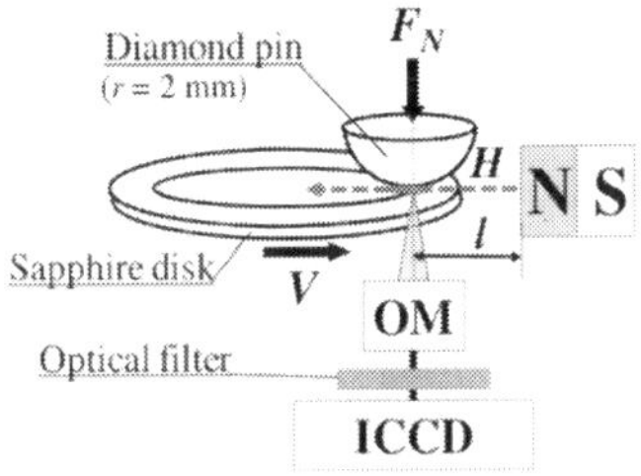

Figure 2 Experimental apparatus.

present paper, the effect of magnetic field on triboplasma generation has been studied in order to understand how the magnetic field acts on the intensity and the distribution of the triboplasma.

2. Experimental procedure

Figure 2 shows the apparatus used for obtaining photographs of the optical emission (tribo-photons) from the sliding contact and its vicinity. A diamond pin with a tip radius r = 2 mm slides against a rotating sapphire (single crystal Al_2O_3) disk with a thickness of 1 mm under the application of a magnetic field H, in which the magnetic flux density is expressed by $B = \mu_0 H$, where μ_0 is the magnetic permeability in vacuum. The photon images were measured using an ICCD camera after magnifying with an optical microscope (OM). By inserting filters, which had transmission wavelength regions of 290-420 nm, 400-800 nm and 800-2500 nm, respectively, UV, visible and IR photon images were obtained since the ICCD camera is sensitive to photon wavelength ranging from 200 to 900 nm. All the sliding experiments were performed at a room temperature of 25°C in ambient air of 44% relative humidity under a normal force F_N = 940 mN and a sliding velocity V = 420 mm/s.

Figure 3 shows the relation between the magnetic flux density, B, of the magnet and the distance, l, from the N-pole or S-pole faces (7×12 mm^2) of the magnet, the length of which was 76 mm. The value of B at the N- and S-poles surface was 0.42 T, but decreases steeply with increasing distance, l, as seen in figure 3, since the magnetic field, H, decreases with increasing l, as expressed by the relation $H = m/(4\pi \mu_0 l^2)$, where m is the strength of the magnetic pole. The distance between the tip of the magnet and the center of the sliding contact was fixed to be l = 1.3 mm, so that B was assumed to be ~0.25 T at the diamond tip, although it may differ from that expressed by figure 3 due to detailed factors related to geometry and materials of the test set-up.

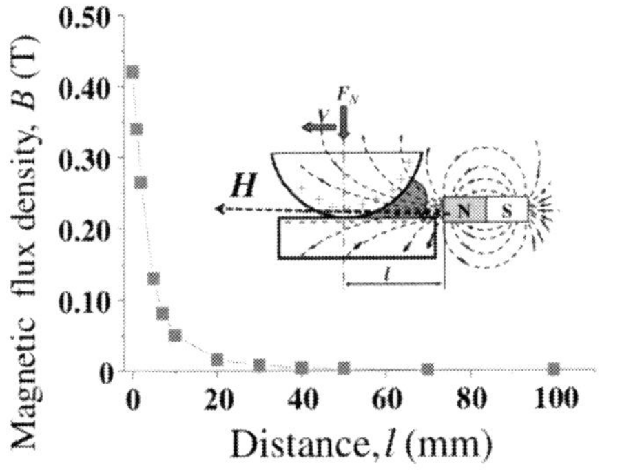

Figure 3 Characteristics of the magnet.

3. Results and discussion

3.1 Total photons

Figure 4 (middle) shows the total photon images measured before applying the magnetic field (a), and under the magnetic field application from N- (b) and S- (c) pole, respectively. The white solid arrow shows the direction of the pin sliding. As seen in figure 4a, a ring-shaped triboplasma with a tail was generated before application of the magnetic field as reported previously [5, 6]. By applying the magnetic field from the N- and S-pole, half of the circular triboplasma almost disappeared, due to the action of the Lorenz force of the magnetic field acting on the moving electrons and ions in the plasma.

Figure 4 (top) shows the *total* photon intensity profiles measured perpendicular to the frictional track at the center of the triboplasma ring from the top to the bottom of the image. Before applying the magnetic field, the photon intensity i.e., the triboplasma intensity at the two positions, C and D in the triboplasma ring were almost the same, as seen in figure 4d. By applying the magnetic field, only the peak height at the position D near the magnet became smaller. Namely, the Lorenz force acted effectively on the plasma at D, but not at C which was only 130 μm further away than D from the magnet.

Figure 4 (bottom) shows the total tribo-photon intensity profile measured from the left to the right along the wear track, which was 1.3 mm distance from the N- and S-pole face of the magnet. It is seen that, under the magnetic field application, the photon intensities at the contact point and the wear track were slightly lower than before. This means that a magnetic flux density of at least 0.42 T at the N- and S- pole can affect the plasma generation at a contact located at a distance of 1.3 mm from the magnet pole.

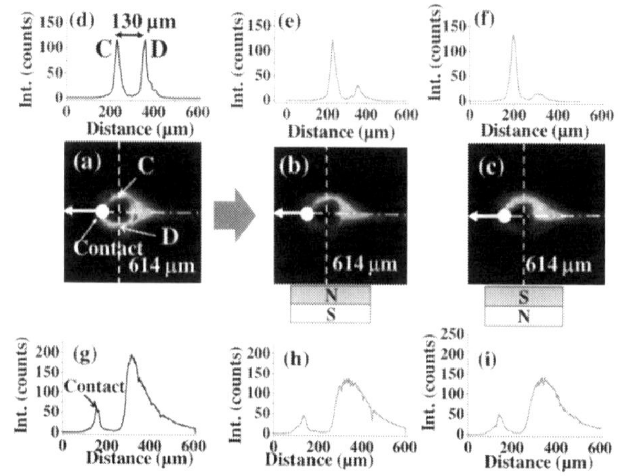

Figure 4 Total tribo-photon images (middle), and line profiles of the photon intensity measured vertical to (top) and along the friction track (bottom).

3.2 UV photons

Figure 5 (middle) and (top) shows the two-dimensional images and the intensity profiles of the UV tribo-photons measured perpendicular to the frictional track, before and under application of magnetic fields from N- and S-poles, respectively. It is seen that by applying the magnetic field, half of the triboplasma ring near the magnet has again completely disappeared. As the UV photons are emitted from the air molecules excited by the electron impact [6], it can be concluded that, at position D, the excited charged particles have been almost completely removed by the Lorenz force. Figure 5 (bottom) shows the photon intensity profiles measured along the wear track. One of the other important features seen in this profile is that some UV photons are definitely emitted from the sliding contact. It suggests that triboplasma is generated even in the extremely narrow gap of the sliding contact. This further suggests that extremely intense electric fields are generated in the small asperity gap of the sliding contact, since the spark voltage increases steeply with the decrease in the narrow electrode distance as seen in the *Paschen* curve [10].

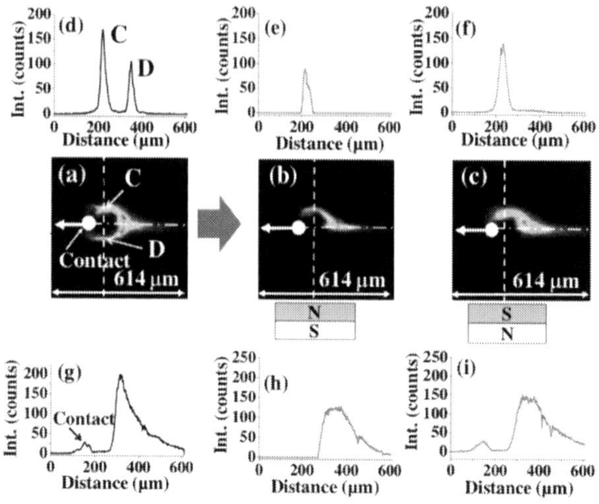

Figure 5 Two-dimensional images of the UV-tribo-photons (middle), and line profiles of the photon intensity measured vertical to (top) and along the friction track (bottom).

3.3 Visible and IR photons

Figure 6 (bottom) and (top) shows the visible tribo-photon images and profiles of the visible photon intensity measured perpendicular to the frictional track at the center of the plasma ring. In contrast to the UV photons, for visible photons the plasma-ring still remains under application of the magnetic field, though the intensity of the photon emission at D located near the magnet decreased greatly. This is probably because visible photons are emitted not only from the ionized plasma gas, but also from surface atoms excited by back-electrons, accelerated by the inverse electric field generated by charges accumulating on the diamond surface by electron bombardment, as reported previously [11].

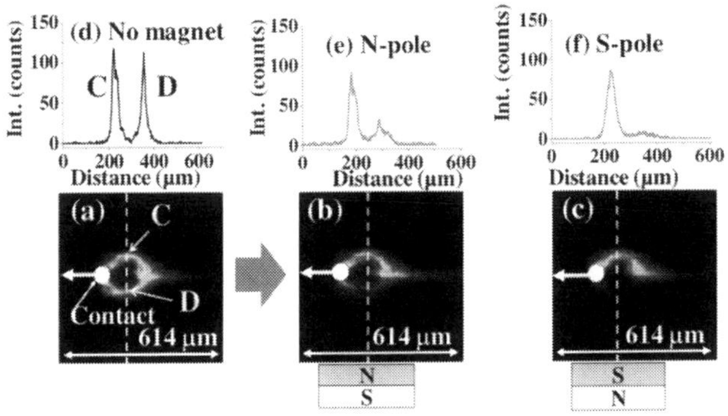

Figure 6 Two-dimensional images of the visible tribo-photons (below), and line profiles of the photon intensity measured vertical to the friction track (top).

Figure 7 (bottom) and (top) shows the IR tribo-photon images before and under application of the magnetic field from the S-pole, and the photon intensity profiles measured perpendicular to the frictional track at the center of the plasma ring. In figure 7c, the photon intensity peak height at C and D is not the same even though no magnetic field was applied. This kind of observation happens sometimes because the tribocharging distribution was not always uniform. However, it is definitely seen that the IR photon intensity at D located near the magnet decreased greatly by applying the magnetic field.

4. Conclusions

The results show that by applying a magnetic field to the sliding contact between diamond and sapphire, the distribution and intensity of the triboplasma can be changed. This is because the electrons and positive ions in the triboplasma, which are moving to the positively and negatively charged sliding surfaces, are subjected to the Lorenz force in the magnetic field. Both magnetic fields from N- and S- poles can change the triboplasma distribution. This means that we can develop a triboplasma control technology by utilising a magnetic field.

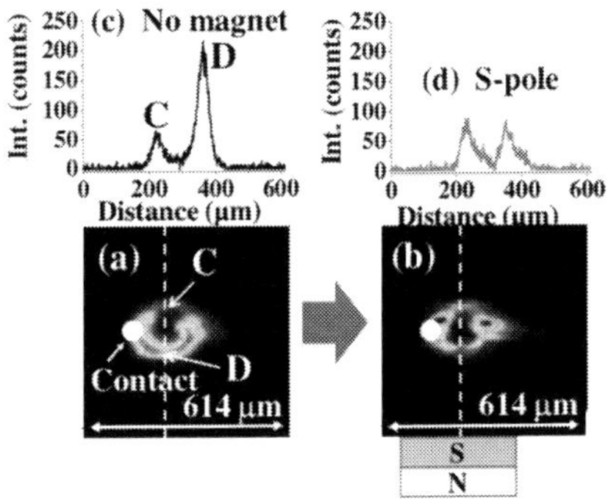

Figure 7 Two-dimensional images of the IR tribo-photons (below), and line profiles of the photon intensity measured vertical to the friction track.

Acknowledgement

This work was supported by JSPS Grant-in-Aid for Scientific Research (A) (20246035).

References

[1] Bowden FP and Tabor D 1954 *Friction and Lubrication of Solids* (Oxford Clarendon Press).
[2] Sakurai T and Sato K 1966 *ASLE Trans.* **9** pp 77-87.
[3] Adachi K and Kato K 2008 *Tribology of Diamond-Like Carbon Films: Fundamentals and Applications,* ed C. Donnet and A Erdemir (New York: Springer) pp 339-361.
[4] Ohta T., Yan J. et al 2009 *Key Eng. Mater.*, **389-390** pp 144-150.
[5] Nakayama K and Nevshupa RA 2002 *J. Phys. D: Appl. Phys.***35** L53-L56.
[6] Nakayama K and Nevshupa RA 2003 *Trans. ASME* **125** pp 780-787.
[7] Nakayama K and Hashimoto H 1996 *Tribol. Int.* 29 pp 385-393.
[8] Nakayama K and Mirza SM 2006 *Tribol. Trans.* **49** pp 17–25.
[9] Nakayama K 2008 *Tribology of Diamond-Like Carbon Films: Fundamentals and Applications,* ed C. Donnet and A Erdemir (New York: Springer) pp 291-310.
[10] von Engel A 1994 *Ionized Gases* (New York, American Institute of Physics) p195.
[11] Nakayama K, 2010 *Tribol. Lett.* **37** pp 215-228.

AUTHOR INDEX

AUTHOR INDEX

Institute of Physics
Dirac House, Temple Back
Bristol BS1 6BE UK